Dingyü Xue
**Linear Algebra and Matrix Computations with MATLAB®**

## Also of Interest

*Fractional-Order Control Systems, Fundamentals and Numerical Implementations*
Dingyü Xue, 2017
ISBN 978-3-11-049999-5, e-ISBN (PDF) 978-3-11-049797-7,
e-ISBN (EPUB) 978-3-11-049719-9

*MATLAB® Programming, Mathematical Problem Solutions*
Dingyü Xue, 2020
ISBN 978-3-11-066356-3, e-ISBN (PDF) 978-3-11-066695-3,
e-ISBN (EPUB) 978-3-11-066370-9

*Calculus Problem Solutions with MATLAB®*
Dingyü Xue, 2020
ISBN 978-3-11-066362-4, e-ISBN (PDF) 978-3-11-066697-7,
e-ISBN (EPUB) 978-3-11-066375-4

*Solving Optimization Problems with MATLAB®*
Dingyü Xue, 2020
ISBN 978-3-11-066364-8, e-ISBN (PDF) 978-3-11-066701-1,
e-ISBN (EPUB) 978-3-11-066369-3

*Differential Equation Solutions with MATLAB®*
Dingyü Xue, 2020
ISBN 978-3-11-067524-5, e-ISBN (PDF) 978-3-11-067525-2,
e-ISBN (EPUB) 978-3-11-067531-3

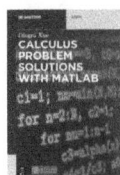

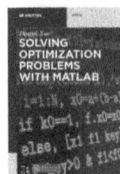

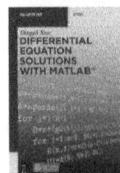

Dingyü Xue

# Linear Algebra and Matrix Computations with MATLAB®

—

清華大学出版社
TSINGHUA UNIVERSITY PRESS

**DE GRUYTER**

**Author**
Prof. Dingyü Xue
School of Information Science and Engineering
Northeastern University
Wenhua Road 3rd Street
110819 Shenyang
China
xuedingyu@mail.neu.edu.cn

MATLAB and Simulink are registered trademarks of The MathWorks, Inc. See www.mathworks.com/trademarks for a list of additional trademarks. The MathWorks Publisher Logo identifies books that contain MATLAB and Simulink content. Used with permission. The MathWorks does not warrant the accuracy of the text or exercises in this book. This book's use or discussion of MATLAB and Simulink software or related products does not constitute endorsement or sponsorship by The MathWorks of a particular use of the MATLAB and Simulink software or related products. For MATLAB® and Simulink® product information, or information on other related products, please contact:

The MathWorks, Inc.
3 Apple Hill Drive
Natick, MA, 01760-2098 USA
Tel: 508-647-700
Fax: 508-647-7001
E-mail: info@mathworks.com
Web: www.mathworks.com

ISBN 978-3-11-066363-1
e-ISBN (PDF) 978-3-11-066699-1
e-ISBN (EPUB) 978-3-11-066371-6

**Library of Congress Control Number: 2019956330**

**Bibliographic information published by the Deutsche Nationalbibliothek**
The Deutsche Nationalbibliothek lists this publication in the Deutsche Nationalbibliografie;
detailed bibliographic data are available on the Internet at http://dnb.dnb.de.

Cover image: Dingyü Xue
Typesetting: VTeX UAB, Lithuania
Printing and binding: CPI books GmbH, Leck

www.degruyter.com

# Preface

Scientific computing is commonly and inevitably encountered in course learning, scientific research and engineering practice for each scientific and engineering student and researcher. For the students and researchers in the disciplines which are not pure mathematics, it is usually not a wise thing to learn thoroughly low-level details of related mathematical problems, and also it is not a simple thing to find solutions of complicated problems by hand. It is an effective way to tackle scientific problems, with high efficiency and in accurate and creative manner, with the most advanced computer tools. This method is especially useful in satisfying the needs for those in the area of science and engineering.

The author had made some effort towards this goal by addressing directly the solution methods for various branches in mathematics in a single book. Such a book, entitled "MATLAB based solutions to advanced applied mathematics", was published first in 2004 by Tsinghua University Press. Several new editions were published afterwards: in 2015, the second edition in English by CRC Press, and in 2018, the fourth Edition in Chinese were published. Based on the latest Chinese edition, a brand new MOOC project was released in 2018,[1] and received significant attention. The number of the registered students was about 14 000 in the first round of the MOOC course, and reached tens of thousands in later rounds. The textbook has been cited tens of thousands times by journal papers, books, and degree theses.

The author has over 30 years of extensive experience of using MATLAB in scientific research and education. Significant amount of materials and first-hand knowledge has been accumulated, which cannot be covered in a single book. A series entitled "Professor Xue Dingyü's Lecture Hall" of such works are scheduled with Tsinghua University Press, and the English editions are included in the DG STEM series with De Gruyter. These books are intended to provide systematic, extensive and deep explorations in scientific computing skills with the use of MATLAB and related tools. The author wants to express his sincere gratitude to his supervisor, Professor Derek Atherton of Sussex University, who first brought him into the paradise of MATLAB.

The MATLAB series is not a simple revision of the existing books. With decades of experience and material accumulation, the idea of "revisiting" is adopted in authoring the series, in contrast to other mathematics and other MATLAB-rich books. The viewpoint of an engineering professor is established and the focus is on solving various applied mathematical problems with tools. Many innovative skills and general-purpose solvers are provided to solve problems with MATLAB, which is not possible by any other existing solvers, so as to better illustrate the applications of computer tools in solving mathematical problems in every mathematics branch. It also helps

---

1 MOOC (in Chinese) address: https://www.icourse163.org/learn/NEU-1002660001

https://doi.org/10.1515/9783110666991-201

the readers broaden their viewpoints in scientific computing, and even in finding innovative solutions by themselves to scientific computing which cannot be solved by any other existing methods.

The first title in the MATLAB series: "MATLAB Programming", can be used as an entry-level textbook or reference book to MATLAB programming, so as to establish a solid foundation and deep understanding for the application of MATLAB in scientific computing. Each subsequent volume tries to cover a branch or topic in mathematical courses. Bearing in mind the "computational thinking" in authoring the series, deep understanding and explorations are made for each mathematics branch involved. These MATLAB books are suitable for the readers who have already learnt the related mathematical courses, and want to revisit the courses to learn how to solve the problems by using computer tools. It can also be used as a companion in synchronizing the learning of related mathematics courses, and viewing the course from a different angle, so that the readers may expand their knowledge in learning the related courses, so as to better learn, understand, and practice the materials in the courses.

This book is the third in the MATLAB series. A brand new viewpoint is introduced for presenting linear algebra in a conventional format, with a focus on the direct solutions of linear algebra and matrix analysis problems employing MATLAB. Matrix input methods are introduced first, followed by matrix analysis methods and matrix transformation problems. The methods of solving matrix equations and computing arbitrary matrix functions are then explored. Finally, modeling and solving in linear algebra applications are addressed.

At the time the books are published, the author wishes to express his sincere gratitude to his wife, Professor Yang Jun. Her love and selfless care over the decades provided the author immense power, which supports his academic research, teaching, and writing.

September 2019                                                                                      Xue Dingyü

# Contents

# 1 Introduction to linear algebra and matrix computation

## 1.1 Matrices and linear equations

The study of linear algebra is originated from linear equations. Linear equations are the most widely used mathematical models in science and technology. In practical applications, more complicated linear, or even nonlinear, equations can be established. For the convenience of further research, the concept of a matrix ought to be introduced to describe linear equations. In this section, several examples are used to demonstrate matrix representations, and also to show the importance of mathematical modeling of linear equations.

### 1.1.1 Matrix representation of tables

In everyday life and in scientific research, various data are described by tables. How can we effectively represent these tables? A different variety of forms can be taken to represent the data in mathematics and in computers. The most effective is the matrix representation.

A matrix is the most important unit in linear algebra, whose mathematical form can be expressed as

$$A = \begin{bmatrix} a_{11} & a_{12} & a_{23} & \cdots & a_{1n} \\ a_{21} & a_{22} & a_{23} & \cdots & a_{2n} \\ \vdots & \vdots & \vdots & \ddots & \vdots \\ a_{m1} & a_{m2} & a_{m3} & \cdots & a_{mn} \end{bmatrix}. \tag{1.1.1}$$

Some examples are given in this section to demonstrate the use of matrices.

**Example 1.1.** Color images can be described in many ways in computers, RGB representation is the most widely used one. Each color is defined as a combination of three primary colors: red, green, and blue. The commonly used eight colors are described in Table 1.1. Use a matrix to present the table.

**Table 1.1:** RGB components of the commonly used colors.

| components | black | blue | green | cyan | red | magenta | yellow | white |
|---|---|---|---|---|---|---|---|---|
| red | 0 | 0 | 0 | 0 | 255 | 255 | 255 | 255 |
| green | 0 | 0 | 255 | 255 | 0 | 0 | 255 | 255 |
| blue | 0 | 255 | 0 | 255 | 0 | 255 | 0 | 255 |

https://doi.org/10.1515/9783110666991-001

**Solutions.** If the rows are defined as the three primary colors, and the columns are assigned for the colors like "black", "blue", …, "white", a $3 \times 8$ table is needed to represent the whole table. The arrangement of the matrix elements is the same as in the original table:

$$A = \begin{bmatrix} 0 & 0 & 0 & 0 & 255 & 255 & 255 & 255 \\ 0 & 0 & 255 & 255 & 0 & 0 & 255 & 255 \\ 0 & 255 & 0 & 255 & 0 & 255 & 0 & 255 \end{bmatrix}.$$

With the mathematical presentation in a matrix, the following statements can be used to input it directly to MATLAB:

```
>> A=[0 0 0 0 255 255 255 255; 0 0 255 255 0 0 255 255
      0 255 0 255 0 255 0 255];
```

**Example 1.2.** Some of the parameters of the eight planets in the solar system are given in Table 1.2, where the relative parameters in the first four columns are the values with respect to the Earth. The unit of semimajor axis is AU (astronomical unit, defined as 149 597 870 700 m $\approx 1.5 \times 10^{11}$ m). The unit of rotation period is day. Represent the table with a matrix.

**Table 1.2:** Some parameters of the eight planets.

| name of planet | relative diameter | relative mass | semimajor axis | orbital period | orbital eccentricity | rotation period | confirmed moons | rings |
|---|---|---|---|---|---|---|---|---|
| Mercury | 0.382 | 0.06 | 0.39 | 0.24 | 0.206 | 58.64 | 0 | no |
| Venus | 0.949 | 0.82 | 0.72 | 0.62 | 0.007 | −243.02 | 0 | no |
| Earth | 1 | 1 | 1 | 1 | 0.017 | 1 | 1 | no |
| Mars | 0.532 | 0.11 | 1.52 | 1.88 | 0.093 | 1.03 | 2 | no |
| Jupiter | 11.209 | 317.8 | 5.20 | 11.86 | 0.048 | 0.41 | 69 | yes |
| Saturn | 9.449 | 95.2 | 9.54 | 29.46 | 0.054 | 0.43 | 62 | yes |
| Uranus | 4.007 | 14.6 | 19.22 | 84.01 | 0.047 | −0.72 | 27 | yes |
| Neptune | 3.883 | 17.2 | 30.06 | 164.8 | 0.009 | 0.67 | 14 | yes |

**Solutions.** It can be seen that most of the data in Table 1.2 are numbers. Besides, the entries of the last column are "yes" and "no", which can be converted to logical numbers, 1 for "yes" and 0 for "no". If we are only interested in the numbers in the table, it is a concise way to represent it with a matrix

$$
A = \begin{bmatrix}
0.382 & 0.06 & 0.39 & 0.24 & 0.206 & 58.64 & 0 & 0 \\
0.949 & 0.82 & 0.72 & 0.62 & 0.007 & -243.02 & 0 & 0 \\
1 & 1 & 1 & 1 & 0.017 & 1 & 1 & 0 \\
0.532 & 0.11 & 1.52 & 1.88 & 0.093 & 1.03 & 2 & 0 \\
11.209 & 317.8 & 5.2 & 11.86 & 0.048 & 0.41 & 69 & 1 \\
9.449 & 95.2 & 9.54 & 29.46 & 0.054 & 0.43 & 62 & 1 \\
4.007 & 14.6 & 19.22 & 84.01 & 0.047 & -0.72 & 27 & 1 \\
3.883 & 17.2 & 30.06 & 164.8 & 0.009 & 0.67 & 14 & 1
\end{bmatrix}.
$$

With the mathematical form of the matrix, the following statements can be used to input it into MATLAB workspace:

```
>> A=[0.382,0.06,0.39,0.24,0.206,58.64,0,0;
    0.949,0.82,0.72,0.62,0.007,-243.02,0,0;
    1,1,1,1,0.017,1,1,0;
    0.532,0.11,1.52,1.88,0.093,1.03,2,0;
    11.209,317.8,5.2,11.86,0.048,0.41,69,1;
    9.449,95.2,9.54,29.46,0.054,0.43,62,1;
    4.007,14.6,19.22,84.01,0.047,-0.72,27,1;
    3.883,17.2,30.06,164.8,0.009,0.67,14,1];
```

The whole table can alternatively be represented by the data type table, see Volume I in this series for details.

**Example 1.3.** The relative data in Example 1.2 are multiples of actual data with respect to the data of the Earth. If the mass of the Earth is $5.965 \times 10^{24}$ kg, compute the masses of the other planets, for instance, the mass of Jupiter.

**Solutions.** From the original table, it can be seen that the relative masses are stored in the second column of the matrix. All the elements in the column can be extracted with $A(:,2)$ command, with the fifth one of which the mass of Jupiter. Therefore, with the following statements, the actual masses of all the eight planets can be computed. The fifth data from the list is the mass of Jupiter, and it is $1.8957 \times 10^{27}$ kg.

```
>> M0=5.965e24; M=A(:,2)*M0; M(5)
```

### 1.1.2 Modeling and solving linear equations

Some examples are given in this section to demonstrate linear equations modeling of some practical problems.

**Example 1.4.** In the fourth to the fifth centuries CE, an ancient Chinese book *The Mathematical Classic of Sun Zi* was published. A famous chick–rabbit cage problem was proposed: "In a cage, there are chicks and rabbits. When they are counted, the number of heads is 35, while the number of feet is 94. How many chicks and rabbits in the cage?"

**Solutions.** There are various methods applicable to solve this problem. If the idea of equations is introduced, one may denote the number of chicks as $x_1$, and the number of rabbits as $x_2$, then the following linear algebraic equations can be established:

$$\begin{cases} x_1 + x_2 = 35, \\ 2x_1 + 4x_2 = 94. \end{cases}$$

If the concept of matrix is introduced, the equations can be expressed as

$$\begin{bmatrix} 1 & 1 \\ 2 & 4 \end{bmatrix} \begin{bmatrix} x_1 \\ x_2 \end{bmatrix} = \begin{bmatrix} 35 \\ 94 \end{bmatrix}.$$

If one denotes the matrices and vectors as

$$A = \begin{bmatrix} 1 & 1 \\ 2 & 4 \end{bmatrix}, \quad x = \begin{bmatrix} x_1 \\ x_2 \end{bmatrix}, \quad B = \begin{bmatrix} 35 \\ 94 \end{bmatrix},$$

the standard form of the linear equations can be obtained as

$$Ax = B. \tag{1.1.2}$$

Linear equations can be solved directly with MATLAB command $x=A\backslash B$, therefore the following statements can be used, and one has $x = [23, 12]^T$. The physical interpretation to the results is as follows: there are 23 chicks and 12 rabbits.

```
>> A=[1 1; 2 4]; B=[35; 94]; x=A\B
```

Besides, symbolic computation commands are also allowed. The following statements can be used to solve the chick–rabbit cage problem, and the same results can be obtained.

```
>> syms x y
   [x y]=solve(x+y==35, 2*x+4*y=94)
```

**Example 1.5.** A beam balancing problem is presented in [19]. Assume that the structure of a beam system is shown in Figure 1.1, where each segment presents a beam, and a circle represents a joint. Assume also that the declining angles of the beams are all 45°. Joint 1 is fixed both horizontally and vertically, while Joint 8 is fixed vertically. Loads are added at Joints 2, 5, and 6. To balance the beam system, write down linear equations at each joint in vertical and horizontal directions. Also, express the equations in matrix form.

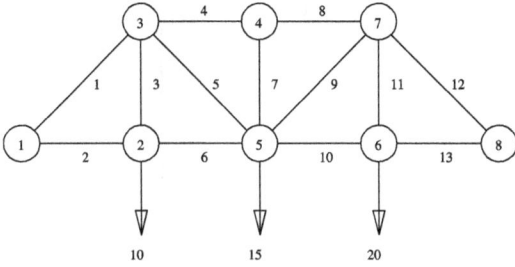

**Figure 1.1:** Planar structure of beams.

**Solutions.** Since Joint 1 is fixed horizontally and vertically, there is no need to write down balancing equations for it. Now let us consider Joint 2 first. It can be seen that in the horizontal direction, there are two forces, one is from beam 2, denoted by $f_2$, the other is from beam 6, denoted as $f_6$. To make the joint balanced, one has $f_2 = f_6$, or alternatively expressed as $f_2 - f_6 = 0$.

In the vertical direction of Joint 2, the force comes from beam 3, i. e., $f_3$, and an external load of 10. Therefore the balancing equation is $f_3 = 10$.

For the convenience of modeling, it is assumed that the forces from the left or up are defined as positive, while those from the right or down are marked as negative.

Now let us consider Joint 3. In the horizontal direction, three forces, $f_1$, $f_4$ and $f_5$, are involved. Since $f_1$ and $f_5$ are declined ones, with an angle of 45°, the actual force should be multiplied with $\cos 45° = 1/\sqrt{2}$. Denoting $\alpha = 1/\sqrt{2}$, the horizontal balancing equation is $\alpha f_1 - f_4 - \alpha f_5 = 0$. In the vertical direction, the balancing equation is $\alpha f_1 + f_3 + \alpha f_5 = 0$.

Similarly, at Joint 4, the horizontal balancing equation is $f_4 - f_8 = 0$, while the vertical one is $f_7 = 0$.

At Joint 5, the two equations are

$$\alpha f_5 + f_6 - \alpha f_9 - f_{10} = 0, \quad f_5 + f_7 + \alpha f_9 = 15.$$

At Joint 6, the two equations are

$$f_{10} - f_{13} = 0, \quad f_{11} = 20.$$

At Joint 7, the two equations are

$$f_8 + \alpha f_9 - \alpha f_{12} = 0, \quad \alpha f_9 + f_{11} + \alpha f_{12} = 0.$$

At Joint 8, since it is fixed vertically, the horizontal equation $\alpha f_{12} + f_{13} = 0$. Summarizing the above analysis, there are 13 linear equations in total. The matrix form of the equations can also be constructed as

$$
\begin{bmatrix}
0 & 1 & 0 & 0 & 0 & -1 & 0 & 0 & 0 & 0 & 0 & 0 & 0 \\
0 & 0 & 1 & 0 & 0 & 0 & 0 & 0 & 0 & 0 & 0 & 0 & 0 \\
\alpha & 0 & 0 & -1 & -\alpha & 0 & 0 & 0 & 0 & 0 & 0 & 0 & 0 \\
\alpha & 0 & 1 & 0 & \alpha & 0 & 0 & 0 & 0 & 0 & 0 & 0 & 0 \\
0 & 0 & 0 & 1 & 0 & 0 & 0 & -1 & 0 & 0 & 0 & 0 & 0 \\
0 & 0 & 0 & 0 & 0 & 0 & 1 & 0 & 0 & 0 & 0 & 0 & 0 \\
0 & 0 & 0 & 0 & \alpha & 1 & 0 & 0 & -\alpha & -1 & 0 & 0 & 0 \\
0 & 0 & 0 & 0 & 1 & 0 & 1 & 0 & \alpha & 0 & 0 & 0 & 0 \\
0 & 0 & 0 & 0 & 0 & 0 & 0 & 0 & 0 & 1 & 0 & 0 & -1 \\
0 & 0 & 0 & 0 & 0 & 0 & 0 & 0 & 0 & 0 & 1 & 0 & 0 \\
0 & 0 & 0 & 0 & 0 & 0 & 0 & 1 & -\alpha & 0 & 0 & -\alpha & 0 \\
0 & 0 & 0 & 0 & 0 & 0 & 0 & 0 & \alpha & 0 & 1 & \alpha & 0 \\
0 & 0 & 0 & 0 & 0 & 0 & 0 & 0 & 0 & 0 & 0 & \alpha & 1
\end{bmatrix}
\begin{bmatrix}
f_1 \\ f_2 \\ f_3 \\ f_4 \\ f_5 \\ f_6 \\ f_7 \\ f_8 \\ f_9 \\ f_{10} \\ f_{11} \\ f_{12} \\ f_{13}
\end{bmatrix}
=
\begin{bmatrix}
0 \\ 10 \\ 0 \\ 0 \\ 0 \\ 0 \\ 0 \\ 15 \\ 0 \\ 20 \\ 0 \\ 0 \\ 0
\end{bmatrix}.
$$

It is rather complicated to solve these equations by hand. Therefore, the best way to solve them is to send the system to computers and let them work for you. Details of the solution process will be fully presented later. Here, the following MATLAB can be considered as a demonstration:

```
>> alpha=1/sqrt(sym(2));
   A=[0  1  0  0  0  -1  0  0  0  0  0  0  0
      0  0  1  0  0  0  0  0  0  0  0  0  0
      alpha  0  0  -1  -alpha  0  0  0  0  0  0  0  0
      alpha  0  1  0  alpha  0  0  0  0  0  0  0  0
      0  0  0  1  0  0  0  -1  0  0  0  0  0
      0  0  0  0  0  0  1  0  0  0  0  0  0
      0  0  0  0  alpha  1  0  0  -alpha  -1  0  0  0
      0  0  0  0  1  0  1  0  alpha  0  0  0  0
      0  0  0  0  0  0  0  0  0  1  0  0  -1
      0  0  0  0  0  0  0  0  0  0  1  0  0
      0  0  0  0  0  0  0  1  -alpha  0  0  -alpha  0
      0  0  0  0  0  0  0  0  alpha  0  1  alpha  0
      0  0  0  0  0  0  0  0  0  0  0  alpha  1];
   B=[0; 10; 0; 0; 0; 0; 0;15; 0; 20; 0; 0; 0];
   f=simplify(A\B)
```

It can be seen that the equations can be solved directly, and the solutions are: $f_1 = -15\sqrt{2}$, $f_2 = 45 - 10\sqrt{2}$, $f_3 = 10$, $f_4 = -20$, $f_5 = 5\sqrt{2}$, $f_6 = 45 - 10\sqrt{2}$, $f_7 = 0$, $f_8 = -20$, $f_9 = 15\sqrt{2} - 10$, $f_{10} = 35 - 5\sqrt{2}$, $f_{11} = 20$, $f_{12} = 10 - 35\sqrt{2}$, $f_{13} = 35 - 5\sqrt{2}$.

In this example, the number of equations equals the number of unknowns, and the solution is unique. Such a system of equations is also known as consistent. In some other cases, the equations can have multiple solutions or none at all (be inconsistent).

**Example 1.6.** Consider again the problem in Example 1.5. If the Joints 1 and 8 are not fixed, three more equations can be formulated. Write down the linear algebraic equations and their matrix form.

**Solutions.** If Joints 1 and 8 are not fixed, the structure cannot be suspended in the air. The supporting forces $s_1$ and $s_2$ should be introduced at Joints 1 and 8, and yet another horizontal force $s_3$ should be introduced at Joint 1, as shown in Figure 1.2.

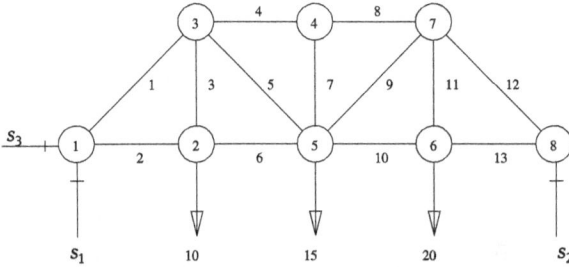

**Figure 1.2:** A modified beam structure.

If Joint 1 is no longer fixed, the horizontal equation becomes $\alpha f_1 + f_2 = s_3$, and the vertical one is $\alpha f_1 = s_1$. If Joint 8 is not fixed in the vertical direction, the corresponding equation becomes $\alpha f_{12} = s_2$. Therefore, the original matrix form is changed to

$$
\begin{bmatrix}
0 & 1 & 0 & 0 & 0 & -1 & 0 & 0 & 0 & 0 & 0 & 0 & 0 \\
0 & 0 & 1 & 0 & 0 & 0 & 0 & 0 & 0 & 0 & 0 & 0 & 0 \\
\alpha & 0 & 0 & -1 & -\alpha & 0 & 0 & 0 & 0 & 0 & 0 & 0 & 0 \\
\alpha & 0 & 1 & 0 & \alpha & 0 & 0 & 0 & 0 & 0 & 0 & 0 & 0 \\
0 & 0 & 0 & 1 & 0 & 0 & 0 & -1 & 0 & 0 & 0 & 0 & 0 \\
0 & 0 & 0 & 0 & 0 & 0 & 1 & 0 & 0 & 0 & 0 & 0 & 0 \\
0 & 0 & 0 & 0 & \alpha & 1 & 0 & 0 & -\alpha & -1 & 0 & 0 & 0 \\
0 & 0 & 0 & 0 & 1 & 0 & 1 & 0 & \alpha & 0 & 0 & 0 & 0 \\
0 & 0 & 0 & 0 & 0 & 0 & 0 & 0 & 0 & 1 & 0 & 0 & -1 \\
0 & 0 & 0 & 0 & 0 & 0 & 0 & 0 & 0 & 0 & 1 & 0 & 0 \\
0 & 0 & 0 & 0 & 0 & 0 & 0 & 1 & -\alpha & 0 & 0 & -\alpha & 0 \\
0 & 0 & 0 & 0 & 0 & 0 & 0 & 0 & \alpha & 0 & 1 & \alpha & 0 \\
0 & 0 & 0 & 0 & 0 & 0 & 0 & 0 & 0 & 0 & 0 & \alpha & 1 \\
\alpha & 1 & 0 & 0 & 0 & 0 & 0 & 0 & 0 & 0 & 0 & 0 & 0 \\
\alpha & 0 & 0 & 0 & 0 & 0 & 0 & 0 & 0 & 0 & 0 & 0 & 0 \\
0 & 0 & 0 & 0 & 0 & 0 & 0 & 0 & 0 & 0 & 0 & \alpha & 0
\end{bmatrix}
\begin{bmatrix}
f_1 \\ f_2 \\ f_3 \\ f_4 \\ f_5 \\ f_6 \\ f_7 \\ f_8 \\ f_9 \\ f_{10} \\ f_{11} \\ f_{12} \\ f_{13}
\end{bmatrix}
=
\begin{bmatrix}
0 \\ 10 \\ 0 \\ 0 \\ 0 \\ 0 \\ 15 \\ 0 \\ 20 \\ 0 \\ 0 \\ 0 \\ s_3 \\ s_1 \\ s_2
\end{bmatrix}.
$$

Since the number of equations is larger than the number of unknowns, $A$ becomes a $16 \times 13$ rectangular matrix. Such a system of equation is also known as overdetermined. If the unknowns $s_1$, $s_2$ and $s_3$ can be appended to the vector $f$, then the new vector $f$ has 16 elements, and the corresponding $A$ matrix is changed to a $16 \times 16$ square matrix:

$$
\begin{bmatrix}
0 & 1 & 0 & 0 & 0 & -1 & 0 & 0 & 0 & 0 & 0 & 0 & 0 & 0 & 0 & 0 \\
0 & 0 & 1 & 0 & 0 & 0 & 0 & 0 & 0 & 0 & 0 & 0 & 0 & 0 & 0 & 0 \\
\alpha & 0 & 0 & -1 & -\alpha & 0 & 0 & 0 & 0 & 0 & 0 & 0 & 0 & 0 & 0 & 0 \\
\alpha & 0 & 1 & 0 & \alpha & 0 & 0 & 0 & 0 & 0 & 0 & 0 & 0 & 0 & 0 & 0 \\
0 & 0 & 0 & 1 & 0 & 0 & 0 & -1 & 0 & 0 & 0 & 0 & 0 & 0 & 0 & 0 \\
0 & 0 & 0 & 0 & 0 & 0 & 1 & 0 & 0 & 0 & 0 & 0 & 0 & 0 & 0 & 0 \\
0 & 0 & 0 & 0 & \alpha & 1 & 0 & 0 & -\alpha & -1 & 0 & 0 & 0 & 0 & 0 & 0 \\
0 & 0 & 0 & 0 & 1 & 0 & 1 & 0 & \alpha & 0 & 0 & 0 & 0 & 0 & 0 & 0 \\
0 & 0 & 0 & 0 & 0 & 0 & 0 & 0 & 1 & 0 & 0 & -1 & 0 & 0 & 0 & 0 \\
0 & 0 & 0 & 0 & 0 & 0 & 0 & 0 & 0 & 1 & 0 & 0 & 0 & 0 & 0 & 0 \\
0 & 0 & 0 & 0 & 0 & 0 & 0 & 1 & -\alpha & 0 & 0 & -\alpha & 0 & 0 & 0 & 0 \\
0 & 0 & 0 & 0 & 0 & 0 & 0 & 0 & \alpha & 0 & 1 & \alpha & 0 & 0 & 0 & 0 \\
0 & 0 & 0 & 0 & 0 & 0 & 0 & 0 & 0 & 0 & 0 & \alpha & 1 & 0 & 0 & 0 \\
\alpha & 1 & 0 & 0 & 0 & 0 & 0 & 0 & 0 & 0 & 0 & 0 & 0 & 0 & 0 & 1 \\
\alpha & 0 & 0 & 0 & 0 & 0 & 0 & 0 & 0 & 0 & 0 & 0 & 0 & 1 & 0 & 0 \\
0 & 0 & 0 & 0 & 0 & 0 & 0 & 0 & 0 & 0 & \alpha & 0 & 0 & 1 & 0
\end{bmatrix}
\begin{bmatrix}
f_1 \\ f_2 \\ f_3 \\ f_4 \\ f_5 \\ f_6 \\ f_7 \\ f_8 \\ f_9 \\ f_{10} \\ f_{11} \\ f_{12} \\ f_{13} \\ s_1 \\ s_2 \\ s_3
\end{bmatrix}
=
\begin{bmatrix}
0 \\ 10 \\ 0 \\ 0 \\ 0 \\ 0 \\ 0 \\ 15 \\ 0 \\ 20 \\ 0 \\ 0 \\ 0 \\ 0 \\ 0 \\ 0
\end{bmatrix}.
$$

The system of equations now is changed to a consistent system. The following statements can be used to solve the new equations:

```
>> A=[A; alpha  1  0  0  0  0  0  0  0  0  0  0
            alpha  0  0  0  0  0  0  0  0  0  0  0
            0  0  0  0  0  0  0  0  0  0  0  alpha  0];
   A=[A zeros(16,3)]; A(14,16)=1; A(15,14)=1; A(16,15)=1;
   B=[B; 0; 0; 0]; x=simplify(inv(A)*B)
```

The forces $f_1$~$f_{13}$ are exactly the same as those in Example 1.5. The three new forces are $s_1 = 15$, $s_2 = 35 - 5\sqrt{2}$, $s_3 = 10\sqrt{2} - 30$.

In contrast to overdetermined systems, there are also underdetermined systems. They are classified as inconsistent, if no solution exists. Later in this book, solving over- and underdetermined systems of equations is also systematically presented.

## 1.2 Introduction of linear algebra

### 1.2.1 Mathematical development of linear algebra

The study of linear algebra originated from the solutions of linear equations, and the concept of determinants. In 1693, German mathematician Gottfried Wilhelm Leibniz

(1646–1716) studied the determinant problems. In 1750, Swiss mathematician Gabriel Cramer (1704–1752) used the concept of determinant to present the explicit solution of linear equations. The method is referred to as Cramer's rule. Later, German mathematician Johann Carl Friedrich Gauss (1777–1855) proposed a method of solving linear equations, known as Gaussian elimination method.

The earliest systematic study on matrix algebra originated in the middle of the 18th century in Europe. In 1844, German mathematician Hermann Günther Grassmann (1809–1877) published a book entitled *"Die Ausdehnungslehre von 1844"* (The Theory of Linear Extension, known in mathematics history as A1. In 1862, A2 was published)[11]. He created a new mathematics branch, called linear algebra, separated it from geometry, and introduced the prototype form of linear algebra. In 1848, British mathematician James Joseph Sylvester (1817–1897) coined the word "matrix". In studying matrix transforms, British mathematician Arthur Cayley (1821–1895) used a letter to indicate the whole matrix. He defined matrix multiplication method, and introduced the concept of inverse matrices.

In 1882, the Ottoman Empire mathematician and general Vidinli Hüseyin Tevfik Paşa (1832–1901) wrote a book entitled *Linear Algebra*. In 1888, Italian mathematician Giuseppe Peano (1858–1932) introduced a more accurate definition of a linear space. In 1900, linear transformation theory of finite-dimensional vector spaces was established. Other branches also appeared in linear algebra. Applications of linear algebra are available in many fields.

Professor Wassily Wassilyevich Leontief (1906–1999) of Harvard University established the input–output theory in economics, to analyze the impact of economics sectors. At the end of 1949, he used data from US Bureau of Labor Statistics to divide US economy into 500 sectors, and set up linear equations with 500 unknowns. Since he was not able to solve the equations on the computers of that time, he simplified the equations into a system with 42 unknowns. He spent months of coding and data inputting, and spent 56 hours in solving the linear equations on the computer[16]. Professor Leontief founded the branch of mathematical modeling in economics, and received the Nobel Prize in Economics in 1973. With the rapid development of computer hardware and software, the handling of large-scale matrices became very easy. They can be used easily in solving the input–output equations of the size mentioned in seconds. The iterative method he used influenced the PageRank algorithm by Google in 1998. It can be used to handle extra large-scale matrix computation problems.

Since matrices are widely used in the fields such as quantum mechanics, special theory of relativity, statistics, control theory, and so on, linear algebra studies helped extend theoretical research into various application fields. The emergence of digital computers also led to the development of algorithms such as Gaussian elimination methods, and high efficiency algorithms in matrix decompositions, such that linear algebra became the ubiquitous tool in system modeling and simulation fields.

### 1.2.2 Numerical linear algebra

Early progress in linear algebra was related to computational methods. They can be regarded as the prototype of numerical linear algebra. It is natural that numerical linear algebra nowadays is a well-developed branch in linear algebra, when computers are widely available. Numerical algebra is a steering force and contributed to a lot of developments in linear algebra and matrix theory.

Around 1950, British physicist Professor James Hardy Wilkinson (1919–1986) at National Physics Laboratory started research on numerical analysis. His works include [25, 26, 27]. Meanwhile, at the American Numerical Institute, American mathematician Professor George Elmer Forsythe (1917–1972) and his coworkers started numerical analysis research, especially the research in numerical linear algebra[6, 7].

American mathematician and computer scientist Cleve Barry Moler (1939–) and coworkers developed the most advanced packages such as EISPACK and LINPACK. In 1979, Professor Moler conceived and developed MATLAB (MATrix LABoratory). He and Jack Little in 1984 cofounded a company named MathWorks, devoting to the development of MATLAB and other related products. MATLAB language has become the top-selected computer language in many fields, and had great impact on many other computer languages. Its creators are very active in the research and engineering applications of linear algebra.

## 1.3 Outline of the book

In this book, we concentrate on the systematic introduction to the MATLAB-based solutions of linear algebra problems.

In Chapter 2, the fundamental concept of a matrix is presented first. Then the methods are presented to input ordinary matrices into MATLAB workspace, and we show how to input special matrices such as identity, random, diagonal, Hankel matrices, and so on, into MATLAB workspace. Also the input of arbitrary symbolic matrices and symbolic function matrices is introduced. Sparse matrices, their conversion, extraction, and graphical representation are also discussed in the chapter. Besides, some fundamental operations such as algebraic computations, complex matrix handling and matrix calculus are illustrated.

In Chapter 3, we focus on the fundamental topics of matrix analysis, including the computation of determinants, traces, ranks, and norms. Besides the topics of vector spaces, inverse matrices, and reduced row echelon form conversions, generalized inverse matrices are also discussed. The computation of characteristic polynomials and eigenvalues is discussed, and also the concepts of matrix polynomials are presented.

In Chapter 4, matrix transformation and decomposition are the main topics. The concept and properties of the similarity transform of matrices are introduced first, followed by the concept of orthogonal matrices. The methods of elementary row and column transformations, and the methods of pivoting in the evaluation of inverse matrices are summarized. The linear equation solution methods such as Gaussian elimination method, triangular factorization, and Cholesky decomposition of symmetrical matrices are introduced. Conversion methods to transform a matrix into companion, diagonal, and Jordan forms are presented. The concepts and computation of singular value decomposition and condition number of ordinary matrices are illustrated. Givens and Householder transformation methods are presented.

In Chapter 5, we concentrate on how to solve various matrix equations with MATLAB. Numerical and analytical solutions of simple linear algebraic equations are introduced, and according to the equation classifications, unique solutions, infinite number of solutions, and least squares solutions are discussed. Also the numerical and analytical solutions of other equations such as Lyapunov and Sylvester equations are discussed, and numerical methods on Riccati equations are presented. Attempts are made to find all possible solutions of various Riccati equations. A general-purpose solver for ordinary nonlinear matrix equations is presented, which is capable of finding unique and multiple solutions of nonlinear matrix equations of any complexity. Polynomial Diophantine equations and Bézout identities are analytically solved.

In Chapter 6, numerical and analytical operations of matrix functions are mainly discussed. The concept of matrix functions is given first, and commonly used matrix exponential functions are studied. Special matrix functions such as logarithmic, square root, and trigonometric functions are introduced. A general-purpose arbitrary matrix function evaluation program is provided, and attempts are also made to compute matrix powers such as $A^k$ and $k^A$.

In Chapter 7, applications of linear algebra in various fields are studied, including equation solution-based applications in electrical engineering, mechanics, and chemistry. In linear control system theory, qualitative analysis of system properties, transformations, and analytical solutions to linear differential equations are mainly summarized. In the field of digital image processing, image compression, scaling, and rotation are presented. In the field of graph theory, topics such as matrix expression of graphs, shortest path problem, and complicated block diagram simplifications are presented. The analytical and numerical solutions of various difference equations are introduced, and Markov chain modeling and computing are studied. In the fields of data fitting and analysis, the applications in linear regression, least-squares curve fitting, and principal component analysis are presented.

If the readers carefully studied the theoretical foundations and MATLAB applications presented in this book, it may also be helpful for them to enhance their ability in solving linear algebra related problems.

## 1.4 Exercises of the chapter

1.1 Convert the following simultaneous equations into matrix form. Find and validate solutions of the equations for

(1) $\begin{cases} x_1 + 2x_2 + 3x_3 = 0, \\ 4x_1 + x_2 - 2x_3 = 0, \\ 3x_1 + 2x_2 + x_3 = 2, \end{cases}$

(2) $\begin{cases} x_1 + x_2 + x_3 + x_4 + x_5 = 1, \\ 3x_1 + 2x_2 + x_3 + x_4 - 3x_5 = 2, \\ x_2 + 2x_3 + 2x_4 + 6x_5 = 3, \\ 5x_1 + 4x_2 + 3x_3 + 3x_4 - x_5 = 4, \\ 4x_2 + 3x_3 - 5x_4 = 12. \end{cases}$

1.2 Consider the given linear equations in matrix form. Write the simultaneous equations for

(1) $\begin{bmatrix} 16 & 2 & 3 & 13 \\ 5 & 11 & 10 & 8 \\ 9 & 7 & 6 & 12 \\ 4 & 14 & 15 & 2 \end{bmatrix} X = \begin{bmatrix} 1 \\ 3 \\ 4 \\ 7 \end{bmatrix};$

(2) $\begin{bmatrix} 2 & 9 & 4 & 12 & 5 & 8 & 6 \\ 12 & 2 & 8 & 7 & 3 & 3 & 7 \\ 3 & 0 & 3 & 5 & 7 & 5 & 10 \\ 3 & 11 & 6 & 6 & 9 & 9 & 1 \\ 11 & 2 & 1 & 4 & 6 & 8 & 7 \\ 5 & -18 & 1 & -9 & 11 & -1 & 18 \\ 26 & -27 & -1 & 0 & -15 & -13 & 18 \end{bmatrix} X = \begin{bmatrix} 1 & 9 \\ 5 & 12 \\ 4 & 12 \\ 10 & 9 \\ 0 & 5 \\ 10 & 18 \\ -20 & 2 \end{bmatrix}.$

# 2 Representation and fundamental computations of matrices

If one wants to solve linear algebra problems with computers, the first thing to do is to input matrices to computers, then matrix computations of any sort can be carried out, and various problems in linear algebra can be tackled. In Section 2.1, fundamental concepts of matrices are presented, and methods of inputting ordinary matrices to MATLAB workspace are explained with examples. In Section 2.2, the input of special matrices such as identity, random, diagonal, and Hankel matrices are presented. In Section 2.3, representations of arbitrary matrices under the symbolic framework are presented. Arbitrary matrices of symbolic functions are also addressed. In Section 2.4, the input, storage, and conversion of sparse matrices are presented, followed by the extracting of nonzero elements and graphical representation of sparse matrices. Fundamental matrix computations are presented in Section 2.5, with simple algebraic computation and manipulation of complex matrices. In Section 2.6, integrals and derivatives of matrix functions are presented.

## 2.1 Input of an ordinary matrix

Matrices are the basic mathematical elements in linear algebra. In this section, mathematical representation of a matrix is presented first, followed by the ways of inputting real and complex matrices in MATLAB.

**Definition 2.1.** A matrix with $n$ rows $m$ columns is also known as an $n \times m$ matrix, whose mathematical expression is

$$
A = \begin{bmatrix}
a_{11} & a_{12} & \cdots & a_{1m} \\
a_{21} & a_{22} & \cdots & a_{2m} \\
\vdots & \vdots & \ddots & \vdots \\
a_{n1} & a_{n2} & \cdots & a_{nm}
\end{bmatrix}. \tag{2.1.1}
$$

**Definition 2.2.** If $A$ is an $n \times m$ real matrix, it is denoted as $A \in \mathscr{R}^{n \times m}$. A complex matrix is denoted as $A \in \mathscr{C}^{n \times m}$.

**Definition 2.3.** An $n \times 1$ matrix is referred to as a column vector, while a $1 \times n$ matrix is also known as a row vector. A $1 \times 1$ matrix is a scalar.

One simple statement can be used to input the whole matrix into MATLAB environment. Simple commands can also be used to analyze behaviors of matrices. Double precision scheme is normally used to represent matrices in MATLAB, while symbolic data type can also be used. The matrix computations under the two data types are different in MATLAB, with the former corresponding to numerical solutions, while the

https://doi.org/10.1515/9783110666991-002

latter to symbolic computations, which may lead to analytical solutions. Examples are used here to demonstrate the input methods of given matrices.

**Example 2.1.** Input the following real matrix into MATLAB environment:

$$A = \begin{bmatrix} -1 & 5 & 4 & 6 \\ 0 & 2 & 4 & -2 \\ 4 & 0 & -2 & 5 \end{bmatrix}.$$

**Solutions.** The whole matrix can be entered into MATLAB environment in a row-by-row fashion, with spaces and commas to separate the elements in the same row, and they are exactly the same in MATLAB. Elements in different rows can be separated with semicolons or carriage returns. The above $A$ matrix can be input into MATLAB with the following statements:

```
>> A=[-1,5,4,6; 0,2,4,-2; 4,0,-2,5]
```

Normally, the default data type of the matrix is under double precision scheme. If one wants to convert it into symbolic data type, sym() function can be called to complete the conversion.

**Example 2.2.** Input the following complex matrix into MATLAB workspace:

$$A = \begin{bmatrix} -1 + 6j & 5 + 3j & 4 + 2j & 6 - 2j \\ j & 2 - j & 4 & -2 - 2j \\ 4 & -j & -2 + 2j & 5 - 2j \end{bmatrix}.$$

**Solutions.** To enter a complex matrix, the notations of i and j should be used. For instance, mathematical 6j term can be expressed as 6i or 6j and can be used directly. Therefore the following statements can be used to input the complex matrix, and the matrix $A$ is stored as a double precision matrix, while $B$ is stored as a symbolic one. To display the two matrices, it can easily be noticed that the formats are completely different, and it is suggested that the readers can input and display the two matrices, and experience the differences.

```
>> A=[-1+6i,5+3i,4+2i,6-2i; 1i,2-1i,4,-2-2i; 4,-1i,-2+2i,5-2i]
   B=sym(A)
```

**Example 2.3.** To input the quantity j in a complex matrix, it is suggested to use 1i or 1j, rather than i or j, otherwise, unpredictable results may be obtained. Sometimes, extra spaces in the statements may lead to errors. For instance, the following commands:

```
>> A=[-1 +6i,5+3i,4+2i,6-2i; 1i,2-1i,4,-2-2i; 4,-1i,-2+2i,5-2i]
```

may produce an error, since there is an extra space after the first element, and the mechanism of MATLAB may understand –1 as the first element, and +6i as the second element, so that five elements are recognized in the first row, while other rows contain four elements each, which may eventually lead to errors.

## 2.2 Input of special matrices

A simple one-by-one element input method is useful, however, in some situations, for instance, if one wants to input a 10 × 20 zero matrix, the one-by-one input method is awkward. Some special functions should be used to input special matrices. In this section, the input of some special matrices is presented.

### 2.2.1 Matrices of zeros, ones, and identity matrices

**Definition 2.4.** If a matrix is such that all its elements are zero, it is referred to as a matrix of zeros, while if all elements in the matrix are ones, it is referred to as a matrix of ones. If all elements on the main diagonal are ones, while others are zeros, the matrix is referred to as an identity matrix. The concept of a typical identity matrix is extended to include $m \times n$ matrices.

Matrices of zeros and ones, as well as identity matrices, can be generated in MATLAB with respectively the following functions:

$A$=zeros$(n)$, $B$=ones$(n)$, $C$=eye$(n)$ % generate $n \times n$ square matrix

$A$=zeros$(m,n)$; $B$=ones$(m,n)$; $C$=eye$(m,n)$ % generate $m \times n$ matrix

$A$=zeros(size$(B)$) % generate a matrix with the same size of $B$

**Example 2.4.** Generate first a 3×8 matrix of zeros, then generate an extended identity matrix of the same size.

**Solutions.** A 3 × 8 matrix $A$ of zeros can be generated first with the following statements, and based on it, an extended identity matrix $B$ with the same size of $A$ can be generated. It can be seen that the input of special matrices is simpler.

```
>> A=zeros(3,8)     % generate a matrix of zeros
   B=eye(size(A))  % generate identity matrix of the same size
```

The two matrices below can be entered into MATLAB workspace

$$A = \begin{bmatrix} 0 & 0 & 0 & 0 & 0 & 0 & 0 & 0 \\ 0 & 0 & 0 & 0 & 0 & 0 & 0 & 0 \\ 0 & 0 & 0 & 0 & 0 & 0 & 0 & 0 \end{bmatrix}, \quad B = \begin{bmatrix} 1 & 0 & 0 & 0 & 0 & 0 & 0 & 0 \\ 0 & 1 & 0 & 0 & 0 & 0 & 0 & 0 \\ 0 & 0 & 1 & 0 & 0 & 0 & 0 & 0 \end{bmatrix}.$$

**Example 2.5.** Functions zeros() and ones() can also be used to input multidimensional arrays. For instance, command zeros(3,4,5) generates a 3×4×5 three-dimensional array, with all its elements being zero.

### 2.2.2 Matrices with random elements

Matrices with random elements can be generated in two major ways. The first is to generate physical random numbers with electronic devices, while the other is to generate random numbers with mathematical formulas. The random numbers generated using the second approach are referred to as pseudorandom numbers. The benefit of pseudorandom numbers is that they can be repeated.

(1) Uniform distribution. If the random numbers in the matrix come from the uniform distribution in the interval $[0, 1]$, the numbers can be generated with the MATLAB function rand(), with the syntaxes

$A$=rand($n$), % generate $n \times n$ uniform distribution matrix

$A$=rand($n,m$), % generate $n \times m$ matrix of standard pseudorandom numbers

Function rand() can be used to generate multidimensional pseudorandom arrays, for instance, $A$=rand(5,4,6) can be used to generate three-dimensional random arrays.

More generally, if uniform random numbers in the interval $(a, b)$ are expected, standard uniform random numbers $V$ in the interval $(0, 1)$ should be generated first, then through the following commands, the expected random matrix $V_1$ can be generated:

$V$=rand($n,m$),      $V_1$=$a + (b - a)*V$

**Example 2.6.** Generate a set of 50 000 uniformly distributed random numbers in the interval $[-2, 3]$, compute their mean and draw the histogram.

**Solutions.** It is not difficult to generate the pseudorandom vector and compute the mean of it. It can be found that the mean obtained is $\bar{v} = 0.495245337225170$, which is close to the theoretical value of 0.5. Besides, the histogram of the data can be obtained, as shown in Figure 2.1. It can be seen that the random numbers in different subintervals are virtually the same.

```
>> N=50000; a=-2; b=3; v=a+(b-a)*rand(1,N);
   m=mean(v)
   c=linspace(-2,5,10); histogram(v,c)
```

As it was pointed out earlier, there are two advantages of the random numbers. First, they can be generated with a mathematical formula, and second, they can be repeated. How can we generate repeatable random numbers? Seeds are needed to

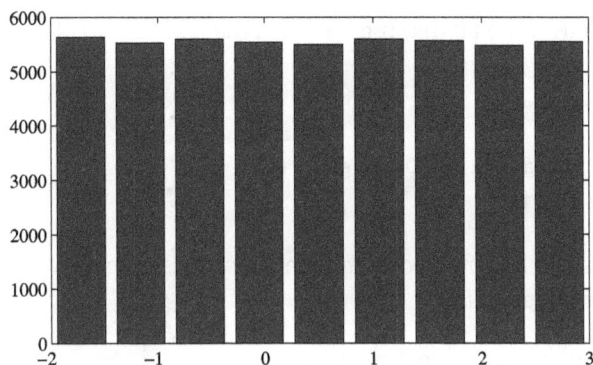

**Figure 2.1:** Histogram of pseudorandom numbers.

generate pseudorandom numbers. If the seeds are the same, the random numbers generated are also the same. MATLAB function rng() can be used to assign the seeds of the pseudorandom numbers. The parameters can be obtained with $s = $ rng, which allows for the relevant information in a random number generator to be retrieved. The variable $s$ can be stored, and next time, if repeated random numbers are to be generated again, $s$ can be loaded with rng(s), then repeated pseudorandom numbers can be generated.

**Example 2.7.** Generate two sets of identical uniform pseudorandom numbers and validate the results.

**Solutions.** If the same command rand() is called twice, the two sets of random numbers generated are definitely different. However, with the following commands, function rand() can be called twice, and the error err between the two sets shows zeros:

```
>> s=rng; a=rand(1,100);
   rng(s); b=rand(1,100); err=a-b
```

(2) Uniform integer matrix. With the MATLAB function randi(), a random integer matrix in the interval $[a, b]$ can be generated with the syntaxes

$A$=randi([a,b],[n,m]),  $B$=randi([a,b],n)

where $a$, $b$ are integers, and $a \leqslant b$, describing the lower and upper bounds of the interval. The former command generates an $n \times m$ rectangular matrix $A$, while the latter produces an $n \times n$ square matrix $B$.

**Example 2.8.** Generate a nonsingular $10 \times 10$ integer matrix of 0's and 1's.

**Solutions.** An infinite loop can be used to generate such a matrix. If a nonsingular matrix is found, break command can be used to terminate the loop. The function rank()

can be used to compute the rank of the matrix. If the rank of the matrix is 10, $A$ is a $10 \times 10$ nonsingular matrix.

```
>> while(1),
      A=randi([0,1],10); if rank(A)==10, break; end,
   end % found a nonsingular matrix
```

Different nonsingular matrices can be obtained each time the code is executed. For instance, the matrix below can be generated

$$
A = \begin{bmatrix}
1 & 0 & 1 & 1 & 0 & 0 & 0 & 1 & 0 & 1 \\
1 & 1 & 0 & 0 & 0 & 1 & 1 & 0 & 1 & 1 \\
1 & 0 & 0 & 1 & 1 & 0 & 1 & 1 & 0 & 0 \\
1 & 1 & 0 & 1 & 0 & 1 & 0 & 1 & 0 & 0 \\
0 & 1 & 0 & 0 & 0 & 1 & 0 & 0 & 0 & 0 \\
0 & 0 & 0 & 1 & 1 & 0 & 0 & 0 & 1 & 0 \\
0 & 0 & 0 & 0 & 1 & 1 & 1 & 0 & 0 & 0 \\
1 & 0 & 0 & 0 & 1 & 1 & 1 & 1 & 1 & 1 \\
0 & 0 & 0 & 0 & 0 & 1 & 0 & 0 & 1 & 1 \\
1 & 1 & 0 & 0 & 0 & 1 & 0 & 0 & 0 & 0
\end{bmatrix}.
$$

(3) Normal pseudorandom numbers. The pseudorandom numbers from the standard $N(0,1)$ normal distribution can be generated with function randn(), and the syntax is the same as that for rand(). The syntax $B$=randn(size($A$)) can also be used to generate a random matrix with the same size as that of matrix $A$.

If normal random numbers from $N(\mu,\sigma^2)$ are expected, the command $V$=randn($n,m$) can be used to generate the standard normal pseudorandom matrix $V$. Command $V_1=\mu + \sigma * V$ can then be used to generate the expected matrix.

In fact, apart from these two types of pseudorandom numbers, random numbers having other distributions can be generated with random() function.

### 2.2.3 Hankel matrix

Hankel matrix is a special matrix named after German mathematician Hermann Hankel (1839–1873), with the format described below.

**Definition 2.5.** The mathematical form of a Hankel matrix is

$$
H = \begin{bmatrix}
c_1 & c_2 & \cdots & c_m \\
c_2 & c_3 & \cdots & c_{m+1} \\
\vdots & \vdots & \ddots & \vdots \\
c_n & c_{n+1} & \cdots & c_{n+m-1}
\end{bmatrix}. \tag{2.2.1}
$$

If $n \to \infty$, an infinite-dimensional Hankel matrix can be defined. Hankel matrix is a symmetric matrix, and the elements on each back diagonal line are identical. MATLAB can only be used to handle finite-dimensional matrices.

Function `hankel()` can be used to generate two types of Hankel matrices

$$H_1 = \texttt{hankel}(c), \quad H_2 = \texttt{hankel}(c,r)$$

where, for given vectors $c$ and $r$, two different Hankel matrices can be generated:

$$H_1 = \begin{bmatrix} c_1 & c_2 & \cdots & c_n \\ c_2 & c_3 & \cdots & 0 \\ \vdots & \vdots & \ddots & \vdots \\ c_n & 0 & \cdots & 0 \end{bmatrix}, \quad H_2 = \begin{bmatrix} c_1 & c_2 & \cdots & \cdot \\ c_2 & c_3 & \cdots & \cdot \\ \vdots & \vdots & \ddots & \vdots \\ c_n & r_2 & \cdots & r_m \end{bmatrix}.$$

The first syntax is simple, and a square matrix can be generated, whose lower triangle elements are zeros. Other elements of the Hankel matrix can be constructed by the property of identical back diagonals. In the second syntax, it is required to have $c_n = r_1$, otherwise a warning message is given, and $r_1$ is removed. The Hankel matrix obtained is given in $H_2$. The upper-right corner element depends on the lengths of vectors $c$ and $r$. If $n = m$, a square matrix can be generated, with the upper-right corner element being $c_n$.

**Example 2.9.** Input the following two Hankel matrices $H_1$ and $H_2$ in MATLAB:

$$H_1 = \begin{bmatrix} 1 & 2 & 3 & 4 & 5 & 6 & 7 \\ 2 & 3 & 4 & 5 & 6 & 7 & 8 \\ 3 & 4 & 5 & 6 & 7 & 8 & 9 \end{bmatrix}, \quad H_2 = \begin{bmatrix} 1 & 2 & 3 \\ 2 & 3 & 0 \\ 3 & 0 & 0 \end{bmatrix}.$$

**Solutions.** Analyzing the given matrix $H_1$, the first column and last row vectors are needed, which are respectively $c=[1,2,3]$, $r=[3,4,5,6,7,8,9]$. The following commands can be used to input the vectors, so as to generate the required matrices. Note that the two vectors have the common element of "3". Matrix $H_2$ is simple.

```
>> c=[1 2 3]; r=[3 4 5 6 7 8 9];
   H1=hankel(c,r), H2=hankel(c) % Hankel matrices input
```

### 2.2.4 Diagonal matrices

Diagonal matrix is a special matrix whose elements on the main diagonal are arbitrary numbers, while other elements are all zero. The mathematical form of diagonal matrix is $\mathrm{diag}(\alpha_1, \alpha_2, \ldots, \alpha_n)$, where the diagonal matrix can be expressed as

$$\mathrm{diag}(\alpha_1, \alpha_2, \ldots, \alpha_n) = \begin{bmatrix} \alpha_1 & & & \\ & \alpha_2 & & \\ & & \ddots & \\ & & & \alpha_n \end{bmatrix}. \tag{2.2.2}$$

MATLAB function diag() can be used to generate a diagonal matrix with the following syntaxes:

A=diag(v), % form diagonal matrix

v=diag(A), % extract diagonal elements

A=diag(v,k), % generate matrix whose kth diagonal is vector v

If v is a matrix, the kth-diagonal elements are extracted.

The diag() function in MATLAB is a very interesting one since it can be used to input diagonal matrices, and extract diagonal elements. Also it can be used to assign a certain subdiagonal matrix. Examples are shown to demonstrate the use of this function.

**Example 2.10.** Input the three matrices into MATLAB workspace

$$A = \begin{bmatrix} 1 & 0 & 0 \\ 0 & 2 & 0 \\ 0 & 0 & 3 \end{bmatrix}, \quad B = \begin{bmatrix} 0 & 0 & 1 & 0 & 0 \\ 0 & 0 & 0 & 2 & 0 \\ 0 & 0 & 0 & 0 & 3 \\ 0 & 0 & 0 & 0 & 0 \\ 0 & 0 & 0 & 0 & 0 \end{bmatrix}, \quad C = \begin{bmatrix} 0 & 0 & 0 & 0 \\ 1 & 0 & 0 & 0 \\ 0 & 2 & 0 & 0 \\ 0 & 0 & 3 & 0 \end{bmatrix}.$$

**Solutions.** The vector $v = [1, 2, 3]$ can be input into MATLAB workspace, and $A$ matrix can be generated with diag(). Now let us consider matrix $B$. Since its second diagonal is vector $v$, $k$ should be set to 2, while in matrix $C$, the first diagonal under the main one is vector $v$, so $k$ should be set to $-1$. Therefore, the following statements can be used to generate the three matrices into MATLAB workspace:

```
>> v=[1 2 3]; A=diag(v), B=diag(v,2), C=diag(v,-1)
```

The following statements can be used to extract the diagonal elements from the above matrices, and column vectors can be created:

```
>> v1=diag(A), v2=diag(B,2), v3=diag(C,-1)
```

**Example 2.11.** An $n \times n$ Jordan matrix is expressed in the form as follows. Write a MATLAB function to construct such a matrix.

$$J = \begin{bmatrix} -\alpha & 1 & 0 & \cdots & 0 \\ 0 & -\alpha & 1 & \cdots & 0 \\ \vdots & \vdots & \vdots & \ddots & \vdots \\ 0 & 0 & 0 & \cdots & -\alpha \end{bmatrix}.$$

**Solutions.** A Jordan matrix can be expressed as the sum of two matrices, one is $\alpha I$, where $I$ is an identity matrix, and the other is a nilpotent matrix, whose first subdiagonal contains ones, while other entries are zeros. This type of matrix will be presented later. With such an idea, it is not difficult to construct an $n \times n$ Jordan matrix.

```
function J=jordan_matrix(alpha,n)
v=ones(1,n-1); J=alpha*eye(n)+diag(v,1);
```

**Definition 2.6.** If $A_1, A_2, \ldots, A_n$ are given matrices, the block-diagonal matrix is mathematically expressed as

$$A = \begin{bmatrix} A_1 & & & \\ & A_2 & & \\ & & \ddots & \\ & & & A_n \end{bmatrix}. \tag{2.2.3}$$

A MATLAB function `diagm()` can be written to establish block-diagonal matrix from given submatrices, with the syntax of $A=\text{diagm}(A_1, A_2, \ldots, A_n)$, where an arbitrary number of submatrices can be used as input arguments.

```
function A=diagm(varargin), A=[];
for i=1:length(varargin), A1=varargin{i}; % for block-diagonal matrix
    [n,m]=size(A); [n1,m1]=size(A1); A(n+1:n+n1,m+1:m+m1)=A1;
end
```

In fact, a MATLAB function `blkdiag()` can also be used to establish block-diagonal matrices, and it is even more powerful, since symbolic or sparse matrices can be handled by the function.

**Example 2.12.** Establish a block diagonal matrix from the known matrices

$$A = \begin{bmatrix} 8 & 1 & 6 \\ 3 & 5 & 7 \end{bmatrix}, \quad B = \begin{bmatrix} 1 & 3 \\ 4 & 2 \end{bmatrix}.$$

**Solutions.** Two ways can be used to construct block-diagonal matrices

```
>> A=[8,1,6; 3,5,7]; B=[1,3; 4,2];
   C=blkdiag(A,B), D=diagm(A,B)
```

and the results are identical.

$$C = D = \begin{bmatrix} 8 & 1 & 6 & 0 & 0 \\ 3 & 5 & 7 & 0 & 0 \\ 0 & 0 & 0 & 1 & 3 \\ 0 & 0 & 0 & 4 & 2 \end{bmatrix}.$$

If matrix $B$ is symbolic, the resulted $C$ is the expected symbolic matrix, while $D$ is a double precision one.

```
>> B=sym([1,3; 4,2]); C=blkdiag(A,B), D=diagm(A,B)
```

### 2.2.5 Hilbert matrix and its inverse

Hilbert matrix is a special matrix, named after German mathematician David Hilbert (1862–1943).

**Definition 2.7.** Hilbert matrices are a class of special matrices whose $(i,j)$th element is $h_{i,j} = 1/(i+j-1)$. The $n \times n$ Hilbert matrix is expressed as

$$
H = \begin{bmatrix}
1 & 1/2 & 1/3 & \cdots & 1/n \\
1/2 & 1/3 & 1/4 & \cdots & 1/(n+1) \\
\vdots & \vdots & \vdots & \ddots & \vdots \\
1/n & 1/(n+1) & 1/(n+2) & \cdots & 1/(2n-1)
\end{bmatrix}.
\tag{2.2.4}
$$

A Hilbert matrix can be generated from $A$=hilb$(n)$, where $n$ is the size of the matrix, so an $n \times n$ square matrix can be generated.

Large-scale Hilbert matrices are usually badly conditioned, and their direct inversion under double precision scheme may lead to a floating point overflow. An alternative function invhilb() is presented in MATLAB to compute an inverse Hilbert matrix, with the syntax of $B$=invhilb$(n)$. Since a Hilbert matrix approaches a singular one, Symbolic Math Toolbox is recommended for dealing with such matrices. When numerical solutions are found, validation process should be invoked.

**Example 2.13.** If a rectangular Hilbert matrix is expected, the most effective way is to generate a square Hilbert matrix, then extract its submatrix. Of course, the following commands can be used to generate directly rectangular Hilbert matrix:

```
>> m=10000; n=5;
   [x,y]=meshgrid(1:m,1:n); H=1./(x+y-1);
```

### 2.2.6 Companion matrices

**Definition 2.8.** Assuming that a monic polynomial is given by

$$
p(s) = s^n + a_1 s^{n-1} + a_2 s^{n-2} + \cdots + a_{n-1}s + a_n,
\tag{2.2.5}
$$

a companion matrix is defined as

$$
A_c = \begin{bmatrix}
-a_1 & -a_2 & \cdots & -a_{n-1} & -a_n \\
1 & 0 & \cdots & 0 & 0 \\
0 & 1 & \cdots & 0 & 0 \\
\vdots & \vdots & \ddots & \vdots & \vdots \\
0 & 0 & \cdots & 1 & 0
\end{bmatrix} .
\tag{2.2.6}
$$

Companion matrices can be generated with $A_c$=compan$(a)$, where $a$ is a vector of polynomial coefficients in the descending order of $s$. A nonmonic polynomial can be converted first to a monic one automatically.

**Example 2.14.** For the given polynomial $P(s) = 2s^4 + 4s^2 + 5s + 6$, write down its companion matrix.

**Solutions.** The polynomial coefficients can be input first as a vector, then a companion matrix can be created with the following statements, and assigned to matrix $A$:

```
>> P=[2 0 4 5 6]; A=compan(P) % companion matrix from polynomial
```

The following companion matrix can be established:

$$
A = \begin{bmatrix}
0 & -2 & -2.5 & -3 \\
1 & 0 & 0 & 0 \\
0 & 1 & 0 & 0 \\
0 & 0 & 1 & 0
\end{bmatrix} .
$$

### 2.2.7 Vandermonde matrices

Vandermonde matrices are special matrices named after French mathematician Alexandre-Théophile Vandermonde (1735–1796).

**Definition 2.9.** For a given vector $c = [c_1, c_2, \ldots, c_n]$, a matrix whose $(i,j)$th element satisfies $v_{i,j} = c_i^{n-j}$, $i,j = 1, 2, \ldots, n$, is referred to as a Vandermonde matrix. Its mathematical form is

$$
V = \begin{bmatrix}
c_1^{n-1} & c_1^{n-2} & \cdots & c_1 & 1 \\
c_2^{n-1} & c_2^{n-2} & \cdots & c_2 & 1 \\
\vdots & \vdots & \ddots & \vdots & \vdots \\
c_n^{n-1} & c_n^{n-2} & \cdots & c_n & 1
\end{bmatrix} .
\tag{2.2.7}
$$

The command $V$=vander$(c)$ can be used to generate a Vandermonde matrix, from the given vector $c$.

**Example 2.15.** Establish the following Vandermonde matrix in MATLAB workspace:

$$V = \begin{bmatrix} 1 & 1 & 1 & 1 & 1 \\ 1 & 2 & 3 & 4 & 5 \\ 1 & 4 & 9 & 16 & 25 \\ 1 & 8 & 27 & 64 & 125 \\ 1 & 16 & 81 & 256 & 625 \end{bmatrix}.$$

**Solutions.** The required matrix is in different form from that in (2.2.7). It is the transpose of the standard Vandermonde matrix. A vector $c=[1,2,3,4,5]$ should be entered into MATLAB first, then a 90° counterclockwise rotation can be used to establish the expected $V$ matrix.

```
>> c=[1,2,3,4,5]; V=vander(c); V=rot90(V) % matrix rotation
```

### 2.2.8 Some commonly used test matrices

**Definition 2.10.** If an $n \times n$ matrix is such that the sums of elements in each row, column and diagonal, as well as back diagonal, are the same constant, the matrix is referred to as a magic matrix.

**Definition 2.11.** If a matrix is such that the elements in the first row and first column are all ones and the rest are recursively computed from $a_{ij} = a_{i,j-1} + a_{i-1,j}$, the matrix is referred to as a Pascal matrix.

**Definition 2.12.** If all the elements in each diagonal are the same, the matrix is referred to as a Toeplitz matrix. If the first row $r$ and first column $c$ are known, the following Toeplitz matrix can be established:

$$T = \begin{bmatrix} c_1 & r_2 & r_3 & \cdots & r_m \\ c_2 & c_1 & r_2 & \cdots & r_{m-1} \\ c_3 & c_2 & c_1 & \cdots & r_{m-2} \\ \vdots & \vdots & \vdots & \ddots & \vdots \\ c_n & c_{n-1} & c_{n-2} & \cdots & \end{bmatrix}. \tag{2.2.8}$$

Pascal matrix is a special matrix named after French mathematician Blaise Pascal (1623–1662), while Toeplitz matrix is named after German mathematician Otto Toeplitz (1881–1940).

Some MATLAB functions are provided in generating the special matrices, for example,

$M=$magic$(n)$, $P=$pascal$(n)$, $T=$toeplitz$(r,c)$.

**Example 2.16.** Generate $4 \times 4$ magic, Pascal and Toeplitz matrices.

**Solutions.** The following MATLAB statements can be used to generate the expected matrices:

```
>> A=magic(4), B=pascal(4), C=toeplitz(1:4)
```

and the results obtained are

$$A = \begin{bmatrix} 16 & 2 & 3 & 13 \\ 5 & 11 & 10 & 8 \\ 9 & 7 & 6 & 12 \\ 4 & 14 & 15 & 1 \end{bmatrix}, \quad B = \begin{bmatrix} 1 & 1 & 1 & 1 \\ 1 & 2 & 3 & 4 \\ 1 & 3 & 6 & 10 \\ 1 & 4 & 10 & 20 \end{bmatrix}, \quad C = \begin{bmatrix} 1 & 2 & 3 & 4 \\ 2 & 1 & 2 & 3 \\ 3 & 2 & 1 & 2 \\ 4 & 3 & 2 & 1 \end{bmatrix}.$$

It can be seen from the upper-triangular part of the Pascal matrix that each back diagonal consists of binomial coefficients.

Wilkinson matrices are a set of testing matrices named after British mathematician James Hardy Wilkinson (1919–1986).

**Definition 2.13.** Wilkinson matrix is a tri-diagonal matrix, whose first and −1 diagonal elements are ones, while in the $(2m+1) \times (2m+1)$ Wilkinson matrix, the main diagonal elements are

$$v = [m, m-1, \ldots, 2, 1, 0, 1, 2, \ldots, m-1, m], \tag{2.2.9}$$

and for the $(2m) \times (2m)$ matrix the main diagonal elements are

$$v = [(m+1)/2, \ldots, 3/2, 1/2, 1/2, 3/2, \ldots, (m+1)/2]. \tag{2.2.10}$$

Wilkinson matrix can be generated with $W$=wilkinson($n$), and if a symbolic Wilkinson matrix is expected, the command sym() can be used to convert it through data type conversion.

**Example 2.17.** Generate respectively the $5 \times 5$ and $6 \times 6$ Wilkinson matrices.

**Solutions.** The following MATLAB statements can be used to generate directly the expected Wilkinson matrices, where the first is a double precision one, while the second is converted to the symbolic form.

```
>> W1=wilkinson(5), W2=sym(wilkinson(6))
```

The results obtained are

$$W_1 = \begin{bmatrix} 2 & 1 & 0 & 0 & 0 \\ 1 & 1 & 1 & 0 & 0 \\ 0 & 1 & 0 & 1 & 0 \\ 0 & 0 & 1 & 1 & 1 \\ 0 & 0 & 0 & 1 & 2 \end{bmatrix}, \quad W_2 = \begin{bmatrix} 5/2 & 1 & 0 & 0 & 0 & 0 \\ 1 & 3/2 & 1 & 0 & 0 & 0 \\ 0 & 1 & 1/2 & 1 & 0 & 0 \\ 0 & 0 & 1 & 1/2 & 1 & 0 \\ 0 & 0 & 0 & 1 & 3/2 & 1 \\ 0 & 0 & 0 & 0 & 1 & 5/2 \end{bmatrix}.$$

A MATLAB function `gallery()` is provided for generating special matrices, with the syntax

`A=gallery(matrix type,n,other parameters)`

where "matrix type" can be selected as `'binomial'`, `'cauchy'`, `'chebspec'`, and many other options can be found with the `doc gallery` command.

## 2.3 Symbolic matrix input

It has been explained that for a given numerical matrix $A$, the $B$=`sym`$(A)$ command can be used to convert it to symbolic data types. Therefore, all the numerical matrices can be converted into symbolic ones, to increase the accuracy in matrix computation. On the other hand, a symbolic matrix $B$ with numbers only can be converted to a double precision $A_1$ with $A_1$=`double`$(B)$.

If there are symbolic variables in the matrix, `double()` function cannot be used in the conversion; an error message will be displayed.

In this section, the creation of arbitrary symbolic matrices and matrix functions is also discussed.

### 2.3.1 Input of special symbolic matrices

Most of the special matrix functions discussed earlier can be used to handle directly symbolic matrices. Alternatively, double precision matrices can be converted to symbolic ones with function `sym()`. For instance, `eye(5)` function can be used to generate a double precision matrix, then `sym()` function can be used to convert it to a symbolic identity matrix. Alternatively, command `eye(sym(5))` can be used.

**Example 2.18.** Generate a companion matrix from the given polynomial

$$P(\lambda) = a_1\lambda^7 + a_2\lambda^6 + a_3\lambda^5 + \cdots + a_6\lambda^2 + a_7\lambda + a_8.$$

**Solutions.** A symbolic vector $a$ can be expressed first with the following commands, then `compan()` function can be used to generate the companion matrix. Since the original polynomial is not monic, automatic conversion is made in `compan()` function to generate the correct companion matrix

```
>> syms a1 a2 a3 a4 a5 a6 a7 a8;
   a=[a1 a2 a3 a4 a5 a6 a7 a8];
   A=compan(a)              % build up the companion matrix
```

and the companion matrix obtained is

$$
A = \begin{bmatrix}
-a_2/a_1 & -a_3/a_1 & -a_4/a_1 & -a_5/a_1 & -a_6/a_1 & -a_7/a_1 & -a_8/a_1 \\
1 & 0 & 0 & 0 & 0 & 0 & 0 \\
0 & 1 & 0 & 0 & 0 & 0 & 0 \\
0 & 0 & 1 & 0 & 0 & 0 & 0 \\
0 & 0 & 0 & 1 & 0 & 0 & 0 \\
0 & 0 & 0 & 0 & 1 & 0 & 0 \\
0 & 0 & 0 & 0 & 0 & 1 & 0
\end{bmatrix}.
$$

### 2.3.2 Generating arbitrary constant matrices

The sym() function in MATLAB can be used to convert a matrix into symbolic form, also it can be used to declare a matrix having arbitrary elements $a_{ij}$, with the syntax

A=sym('a%d%d',[n,m]), % where %d%d represents double subscripts

The above sentence can be used to generate an arbitrary $n \times m$ matrix whose elements are $a_{ij}, i = 1, 2, \ldots, n, j = 1, 2, \ldots, m$. If $m$ is not assigned, an arbitrary $n \times n$ square matrix can be generated.

Similarly, the following commands can be used to generate arbitrary row vector $v_1$ and column vector $v_2$:

v₁=sym('a',[1,n]), v₂=sym('a',[n,1]), % first method

v₁=sym('a%d',[1,n]), v₂=sym('a%d',[n,1]), % second method

**Example 2.19.** Generate the following matrices and vector in MATLAB:

$$
A = \begin{bmatrix}
a_{11} & a_{12} & a_{13} & a_{14} \\
a_{21} & a_{22} & a_{23} & a_{24} \\
a_{31} & a_{32} & a_{33} & a_{34} \\
a_{41} & a_{42} & a_{43} & a_{44}
\end{bmatrix}, \quad
B = \begin{bmatrix}
f_{11} & f_{12} \\
f_{21} & f_{22} \\
f_{31} & f_{32} \\
f_{41} & f_{42}
\end{bmatrix}, \quad
v = \begin{bmatrix}
v_1 \\
v_2 \\
v_3 \\
v_4
\end{bmatrix}.
$$

**Solutions.** It can be seen that double subscript elements are required in matrices $A$ and $B$, therefore, the string '%d%d' should be used, and the variable names $a$ and $f$ should be used for them. The following commands can be used. On the other hand, the input of vector $v$ is simpler and more straightforward:

```
>> A=sym('a%d%d',4), B=sym('f%d%d',[4,2]), v=sym('v',[4,1])
```

If one wants to further declare certain properties, the functions assume() and assumeAlso() can be used. For instance, the following statements can be employed to further assign the properties of the matrix

```
>> assume(A,'real'); assumeAlso(A,'integer') % assign other properties
```

where other properties such as 'integer', 'rational' and 'positive' can also be assigned. By default a matrix is complex.

**Example 2.20.** Generate again the companion matrix in Example 2.18.

**Solutions.** It is not necessary to declare the vector elements one-by-one in Example 2.18, an arbitrary vector with the command $a$=sym('a',[1,8]) can be used. Even larger matrices can be generated easily.

```
>> a=sym('a',[1,20]); A=compan(a) % set a companion matrix
```

**Example 2.21.** Generate a $3 \times 6$ arbitrary matrix with real rational elements.

**Solutions.** It is not difficult to generate the needed matrix by

```
>> A=sym('a%d%d',[3,6]);
   assume(A,'real'), assumeAlso(A,'rational')
```

### 2.3.3 Arbitrary matrix function input

Function sym() can be used to generate an arbitrary constant matrix, and based on the function, a MATLAB function can be written to generate matrix functions. For instance, a matrix function $M = \{m_{ij}(x,y)\}$ can be generated.

```
function A=any_matrix(nn,sA,varargin) % generate arbitrary matrix
v=varargin; n=nn(1); if length(nn)==1, m=n; else, m=nn(2); end
s=''; k=length(v); K=0; if n==1 || m==1, K=1; end
if k>0, s='('; for i=1:k, s=[s ',' char(v{i})]; end
s(2)=[]; s=[s ')']; end
for i=1:n, for j=1:m % declare elements one-by-one
    if K==0, str=[sA int2str(i),int2str(j)];
    else, str=[sA int2str(i*j)]; end
    eval(['syms ' str s]); eval(['A(i,j)=' str ';']); % matrix elements
end, end
```

The syntax of the function is

$$A(x_1,x_2,\dots,x_k)=\text{any\_matrix}([n,m],'a',x_1,x_2,\dots,x_k)$$

where $n$ and $m$ are numbers of rows and columns. If only one argument $n$ is used, a square matrix can be generated. The arguments $x_1, x_2, \dots, x_k$ are symbolic variables, and letter 'a' can also be used for the others. The commands above generate the following matrix:

$$A = \begin{bmatrix} a_{11}(x_1, x_2, \ldots, x_k) & \cdots & a_{1m}(x_1, x_2, \ldots, x_k) \\ \vdots & \ddots & \vdots \\ a_{n1}(x_1, x_2, \ldots, x_k) & \cdots & a_{nm}(x_1, x_2, \ldots, x_k) \end{bmatrix}.$$

**Example 2.22.** If the symbolic variables $x$, $y$ and $t$ are declared, the following statements can be used to generate matrix functions:

```
>> syms x y t; clear v X
   v(x,y)=any_matrix([3,1],'a',x,y), X(t)=any_matrix(3,'m',t)
```

The generated arbitrary matrix and vector functions are

$$v(x,y) = \begin{bmatrix} a_1(x,y) \\ a_2(x,y) \\ a_3(x,y) \end{bmatrix}, \quad X(t) = \begin{bmatrix} m_{11}(t) & m_{12}(t) & m_{13}(t) \\ m_{21}(t) & m_{22}(t) & m_{23}(t) \\ m_{31}(t) & m_{32}(t) & m_{33}(t) \end{bmatrix}.$$

Note that the clear command is essential to avoid errors, since before executing the remaining commands, the variables $v$ and $X$ may have been defined as other functions. With clear command, they can be redefined as the new matrix functions.

## 2.4 Input of sparse matrices

In many applications, large-scale matrices are often used, most of whose elements are zeros, with very few nonzero elements. Such matrices are referred to as sparse matrices. If suitable algorithms are adopted, the efficiency of sparse matrix computation is higher than with ordinary matrices. Sparse matrix input is allowed in MATLAB, and they are very useful in certain applications.

Function $A$=sparse($p$,$q$,$w$) can be used to input a matrix into MATLAB, where $p$ and $q$ are the vectors storing the row and column numbers of nonzero elements, while $w$ is the vector storing the nonzero elements. The lengths of the three vectors are the same, otherwise error information is given.

For the double precision data structure, storing an element requires 8 bytes. Therefore an $n \times n$ square matrix requires $8n^2$ bytes to store it. For sparse matrices, each nonzero element requires 8 bytes to store the number, and an extra of 16 bytes are used to store the row and column positions. Therefore a total of $24m$ bytes are required to store all the nonzero elements, where $m$ is the number of nonzero elements. It can be seen that when $m = n^2/3$, storage requirements for double precision and sparse matrices are identical, meaning that if more than $n^2/3$ elements are zero, it is more economical to store the matrix in the sparse matrix format.

Regular matrix **B** and sparse matrix **A** can be converted, also one can judge if a matrix **A** sparse or not. The following functions can be used for the conversion and judgement:

**B**=full(**A**),  **A**=sparse(**B**), key=issparse(**A**)

The number of nonzero elements in regular and sparse matrices can directly be measured with $n$=nnz(**A**). If one wants to extract all the nonzero elements, the command $v$=nonzeros(**A**) can be used, and the elements are arranged according to their columnwise positions.

The positions of the nonzero elements in a sparse matrix **A** can be displayed with the command spy(**A**). The positions can also be displayed even if **A** is not a sparse matrix. Matrix **A** can also be visualized with the function spy().

**Example 2.23.** If the simple beam structure in Example 1.5 is represented with a sparse matrix, analyze the sparsity of the matrix.

**Solutions.** A double precision data type can be used to represent the matrix. The following statements can be used to input the original matrix again, and it can be seen that the number of nonzero elements is 30. If the matrix is converted into a sparse matrix, the distribution of nonzero elements can be displayed as shown in Figure 2.2. The nonzero elements are denoted by dots, and those left blank are zeros.

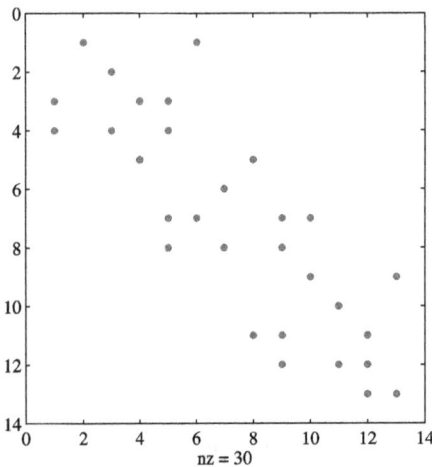

**Figure 2.2:** Distribution of nonzero elements.

```
>> alpha=1/sqrt(2);
   A=[0   1  0  0  0  -1  0  0  0  0  0  0  0
      0  0   1  0  0  0  0  0  0  0  0  0  0
      alpha  0  0  -1  -alpha  0  0  0  0  0  0  0  0
```

```
 alpha  0  1  0  alpha  0  0  0  0  0  0  0  0
 0  0  0  1  0  0  0  -1  0  0  0  0  0
 0  0  0  0  0  0  1  0  0  0  0  0  0
 0  0  0  0  alpha  1  0  0  -alpha  -1  0  0  0
 0  0  0  0  1  0  1  0  alpha  0  0  0  0
 0  0  0  0  0  0  0  0  0  1  0  0  -1
 0  0  0  0  0  0  0  0  0  0  1  0  0
 0  0  0  0  0  0  0  1  -alpha  0  0  -alpha  0
 0  0  0  0  0  0  0  0  alpha  0  1  alpha  0
 0  0  0  0  0  0  0  0  0  0  0  alpha  1];
nnz(A), B=sparse(A), spy(B)
```

The command whos can be used to display the information about the storage of the two variables:

```
>> whos A B    % display results are as follows
```

| Name | Size | Bytes | Class | Attributes |
|------|------|-------|-------|------------|
| A | 13x13 | 1352 | double | |
| B | 13x13 | 592 | double | sparse |

It can be seen that the sparsity in the example is not very high. It can be expected that in complex beam structures, for instance, if there are 1000 beams, the established matrices should be represented in the sparse matrix format, where storage space can be significantly reduced.

Some special sparse matrices are supported in MATLAB. For instance, speye() function can be used to input an identity matrix, while functions sprandn() and sprandsym() can be used to generate random sparse matrices, whose elements are normally distributed. The latter generates a symmetric matrix. The syntaxes of the functions are

$E$=speye([$n,m$]),  $A$=sprandn($n,m$,sparsity),  $B$=sprandn($n$,sparsity)

**Example 2.24.** Generate a $50\,000 \times 50\,000$ random matrix with sparsity of 0.0002%, and observe the distribution of nonzero elements. Convert the matrix into a regular one and test the storage.

**Solutions.** The following statements can be used to generate the required sparse matrix, and the 5000 nonzero elements are illustrated as shown in Figure 2.3. If one tries to convert the original matrix into a regular one, the error message of "Requested $50\,000\times50\,000$ (18.6 GB) array exceeds maximum array size preference" is shown, since such a large regular matrix cannot be stored in MATLAB. With the sparse matrix format, even if the sparsity is increased to 1%, the matrix can still be stored.

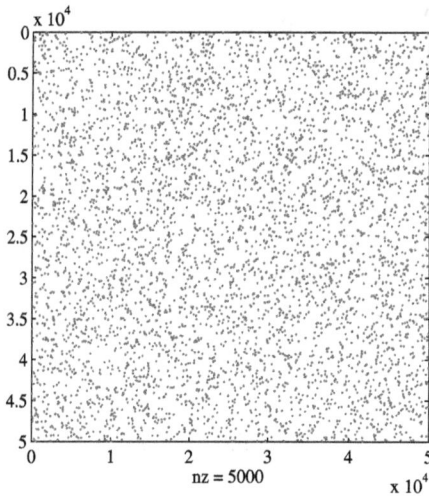

**Figure 2.3:** Distribution of nonzero elements in the sparse matrix.

```
>> A=sprandn(50000,50000,0.0002/100); spy(A), nnz(A)
   whos A, B=full(A)
```

| Name | Size | Bytes | Class | Attributes |
|------|------|-------|-------|------------|
| A | 50000x50000 | 480008 | double | sparse |

Efficient sparse matrix multiplication is still a challenging problem. If the problems cannot be solved effectively, the sparse matrices should first be converted to regular matrices, and the benefits in sparse matrices cannot be used to reduced the computation loads.

## 2.5 Fundamental matrix computations

Matrix computations are fundamental in linear algebra. In this section, complex matrix processing is introduced first, then transposition and rotation of matrices are introduced, Kronecker sum and product are presented, and algebraic computation of matrices is addressed.

### 2.5.1 Handling complex matrices

MATLAB can be used to directly represent complex matrices. Assuming that a complex matrix is given in **Z**, the following simple functions can be used for conversion:

(1) Complex conjugate matrix is obtained with the command $Z_1$=conj(Z).
(2) Real and imaginary parts are extracted with R=real(Z),I=imag(Z).
(3) Magnitude and phase can be extracted with A=abs(Z),P=angle(Z), where the phase is in radians.

In fact, variable Z is not restricted to matrices, it can also be a multidimensional array or a symbolic expression.

**Example 2.25.** For the following complex matrix A, find its real and imaginary parts, and compute its complex conjugate:

$$A = \begin{bmatrix} 1+9j & 2+8j & 3+7j \\ 4+6j & 5+5j & 6+4j \\ 7+3j & 8+2j & j \end{bmatrix}.$$

**Solutions.** The complex matrix can be input into MATLAB environment, and the related matrices can be obtained with

```
>> A=[1+9i,2+8i,3+7j; 4+6j 5+5i,6+4i; 7+3i,8+2j 1i];
   R=real(A), I=imag(A), C=conj(A)
```

The real and imaginary parts, as well as complex conjugate, are respectively:

$$R = \begin{bmatrix} 1 & 2 & 3 \\ 4 & 5 & 6 \\ 7 & 8 & 0 \end{bmatrix}, \quad I = \begin{bmatrix} 9 & 8 & 7 \\ 6 & 5 & 4 \\ 3 & 2 & 1 \end{bmatrix}, \quad C = \begin{bmatrix} 1-9j & 2-8j & 3-7j \\ 4-6j & 5-5j & 6-4j \\ 7-3j & 8-2j & -j \end{bmatrix}.$$

### 2.5.2 Transposition and rotation of matrices

In matrix manipulations, matrix transposition, flipping and rotation are involved. The manipulations are summarized below:
   (1) Matrix transpose. Two types of matrix transposes are supported:

**Definition 2.14.** Assuming that $A \in \mathscr{C}^{n \times m}$, the elements in a transpose matrix B are defined as $b_{ji} = a_{ij}$, $i = 1, \ldots, n$, $j = 1, \ldots, m$, such that $B \in \mathscr{C}^{m \times n}$, denoted as $B = A^T$. This transpose is also known as a direct transpose.

**Definition 2.15.** If a matrix A contains complex elements, the elements in a transpose matrix B may also be defined as $b_{ji} = a_{ij}^H$, $i = 1, \ldots, n$, $j = 1, \ldots, m$, i. e., we compute the transpose first, then take the complex conjugate of the resulting elements. This type of transpose is also known as Hermitian transpose, denoted as $B = A^H$.

MATLAB command B=A' can be use to compute the Hermitian transpose of A, and the direct transpose can be evaluated with C=A.'.

**Definition 2.16.** If a complex matrix satisfies $A = A^H$, then $A$ is referred to as a Hermitian matrix.

**Definition 2.17.** If a complex matrix is such that $A = -A^H$, then $A$ is also known as a skew-Hermitian matrix.

**Theorem 2.1.** *The transposes of the product of complex matrices $A$ and $B$ satisfy*

$$(AB)^T = B^T A^T, \quad (AB)^H = B^H A^H \qquad (2.5.1)$$

**Example 2.26.** Consider the complex matrix $B$ in Example 2.25. Compute its direct and Hermitian transposes.

**Solutions.** After we input matrix $B$ into MATLAB environment, the two transposes can be obtained with the following statements:

```
>> B=[1+9i,2+8i,3+7j; 4+6j 5+5i,6+4i; 7+3i,8+2j 1i];
   B1=B', B2=B.' % two transposes
```

The obtained Hermitian and direct transposes are

$$B_1 = \begin{bmatrix} 1-9j & 4-6j & 7-3j \\ 2-8j & 5-5j & 8-2j \\ 3-7j & 6-4j & -1j \end{bmatrix}, \quad B_2 = \begin{bmatrix} 1+9j & 4+6j & 7+3j \\ 2+8j & 5+5j & 8+2j \\ 3+7j & 6+4j & 1j \end{bmatrix}.$$

(2) **Matrix flips.** A set of special MATLAB commands is provided for matrix flipping, where left–right flipping can be made with $B=\texttt{fliplr}(A)$, matrix $A$ can be left–right flipped and assigned to $B$, i. e., $b_{ij} = a_{i,n+1-j}$. It can be seen that the function is equivalent to $B=A(:,\texttt{end}:-1:1)$. The command $C=\texttt{flipud}(A)$ can be used to up–down flip matrix $A$ and assign the result to $C$, i. e., $c_{ij} = a_{m+1-i,j}$. The up–down flipping is equivalent to $C=A(\texttt{end}:-1:1,1)$.

Similar to left–right flipping command $B=A(:,\texttt{end}:-1:1)$, flipping can also be made such that any local arrangement or any row and column assignment are achieved.

**Example 2.27.** Make a random row arrangement to the following matrix:

$$A = \begin{bmatrix} 6 & 1 & 1 & 2 & 5 \\ 6 & 3 & 3 & 5 & 0 \\ 2 & 0 & 4 & 2 & 4 \end{bmatrix}.$$

**Solutions.** With the function call $A=\texttt{randperm}(n)$, the integers $1, 2, \ldots, n$ can be arranged in a random manner, to rearrange the rows randomly in matrix $A$.

```
>> A=[6,1,1,2,5; 6,3,3,5,0; 2,0,4,2,4];
   ii=randperm(3), B=A(ii,:)
```

The random vector obtained is $[2, 1, 3]$, and the final result is

$$B = \begin{bmatrix} 6 & 3 & 3 & 5 & 0 \\ 6 & 1 & 1 & 2 & 5 \\ 2 & 0 & 4 & 2 & 4 \end{bmatrix}.$$

(3) Matrix rotation. MATLAB function $D$=rot90($A$) can be used to rotate matrix $A$ by $90°$ in the counterclockwise direction, and then assign to matrix $D$, i. e., $d_{ij} = a_{j,n+1-i}$. Function $E$=rot90($A$,$k$) can be used to rotate matrix $A$ by $90k°$ in the counterclockwise direction, and assign to $E$, where $k$ is an integer.

**Example 2.28.** For the matrix $A$ below, rotate it by $90°$ in the clockwise direction and obtain the new matrix $B$ as follows:

$$A = \begin{bmatrix} 1 & 2 & 3 \\ 4 & 5 & 6 \\ 7 & 8 & 0 \end{bmatrix}, \quad B = \begin{bmatrix} 7 & 4 & 1 \\ 8 & 5 & 2 \\ 0 & 6 & 3 \end{bmatrix}.$$

**Solutions.** The standard rot90() function can be used to rotate the matrix in the counterclockwise direction. To rotate the matrix by $90°$ in the clockwise direction, there are two ways: the first is to use rot90() and let $k = -1$, while the other is to let $k = 3$, i. e., rotate by $270°$ in the counterclockwise direction. The following statements can be used to implement the two methods, with $B_1 = B_2$.

```
>> A=[1 2 3; 4 5 6; 7 8 0]; B1=rot90(A,-1), B2=rot90(A,3)
```

### 2.5.3 Algebraic computation of matrices

Algebraic computations of matrices are the most widely used matrix handling tasks in scientific computation. In this section, definitions of algebraic computations are introduced, and MATLAB implementation of algebraic computation is illustrated.

**Definition 2.18.** Finitely many addition, subtraction, product, division and power computations between variables are referred to as algebraic computations.

The following algebraic computations are supported in MATLAB:

(1) **Addition and subtraction.** Assuming that there are two matrices, $A$ and $B$, in MATLAB workspace, the commands $C = A + B$ and $C = A - B$ can be used to evaluate matrix sum and difference. If the sizes of $A$ and $B$ are the same, the sums or differences of the corresponding terms in matrices $A$ and $B$ can be computed, and the result is returned in variable $C$.

Two special cases are considered in MATLAB:

(a) If one of the matrices is a scalar, it can be added or subtracted from all the elements of the other matrix.

(b) If $A \in \mathscr{C}^{n\times m}$ and $B$ is an $n\times 1$ column vector, or a $1\times m$ row vector, earlier versions of MATLAB give error messages, while in the new versions, the vector can be added to or subtracted from all the rows or columns of the other matrix.

In other situations, error messages will automatically be given, indicating that the two matrices do not match.

**Example 2.29.** A matrix and a vector are given below. What is the sum $A + B$ if

$$A = \begin{bmatrix} 5 \\ 6 \end{bmatrix} \quad \text{and} \quad B = \begin{bmatrix} 1 & 2 \\ 3 & 4 \end{bmatrix}?$$

**Solutions.** The two quantities cannot be added up in mathematics. In earlier versions of MATLAB, error messages are given. The following attempts can be made to evaluate the sum and difference of the matrices in the new versions:

```
>> A=[5;6]; B=[1 2; 3 4]; C=A+B, D=B-A'
```

An alternative "addition" is supported in MATLAB, if $A$ is a column vector: it can be added to all the columns of matrix $B$. The matrix sum obtained is given below. Also, since $A^T$ is a row vector, matrix $D$ equals to the matrix, where from each row of $B$ one subtracts the row vector $A^T$:

$$C = \begin{bmatrix} 6 & 7 \\ 9 & 10 \end{bmatrix}, \quad D = \begin{bmatrix} -4 & -4 \\ -2 & -2 \end{bmatrix}.$$

(2) Matrix multiplication. The product of two matrices is defined below.

**Definition 2.19.** Given two matrices $A$ and $B$, where the number of columns of $A$ and the number of rows of $B$ are the same, or at least one of them is a scalar, the matrices $A$ and $B$ are compatible.

**Definition 2.20.** Assuming that matrices $A \in \mathscr{C}^{n\times m}$ and $B \in \mathscr{C}^{m\times r}$ are known, the product $C = AB \in \mathscr{C}^{n\times r}$ satisfies

$$c_{ij} = \sum_{k=1}^{m} a_{ik} b_{kj}, \quad \text{where } i = 1, 2, \ldots, n, \ j = 1, 2, \ldots, r. \tag{2.5.2}$$

**Definition 2.21.** If one of the two matrices $A$ and $B$ is a scalar, then $AB$ equals to a matrix, elements of which are the products of the scalar and the other matrix elements.

Multiplication of two matrices $A$ and $B$ can easily be obtained with MATLAB command $C=A*B$. It is not necessary to assign the sizes of the matrices $A$ and $B$. If matrices $A$ and $B$ are incompatible, an error is displayed, otherwise, the product of the two matrices can be found and returned in $C$.

(3) Inner products. The inner product of two vectors can be expressed as a scalar defined below.

**Definition 2.22.** For two column vectors $a$, $b$ of equal lengths, the inner product is defined as $\langle a, b \rangle = a^T b$.

**Theorem 2.2.** *Inner products satisfy the following properties:*

$$\langle a, b \rangle = \langle b, a \rangle, \quad \langle \lambda a, b \rangle = \lambda \langle a, b \rangle, \quad \langle a, b + c \rangle = \langle a, b \rangle + \langle a, c \rangle. \tag{2.5.3}$$

**Theorem 2.3.** *One has $\langle a, a \rangle = 0$ if and only if $a \equiv 0$.*

If there exist two vectors $a$ and $b$ in MATLAB workspace, and they are of equal size, the inner product can be evaluated with c=a(:).'*b(:). Furthermore, if the vectors are not column vectors, the inner product can also be obtained.

(4) Left division. The operator "\" is provided in MATLAB for evaluating the left division of two matrices, such that $A \backslash B$ is the solution $X$ of the equation $AX = B$. If $A$ is a nonsingular square matrix, then $X = A^{-1}B$. If $A$ is not a square matrix, $X = A \backslash B$ stands for the least-squares solution $X$ of the equation $AX = B$.

(5) Right division. The operator "/" is also provided in MATLAB, to indicate the right division of two matrices, which is equivalent to finding the solution of equation $XA = B$. If $A$ is a nonsingular square matrix, $B/A$ equals to $BA^{-1}$. More precisely, B/A=(A'\B')'.

(6) Matrix power. A power of the square matrix $A$ is mathematically expressed as $A^x$. If $x$ is a positive integer, the power expression $A^x$ can be obtained by multiplying the $A$ matrix $x$ times. If $x$ is a negative integer, we can multiply the $A$ matrix $-x$ times, and compute the inverse of the result. If $x$ is a fraction, such as $x = n/m$, where $n$ and $m$ are both integers, matrix $A$ can be multiplied $n$ times, then we can compute its $m$th roots. In MATLAB, the matrix power can be evaluated as F=A^x.

(7) Matrix roots. It is known in mathematics that the $m$th root of $A$ is defined to be a matrix such that, by multiplying it $m$ times, we restore the original $A$ matrix. There should be $m$ different $m$th roots of a matrix. Considering $\sqrt[3]{-1}$, for instance, one cubic root is $-1$, the other roots can be obtained by rotating that root by $120°$ to find the second root, and further rotating the result by $120°$, the third root can be found. This rotation by $120°$ can be realized as a multiplication of the results by a scalar $\delta = e^{2\pi j/3}$.

**Theorem 2.4.** *For the given matrix $A$, if one of the $m$th root matrices is $A_0$, the other $m-1$ roots can be obtained as $A_0 e^{2k\pi j/m}$, where $k = 1, 2, \ldots, m - 1$.*

One of the $m$th roots can be found with MATLAB command A^(1/m).

**Example 2.30.** Compute all the cubic roots of matrix $A$, and validate the results if

$$A = \begin{bmatrix} 1 & 2 & 3 \\ 4 & 5 & 6 \\ 7 & 8 & 0 \end{bmatrix}.$$

**Solutions.** One of the cubic roots can be evaluated directly with the "^" operator

```
>> A=[1,2,3; 4,5,6; 7,8,0]; C=A^(1/3),
   e=norm(A-C^3) % find cubic roots and validate the results
```

the result is obtained below, with the norm of error of $e = 1.0145 \times 10^{-14}$. It can be seen that the result is rather accurate:

$$C = \begin{bmatrix} 0.7718 + j0.6538 & 0.4869 - j0.0159 & 0.1764 - j0.2887 \\ 0.8885 - j0.0726 & 1.4473 + j0.4794 & 0.5233 - j0.4959 \\ 0.4685 - j0.6465 & 0.6693 - j0.6748 & 1.3379 + j1.0488 \end{bmatrix}.$$

In fact, there should be three cubic roots altogether for a given matrix, while only one is obtained above. The other two can be constructed by rotating the result twice, that is, by multiplying it by the scalars $Ce^{j2\pi/3}$ and $Ce^{j4\pi/3}$.

```
>> j1=exp(sqrt(-1)*2*pi/3);
   A1=C*j1, A2=C*j1^2      % rotating to find the two other roots
   e1=norm(A-A1^3), e2=norm(A-A2^3)   % root validations
```

The other two root matrices can also be found, with $10^{-14}$-level errors:

$$A_1 = \begin{bmatrix} -0.9521 + j0.3415 & -0.2297 + j0.4296 & 0.1618 + j0.2971 \\ -0.3814 + j0.8058 & -1.1388 + j1.0137 & 0.1678 + j0.7011 \\ 0.3256 + j0.7289 & 0.2497 + j0.9170 & -1.5772 + j0.6343 \end{bmatrix},$$

$$A_2 = \begin{bmatrix} 0.1803 - j0.9953 & -0.2572 - j0.4137 & -0.3382 - j0.0084 \\ -0.5071 - j0.7332 & -0.3085 - j1.4931 & -0.6911 - j0.2052 \\ -0.7941 - j0.0825 & -0.9190 - j0.2422 & 0.2393 - j1.6831 \end{bmatrix}.$$

Variable precision framework can be used to find the cubic roots, with the errors as low as $7.2211 \times 10^{-39}$, and they are much more accurate than those obtained under the double precision framework.

```
>> A=sym([1,2,3; 4,5,6; 7,8,0]); C=A^(sym(1/3));
   C=vpa(C); norm(C^3-A) % high precision solutions
```

**Example 2.31.** The inverse matrix of $B$ is mathematically denoted as $B^{-1}$, and it can be evaluated with the function inv($B$). Find the $(-1)$th power of the complex matrix $B$ in Example 2.25, and see whether it is the inverse matrix of $B$.

**Solutions.** The matrices are computed under the symbolic framework as

```
>> B=[1+9i,2+8i,3+7j; 4+6j 5+5i,6+4i; 7+3i,8+2j 1i];
   B=sym(B); B1=B^(-1), B2=inv(B), C=B1*B
```

It can be seen that the two matrices obtained are exactly the same, indicating that the inverse matrix is indeed the "reciprocal" of the matrix:

$$B_1 = B_2 = \begin{bmatrix} 13/18 - 5j/6 & -10/9 + j/3 & -1/9 \\ -7/9 + 2j/3 & 19/18 - j/6 & 2/9 \\ -1/9 & 2/9 & -1/9 \end{bmatrix}, \quad C = \begin{bmatrix} 1 & 0 & 0 \\ 0 & 1 & 0 \\ 0 & 0 & 1 \end{bmatrix}.$$

### 2.5.4 Kronecker products and sums

Kronecker products and sums are named after German mathematician Leopold Kronecker (1823–1891). They are useful in finding the solutions of linear algebraic equations, as will be shown later.

**Definition 2.23.** The Kronecker product of matrices $A$ and $B$ is defined as

$$C = A \otimes B = \begin{bmatrix} a_{11}B & \cdots & a_{1m}B \\ \vdots & \ddots & \vdots \\ a_{n1}B & \cdots & a_{nm}B \end{bmatrix}. \tag{2.5.4}$$

**Definition 2.24.** The Kronecker sum of matrices $A$ and $B$ is defined as

$$D = A \oplus B = \begin{bmatrix} a_{11} + B & \cdots & a_{1m} + B \\ \vdots & \ddots & \vdots \\ a_{n1} + B & \cdots & a_{nm} + B \end{bmatrix}. \tag{2.5.5}$$

**Theorem 2.5.** *If the sizes of $A$ and $B$ are the same, the Kronecker product satisfies the distributive law:*

$$(A + B) \otimes C = A \otimes C + B \otimes C, \tag{2.5.6}$$

$$C \otimes (A + B) = C \otimes A + C \otimes B. \tag{2.5.7}$$

**Theorem 2.6.** *The associative law and transpose of Kronecker product are given by*

$$A(B \otimes C) = (A \otimes B)C, \quad (A \otimes B)^{\mathrm{T}} = B^{\mathrm{T}} \otimes A^{\mathrm{T}}. \tag{2.5.8}$$

**Theorem 2.7.** *If the sizes of the following matrices are compatible, then*

$$(A \otimes B)(C \otimes D) = (AC) \otimes (BD). \tag{2.5.9}$$

If the $\otimes$ operator is changed to $\oplus$, the above theorems also hold.

Unlike the normal sum and product of matrices, Kronecker sum and product do not require compatibility of the two matrices. Besides, Kronecker sum and product do not satisfy the commutative law.

The function `kron()` in MATLAB can be used to evaluate directly the Kronecker product $A \otimes B$ of two matrices with `C=kron(A,B)`.

Similar to the function `kron()`, a Kronecker sum function `kronsum()` can be written as

```
function C=kronsum(A,B)
[ma,na]=size(A); [mb,nb]=size(B);
A=reshape(A,[1 ma 1 na]); B=reshape(B,[mb 1 nb 1]);
C=reshape(bsxfun(@plus,A,B),[ma*mb na*nb]);
```

**Example 2.32.** Compute the Kronecker product and sum of matrices $A$ and $B$ if

$$A = \begin{bmatrix} -2 & 2 \\ 0 & -1 \end{bmatrix}, \quad B = \begin{bmatrix} -2 & -1 & 1 \\ 0 & 1 & -2 \end{bmatrix}.$$

**Solutions.** The two matrices can be input into MATLAB first, then function kron() can be called to compute directly $A \otimes B$ and $B \otimes A$

```
>> A=[-2,2; 0,-1]; B=[-2,-1,1; 0,1,-2];
   kron(A,B), kron(B,A)
```

The results obtained are given below. It can be seen that they are different, meaning the Kronecker product does not satisfy the commutative law:

$$A \otimes B = \begin{bmatrix} 4 & 2 & -2 & -4 & -2 & 2 \\ 0 & -2 & 4 & 0 & 2 & -4 \\ 0 & 0 & 0 & 2 & 1 & -1 \\ 0 & 0 & 0 & 0 & -1 & 2 \end{bmatrix}, \quad B \otimes A = \begin{bmatrix} 4 & -4 & 2 & -2 & -2 & 2 \\ 0 & 2 & 0 & 1 & 0 & -1 \\ 0 & 0 & -2 & 2 & 4 & -4 \\ 0 & 0 & 0 & -1 & 0 & 2 \end{bmatrix}.$$

The following statements can be used to evaluate $A \oplus B$ and $B \oplus A$:

```
>> kronsum(A,B), kronsum(B,A)
```

and the results are given below, indicating that the Kronecker sum does not satisfy the commutative law either:

$$A \oplus B = \begin{bmatrix} -4 & -3 & -1 & 0 & 1 & 3 \\ -2 & -1 & -4 & 2 & 3 & 0 \\ -2 & -1 & 1 & -3 & -2 & 0 \\ 0 & 1 & -2 & -1 & 0 & -3 \end{bmatrix}, \quad B \oplus A = \begin{bmatrix} -4 & 0 & -3 & 1 & -1 & 3 \\ -2 & -3 & -1 & -2 & 1 & 0 \\ -2 & 2 & -1 & 3 & -4 & 0 \\ 0 & -1 & 1 & 0 & -2 & -3 \end{bmatrix}.$$

## 2.6 Calculus computations with matrices

The definitions of the derivatives and integrals of matrix functions are proposed first, and based on them, derivatives of complicated matrix functions are computed. Jacobian and Hessian matrices are also presented.

## 2.6.1 Matrix derivatives

**Definition 2.25.** If matrix $A(t)$ is a function of $t$, the derivative matrix of $A(t)$ with respect to $t$ is defined as a matrix of elementwise derivatives of $A(t)$, defined as

$$\frac{dA(t)}{dt} = \left[ \frac{d}{dt} a_{ij}(t) \right]. \tag{2.6.1}$$

If matrix $A(t)$ is a function of $t$, then $dA(t)/dt$ can be obtained directly with `diff()` function.

**Example 2.33.** Compute the derivative of the following matrix:

$$A(t) = \begin{bmatrix} t^2/2+1 & t & t^2/2 \\ t & 1 & t \\ -t^2/2 & -t & 1-t^2/2 \end{bmatrix} e^{-t}.$$

**Solutions.** The matrix function can be entered into MATLAB first, then the derivative matrix can be computed directly. Since the symbolic function data format is used here, in order to avoid conflict in the existing matrices, `clear` command should be used first to remove the $A$ matrix in MATLAB workspace, and then the matrix function can be entered.

```
>> syms t; clear A
   A(t)=[t^2/2+1,t,t^2/2; t,1,t; -t^2/2,-t,1-t^2/2]*exp(-t);
   A1=simplify(diff(A,t))
```

The derivative matrix obtained is

$$A_1(t) = \begin{bmatrix} -t^2/2+t-1 & 1-t & -t(t-2)/2 \\ 1-t & -1 & 1-t \\ t(t-2)/2 & t-1 & t^2/2-t-1 \end{bmatrix} e^{-t}.$$

**Theorem 2.8.** *Assuming that $A(t)$ and $B(t)$ are differentiable,*

$$\frac{d}{dt}[A(t) + B(t)] = \frac{dA(t)}{dt} + \frac{dB(t)}{dt}, \tag{2.6.2}$$

$$\frac{d}{dt}[A(t)B(t)] = A(t)\frac{dB(t)}{dt} + \frac{dA(t)}{dt}B(t). \tag{2.6.3}$$

**Example 2.34.** Consider the matrix $A(t)$ in Example 2.33, and matrix $B(t)$ given below. Compute the derivative of $A(t)B(t)$ with respect to $t$ if

$$B(t) = \begin{bmatrix} e^{-t} - e^{-2t} + e^{-3t} & e^{-t} - e^{-2t} & e^{-t} - e^{-3t} \\ 2e^{-2t} - e^{-t} - e^{-3t} & 2e^{-2t} - e^{-t} & e^{-3t} - e^{-t} \\ e^{-t} - e^{-2t} & e^{-t} - e^{-2t} & e^{-t} \end{bmatrix}.$$

**Solutions.** Two methods can be used to compute the derivative of $A(t)B(t)$. In the first method, $A(t)B(t)$ can be evaluated, then function diff() can be used to find the derivative. In the other method, (2.6.3) can be used directly. It can be seen that the results with the two methods are identical. It can be seen that for complicated matrix functions, there is no need to use the indirect method such as that in (2.6.3). Function diff() can be called directly to compute the derivatives. Since the results are too complicated, they are not given here.

```
>> syms t; clear A B
   A(t)=[t^2/2+1,t,t^2/2; t,1,t; -t^2/2,-t,1-t^2/2]*exp(-t);
   B(t)=[exp(-t)-exp(-2*t)+exp(-3*t),...
             exp(-t)-exp(-2*t),exp(-t)-exp(-3*t);
         2*exp(-2*t)-exp(-t)-exp(-3*t),...
             2*exp(-2*t)-exp(-t),exp(-3*t)-exp(-t);
         exp(-t)-exp(-2*t), exp(-t)-exp(-2*t), exp(-t)];
   A1=simplify(A*diff(B)+diff(A)*B)
   A2=simplify(diff(A*B)), simplify(A1-A2)
```

### 2.6.2 Integrals of matrix functions

**Definition 2.26.** The indefinite integral of a matrix function $A(t)$ with respect to $t$ can be computed from

$$\int_{t_0}^{t_n} A(t)\mathrm{d}t = \left\{ \int_{t_0}^{t_n} a_{ij}(t)\mathrm{d}t \right\}. \tag{2.6.4}$$

The int() function provided in the Symbolic Math Toolbox of MATLAB can be used directly to compute the indefinite, definite, and improper integrals. The evaluation of integrals is illustrated through the following example.

**Example 2.35.** Compute the integral of the result in Example 2.33, and see whether the original matrix can be restored.

**Solutions.** Input the matrix function $A(t)$ first, then the derivative of the matrix can be obtained. Taking the integral of the result, the original matrix can be restored. In fact, the result should be $A(t) + C$, where $C$ is an arbitrary constant matrix.

```
>> syms t; clear A
   A(t)=[t^2/2+1,t,t^2/2; t,1,t; -t^2/2,-t,1-t^2/2]*exp(-t);
   A1=simplify(diff(A,t)); A2=simplify(int(A1))
```

### 2.6.3 Jacobian matrices of vector functions

**Definition 2.27.** Assuming that there are $m$ functions with $n$ independent variables

$$\begin{cases} y_1 = f_1(x_1, x_2, \ldots, x_n), \\ y_2 = f_2(x_1, x_2, \ldots, x_n), \\ \vdots \\ y_m = f_m(x_1, x_2, \ldots, x_n), \end{cases} \tag{2.6.5}$$

the first-order partial derivatives of each $y_i$ with respect to each $x_j$ form the following Jacobian matrix

$$J = \begin{bmatrix} \partial y_1/\partial x_1 & \partial y_1/\partial x_2 & \cdots & \partial y_1/\partial x_n \\ \partial y_2/\partial x_1 & \partial y_2/\partial x_2 & \cdots & \partial y_2/\partial x_n \\ \vdots & \vdots & \ddots & \vdots \\ \partial y_m/\partial x_1 & \partial y_m/\partial x_2 & \cdots & \partial y_m/\partial x_n \end{bmatrix}. \tag{2.6.6}$$

A Jacobian matrix is named after German mathematician Carl Gustav Jacob Jacobi (1804–1851). The matrix is also known as the gradient matrix. Jacobian matrix can be computed directly with `jacobian()` function provided in the Symbolic Math Toolbox of MATLAB, with the syntax `J=jacobian(y,x)`, where $x$ is a vector of independent variables, while $y$ is a vector of functions.

**Example 2.36.** The conversion formulas from spherical to Cartesian coordinates are $x = r \sin\theta \cos\phi$, $y = r \sin\theta \sin\phi$, and $z = r \cos\theta$. Compute the Jacobian matrix of the functions $[x, y, z]$ with respect to the independent variables $[r, \theta, \phi]$.

**Solutions.** The three symbolic functions should be declared first, and from them the Jacobian matrix can be computed with the following statements:

```
>> syms r theta phi; % declare symbolic variables
   x=r*sin(theta)*cos(phi); y=r*sin(theta)*sin(phi); z=r*cos(theta);
   J=jacobian([x; y; z],[r theta phi]) % compute Jacobian matrix
```

The Jacobian matrix can be obtained as

$$J = \begin{bmatrix} \sin\theta\cos\phi & r\cos\theta\cos\phi & -r\sin\theta\sin\phi \\ \sin\theta\sin\phi & r\cos\theta\sin\phi & r\sin\theta\cos\phi \\ \cos\theta & -r\sin\theta & 0 \end{bmatrix}.$$

### 2.6.4 Hessian matrices

**Definition 2.28.** For a scalar function $f(x_1, x_2, \ldots, x_n)$ with $n$ independent variables, the Hessian matrix is defined as

$$H = \begin{bmatrix} \partial^2 f/\partial x_1^2 & \partial^2 f/\partial x_1 \partial x_2 & \cdots & \partial^2 f/\partial x_1 \partial x_n \\ \partial^2 f/\partial x_2 \partial x_1 & \partial^2 f/\partial x_2^2 & \cdots & \partial^2 f/\partial x_2 \partial x_n \\ \vdots & \vdots & \ddots & \vdots \\ \partial^2 f/\partial x_n \partial x_1 & \partial^2 f/\partial x_n \partial x_2 & \cdots & \partial^2 f/\partial x_n^2 \end{bmatrix}. \tag{2.6.7}$$

Hessian matrix is named after German mathematician Ludwig Otto Hesse (1811–1874). It contains, in fact, the second-order derivatives of a scalar function $f(x,y)$. A MATLAB function hessian() is provided for finding the Hessian matrices, with the syntax $H$=hessian$(f,x)$, where $x = [x_1, x_2, \ldots, x_n]$.

In the earlier versions of MATLAB, hessian() function was not provided. The nested call of $H$=jacobian(jacobian$(f,x),x)$ function could be used.

**Example 2.37.** Compute the Hessian matrix for the two-dimensional function

$$f(x,y) = (x^2 - 2x)e^{-x^2-y^2-xy}.$$

**Solutions.** The following MATLAB commands can be used to evaluate Hessian matrix of the given function:

```
>> syms x y; f=(x^2-2*x)*exp(-x^2-y^2-x*y);
   H=simplify(hessian(f,[x,y]))
   H1=simplify(hessian(f,[x,y])/exp(-x^2-y^2-x*y))
```

and the result obtained is

$$H_1 = e^{-x^2-y^2-xy} \begin{bmatrix} 4x - 2(2x-2)(2x+y) - 2x^2 - (2x-x^2)(2x+y)^2 + 2 \\ 2x - (2x-2)(x+2y) - x^2 - (2x-x^2)(x+2y)(2x+y) \end{bmatrix}$$

$$\begin{matrix} 2x - (2x-2)(x+2y) - x^2 - (2x-x^2)(x+2y)(2x+y) \\ x(x-2)(x^2+4xy+4y^2-2) \end{matrix} \Bigg].$$

## 2.7 Exercises

2.1 Input the following matrices $A$ and $B$ into MATLAB workspace:

$$A = \begin{bmatrix} 1 & 2 & 3 & 4 \\ 4 & 3 & 2 & 1 \\ 2 & 3 & 4 & 1 \\ 3 & 2 & 4 & 1 \end{bmatrix}, \quad B = \begin{bmatrix} 1+4j & 2+3j & 3+2j & 4+1j \\ 4+1j & 3+2j & 2+3j & 1+4j \\ 2+3j & 3+2j & 4+1j & 1+4j \\ 3+2j & 2+3j & 4+1j & 1+4j \end{bmatrix}.$$

For the $4 \times 4$ matrix, if command $A(5,6) = 5$ is given, what is the result?

2.2 Generate a diagonal matrix with diagonal elements $a_1, a_2, \ldots, a_{12}$.

2.3 Can you judge from the format in display, which matrix is a double precision one, and which is a symbolic one? For two matrices, if $A$ is a double precision one, while $B$ is a symbolic one, what is the data type of their product $C=A*B$? Please validate the results through examples.

2.4 Generate a set of 30 000 standard normal pseudorandom numbers. Find the mean and variance from the generated sample, then draw the histograms of its distribution.

2.5 Jordan matrix is a widely used practical matrix, whose general form is

$$
J = \begin{bmatrix} -\alpha & 1 & 0 & \cdots & 0 \\ 0 & -\alpha & 1 & \cdots & 0 \\ \vdots & \vdots & \vdots & \ddots & \vdots \\ 0 & 0 & 0 & \cdots & -\alpha \end{bmatrix}, \quad \text{e. g.,} \quad J_1 = \begin{bmatrix} -5 & 1 & 0 & 0 & 0 \\ 0 & -5 & 1 & 0 & 0 \\ 0 & 0 & -5 & 1 & 0 \\ 0 & 0 & 0 & -5 & 1 \\ 0 & 0 & 0 & 0 & -5 \end{bmatrix}.
$$

Function diag() can be used to construct matrix $J_1$.

2.6 For a given vector $c = [-4, -3, -2, -1, 0, 1, 2, 3, 4]$, generate from it Hankel, Vandermonde, and companion matrices.

2.7 Generate a $9 \times 9$ magic matrix. Observe the trends in the numbers $1 \to 2 \to \cdots \to 80 \to 81$.

2.8 Generate a random $15 \times 15$ integer matrix with elements 0's and 1's, whose determinant is 1.

2.9 Try to generate the following $20 \times 20$ matrix without loops:

$$
A = \begin{bmatrix} x & a & a & \cdots & a \\ a & x & a & \cdots & a \\ a & a & x & \cdots & a \\ \vdots & \vdots & \vdots & \ddots & a \\ a & a & a & \cdots & x \end{bmatrix}.
$$

2.10 Nilpotent matrix is a category of special matrices, whose first subdiagonal elements are all 1's, and the rest of the elements are 0's. Validate that for nilpotent matrices, $H_n^i = O$ holds for all $i \geqslant n$.

2.11 Enter the following matrices $A$ and $B$ into MATLAB workspace, and convert them into symbolic ones:

$$
A = \begin{bmatrix} 5 & 7 & 6 & 5 & 1 & 6 & 5 \\ 2 & 3 & 1 & 0 & 0 & 1 & 4 \\ 6 & 4 & 2 & 0 & 6 & 4 & 4 \\ 3 & 9 & 6 & 3 & 6 & 6 & 2 \\ 10 & 7 & 6 & 0 & 0 & 7 & 7 \\ 7 & 2 & 4 & 4 & 0 & 7 & 7 \\ 4 & 8 & 6 & 7 & 2 & 1 & 7 \end{bmatrix}, \quad B = \begin{bmatrix} 3 & 5 & 5 & 0 & 1 & 2 & 3 \\ 3 & 2 & 5 & 4 & 6 & 2 & 5 \\ 1 & 2 & 1 & 1 & 3 & 4 & 6 \\ 3 & 5 & 1 & 5 & 2 & 1 & 2 \\ 4 & 1 & 0 & 1 & 2 & 0 & 1 \\ -3 & -4 & -7 & 3 & 7 & 8 & 12 \\ 1 & -10 & 7 & -6 & 8 & 1 & 5 \end{bmatrix}.
$$

2.12 For the complex matrix $A$, compute the transposes $A^{\mathrm{T}}$ and $A^{\mathrm{H}}$ if

$$
A = \begin{bmatrix} 4+3j & 6+5j & 3+3j & 1+2j & 4+j \\ 6+4j & 5+5j & 1+2j & 2j & 5 \\ 1 & 4 & 5+3j & 5+j & 4+6j \end{bmatrix}.
$$

2.13 Generate a $5 \times 5$ square matrix and find all its fifth roots.

2.14 Generate a $9 \times 9$ square matrix. Make random orders of the second to seventh columns, while the other columns should be left unchanged. Observe through statements and see whether the expected results can be achieved.

2.15 For the square matrix $A$ given below, find its inverse matrix with the matrix power method, and also find $A^{-5}$ if

$$A = \begin{bmatrix} 1 & 0 & 6 & 1 \\ 3 & 6 & 1 & 2 \\ 5 & 5 & 5 & 1 \\ 5 & 1 & 6 & 3 \end{bmatrix}.$$

2.16 Compute $A \otimes B$ and $B \otimes A$, and see whether the two matrices are identical if

$$A = \begin{bmatrix} -1 & 2 & 2 & 1 \\ -1 & 2 & 1 & 0 \\ 2 & 1 & 1 & 0 \\ 1 & 0 & 2 & 0 \end{bmatrix}, \quad B = \begin{bmatrix} 3 & 0 & 3 \\ 3 & 2 & 2 \\ 3 & 1 & 1 \end{bmatrix}.$$

2.17 For the $A$ and $B$ matrices in Example 2.16, compute $A \oplus B$ and $B \oplus A$, and see whether they are identical.

2.18 For the matrices given below, validate that

$$A_1 \otimes B_1 + A_2 \otimes B_2 \neq (A_1 + A_2) \otimes (B_1 + B_2),$$

$$A_1 = \begin{bmatrix} 2 & 0 \\ 2 & 2 \end{bmatrix}, \quad A_2 = \begin{bmatrix} 1 & 0 \\ 0 & 1 \end{bmatrix},$$

$$B_1 = \begin{bmatrix} 2 & 2 & 1 & 1 \\ 2 & 1 & 2 & 0 \\ 0 & 2 & 2 & 2 \\ 2 & 0 & 2 & 2 \end{bmatrix}, \quad B_2 = \begin{bmatrix} 2 & 1 & 0 & 2 \\ 2 & 0 & 0 & 0 \\ 2 & 2 & 0 & 2 \\ 1 & 0 & 2 & 0 \end{bmatrix}.$$

2.19 Compute $d[A^2(t)B^3(t)]/dt$ if

$$A(t) = \begin{bmatrix} t^2/2 + 1 & t & t^2/2 \\ t & 1 & t \\ -t^2/2 & -t & 1 - t^2/2 \end{bmatrix} e^{-t},$$

$$B(t) = \begin{bmatrix} e^{-t} - e^{-2t} + e^{-3t} & e^{-t} - e^{-2t} & e^{-t} - e^{-3t} \\ 2e^{-2t} - e^{-t} - e^{-3t} & 2e^{-2t} - e^{-t} & e^{-3t} - e^{-t} \\ e^{-t} - e^{-2t} & e^{-t} - e^{-2t} & e^{-t} \end{bmatrix}.$$

2.20 Compute the Jacobian matrix of

$$f(x,y,z) = \begin{bmatrix} 3x + e^y z \\ x^3 + y^2 \sin z \end{bmatrix}.$$

2.21 For the scalar $f(x,y,z) = 3x + e^y z + x^3 + y^2 \sin z$ with three independent variables, compute its Hessian matrix.

# 3 Fundamental analysis of matrices

Matrix analysis is the subject dealing with matrices and their algebraic properties. The fundamental issues are algebraic computation, linear transforms, eigenvalues and eigenvectors, and also nonlinear matrix functions.

Solving certain matrix analysis problems is the main focus of this chapter. In Section 3.1, matrix determinant computation is presented, with algebraic cofactor methods and their MATLAB implementations. Cramer's rule for solving linear equations is presented, and MATLAB implementation of determinant computation is demonstrated. In Section 3.2, some fundamental questions about matrices such as computing their traces, checking for linear independence and ranks, as well as finding norms of vectors and matrices, are presented. The concept of a vector space is introduced. In Section 3.3, inverse matrix concepts and computations are presented. Reduced row echelon form conversions are presented in MATLAB, and generalized inverse is computed. Section 3.4 presents the concepts and computations of characteristic polynomials and eigenvalues; also the ideas and computations of generalized eigenvalues are introduced. In Section 3.5, matrix polynomial computation and polynomial conversion are introduced.

## 3.1 Determinants

Matrix determinant is a very important concept in matrix analysis, especially since it is an important tool in solving linear algebraic equations. The name "determinant" means that it can be used to determine whether the linear equations have unique solutions. Of course, with the development of linear algebra theory, the importance of determinants becomes smaller and smaller[1].

In this section, the determinants of the matrices will be evaluated with MATLAB, the determinants of matrices of any size can be obtained, and the Cramer's rule for solving linear algebraic equations will be demonstrated.

### 3.1.1 Determinant definition and properties

**Definition 3.1.** The determinant of an $n \times n$ matrix $A = \{a_{ij}\}$ is defined as

$$D = |A| = \det(A) = \sum (-1)^k a_{1k_1} a_{2k_2} \cdots a_{nk_n}, \tag{3.1.1}$$

where $k_1, k_2, \ldots, k_n$ is a permutation of the sequence of $1, 2, \ldots, n$, having $k$ exchanges; $\Sigma$ represents the sum over all the permutations $k_1, k_2, \ldots, k_n$.

It is worth mentioning that determinants exist only for $n \times n$ square matrices. There are no determinants for nonsquare matrices. The determinant of a matrix is a scalar.

https://doi.org/10.1515/9783110666991-003

**Definition 3.2.** The matrices with zero determinants are referred to as singular matrices.

Some of the properties of determinants are listed without proofs:

**Theorem 3.1.** *For square matrices $A$ and $B$, $\det(AB) = \det(A)\det(B)$.*

**Theorem 3.2.** *For a nonsingular square matrix $A$, $\det(A^{-1}) = 1/\det(A)$.*

**Theorem 3.3.** *The determinants of a square matrix and its transpose are identical, $\det(A^T) = \det(A)$.*

**Theorem 3.4.** *If the whole row (or column) of the matrix is multiplied by $k$, then the determinant of the new matrix is $k\det(A)$.*

**Theorem 3.5.** *If the elements in one row (or column) are all zeros, the determinant is zero, and the matrix is singular.*

**Theorem 3.6.** *If two rows (or columns) of a matrix are swapped, the determinant of the new matrix is $-\det(A)$.*

**Theorem 3.7.** *If all the elements of a row (or column) can be written as sums of two quantities,*

$$
A = \begin{bmatrix}
a_{11} & a_{12} & \cdots & a_{1n} \\
\vdots & \vdots & \ddots & \vdots \\
a_{i1} + a'_{i1} & a_{i2} + a'_{i2} & \cdots & a_{in} + a'_{in} \\
\vdots & \vdots & \ddots & \vdots \\
a_{n1} & a_{n2} & \cdots & a_{nn}
\end{bmatrix},
\tag{3.1.2}
$$

*then the determinant of the matrix satisfies*

$$
\det(A) = \begin{vmatrix}
a_{11} & a_{12} & \cdots & a_{1n} \\
\vdots & \vdots & \ddots & \vdots \\
a_{i1} & a_{i2} & \cdots & a_{in} \\
\vdots & \vdots & \ddots & \vdots \\
a_{n1} & a_{n2} & \cdots & a_{nn}
\end{vmatrix}
+
\begin{vmatrix}
a_{11} & a_{12} & \cdots & a_{1n} \\
\vdots & \vdots & \ddots & \vdots \\
a'_{i1} & a'_{i2} & \cdots & a'_{in} \\
\vdots & \vdots & \ddots & \vdots \\
a_{n1} & a_{n2} & \cdots & a_{nn}
\end{vmatrix}.
\tag{3.1.3}
$$

**Theorem 3.8.** *If the elements of a whole row (column) is multiplied by a scalar, and added to another row, the determinant of the new matrix is the same.*

### 3.1.2 Determinants of small-scale matrices

A $1 \times 1$ matrix is a scalar, and the determinant of this matrix is that matrix element. For $2 \times 2$ and $3 \times 3$ matrices, the computation of the determinants is demonstrated through examples. The algebraic cofactor method for finding the determinants of large-scale matrices is illustrated through examples, and limitations are also demonstrated.

**Example 3.1.** Compute the determinants of $2 \times 2$ and $3 \times 3$ matrices and explain the results.

**Solutions.** For a $2 \times 2$ matrix, the products of two diagonal elements can be obtained, and the determinant equals to the difference of the products:

$$\begin{vmatrix} a_{11} & a_{12} \\ a_{21} & a_{22} \end{vmatrix} = a_{11}a_{22} - a_{12}a_{21}.$$

For a $3 \times 3$ matrix, the matrix can be extended into a $3 \times 5$ matrix

$$\begin{bmatrix} a_{11} & a_{12} & a_{13} \\ a_{21} & a_{22} & a_{23} \\ a_{31} & a_{32} & a_{33} \end{bmatrix} \Rightarrow \begin{bmatrix} a_{11} & a_{12} & a_{13} & a_{11} & a_{12} \\ a_{21} & a_{22} & a_{23} & a_{21} & a_{22} \\ a_{31} & a_{32} & a_{33} & a_{31} & a_{32} \end{bmatrix}.$$

The products of three diagonals and three back diagonals can be obtained. The determinant of the $3 \times 3$ matrix can be obtained as

$$\begin{vmatrix} a_{11} & a_{12} & a_{13} \\ a_{21} & a_{22} & a_{23} \\ a_{31} & a_{32} & a_{33} \end{vmatrix} = a_{11}a_{22}a_{33} + a_{12}a_{23}a_{31} + a_{13}a_{21}a_{32} - a_{31}a_{22}a_{13} - a_{32}a_{23}a_{11} - a_{33}a_{21}a_{12}.$$

The method presented above cannot be used to compute determinants of large-scale matrices. In traditional linear algebra courses, the algebraic cofactor method is usually presented.

**Definition 3.3.** For an $n \times n$ matrix $A$, if the elements in the $i$th row and $j$th column are removed, the determinant of the remaining $(n-1) \times (n-1)$ matrix is referred to as the $(i, j)$th minor, denoted as $M_{ij}$. The minor is multiplied by $(-1)^{i+j}$, then $A_{ij} = (-1)^{i+j}M_{ij}$ is referred to as the $(i, j)$th algebraic cofactor.

**Theorem 3.9.** *The determinant of a matrix $A$ can be computed from any row (say, the kth row) as*

$$\det(A) = a_{k1}A_{k1} + a_{k2}A_{k2} + \cdots + a_{kn}A_{kn}, \tag{3.1.4}$$

*or expanded using the mth column as*

$$\det(A) = a_{1m}A_{1m} + a_{2m}A_{2m} + \cdots + a_{nm}A_{nm}, \tag{3.1.5}$$

*where $1 \leqslant k, m \leqslant n$. The algorithm is referred to as the algebraic cofactor method.*

Based on the algorithm presented above, selecting $k = 1$ in (3.1.4), the determinant can be evaluated with the algebraic cofactor method, where matrix $A_2$ is the algebraic cofactor matrix. It can be seen that only simple multiplications are involved, the precision of the algorithm is relatively high. A recursive structure is used in the function such that when the size is high, the computation load may be very high.

```
function d=det1(A)
[n,m]=size(A);
if n==m
   if n==1; d=A;
   elseif n==2, d=A(1,1)*A(2,2)-A(1,2)*A(2,1);
   else, d=0; A1=A; A1(1,:)=[];
      for i=1:n
         A2=A1; A2(:,i)=[]; d=d+A(1,i)*(-1)^(1+i)*det1(A2);
      end, end
else, error('A rectangular matrix cannot be handled.'); end
```

**Example 3.2.** Compute the determinant of the $4 \times 4$ magic matrix

$$A = \begin{bmatrix} 16 & 2 & 3 & 13 \\ 5 & 11 & 10 & 8 \\ 9 & 7 & 6 & 12 \\ 4 & 14 & 15 & 1 \end{bmatrix}.$$

**Solutions.** The function det1() can be used directly and it can be found that the determinant obtained is zero.

```
>> A=[16 2 3 13; 5 11 10 8; 9 7 6 12; 4 14 15 1];
   det1(A)
```

**Example 3.3.** Compute the determinant of the $10 \times 10$ magic matrix and measure the time consumption.

**Solutions.** The determinant of the matrix can be obtained easily with the following statements, the elapsed time is 7.92 s, and the determinant is zero.

```
>> A=magic(10); tic, d=det1(A), toc
```

For large-scale matrices, the algebraic cofactor method is quite time-consuming. For the determinant of a $11 \times 11$ matrix, the elapsed time is 85 s.

   If a matrix has many zero elements, the following MATLAB function can be written. The row (or column) containing the smallest number of zeros is selected automatically such that in the algebraic cofactor computation, when a matrix element is zero, the term in the cofactor computation is bypassed, so as to reduce the total time needed. Based on the algorithm, the following MATLAB function can be written:

```
function d=det2(A)
[n,m]=size(A);
if n==m
```

```
if n==1; d=A;
else, [n1,ix]=nnzc(A,n); [n2,iy]=nnzc(A.',n);
    if n1>n2, ix=iy; else, A=A.'; end
    d=0; A1=A; A1(ix,:)=[];
    for i=1:n, if A(ix,i)~=0,
        A2=A1; A2(:,i)=[]; d=d+A(ix,i)*(-1)^(ix+i)*det2(A2);
    end, end, end
else, error('A rectangular matrix cannot be handled.'); end
function [n0,ix]=nnzc(A,n)
n0=n; ix=1;
for i=1:n, n1=nnz(A(:,i)); if n1<n0, n0=n1; ix=i; end, end
```

A subfunction nnzc() is written to locate the column containing the least number of zeros. The number of nonzero elements is $n_0$, and the column is $i_x$. This subfunction can also be used to find the number of rows containing the least number of zeros. The matrix $A^T$ can be called instead to find the row numbers.

**Example 3.4.** Compute the determinant of the following $11 \times 11$ matrix.

$$A = \begin{bmatrix} 1 & -1 & 0 & 0 & 1 & 1 & 0 & -1 & 0 & 1 & -1 \\ 1 & -1 & 1 & 0 & -1 & -1 & -1 & -1 & 1 & 1 & 0 \\ 0 & 0 & -1 & 0 & -1 & 1 & -1 & 0 & 1 & 1 & 0 \\ 0 & 0 & -1 & 0 & -1 & 1 & 1 & -1 & 0 & 0 & 0 \\ 1 & -1 & -1 & 0 & 1 & 1 & 0 & 0 & 0 & -1 & 1 \\ 1 & 1 & 0 & -1 & 0 & 1 & 1 & 1 & 1 & 0 & 0 \\ 1 & 0 & 1 & 1 & -1 & 0 & -1 & 1 & 1 & 0 & -1 \\ 0 & -1 & -1 & -1 & 1 & 0 & -1 & -1 & -1 & 0 & 0 \\ 1 & -1 & 1 & 0 & 1 & -1 & 1 & 1 & 1 & 0 & 0 \\ 1 & 1 & 1 & 0 & 0 & 0 & 0 & -1 & 1 & 1 & 1 \\ -1 & -1 & -1 & 0 & -1 & 1 & -1 & 1 & -1 & 1 & 0 \end{bmatrix}.$$

**Solutions.** The matrix can be entered into MATLAB environment, and the two functions can be called to compute the determinants, and the results are the same, with the value of 510. The time needed is different: for det1(), the time consumption is measured as 82.65 s, while the time for det2() is reduced to 2.75 s. It can be seen that the efficiency of the function is significantly increased. Normally, the more nonzero elements a matrix has, the less time is needed.

```
>> A=[1,-1,0,0,1,1,0,-1,0,1,-1; 1,-1,1,0,-1,-1,-1,-1,1,1,0;
      0,0,-1,0,-1,1,-1,0,1,1,0; 0,0,-1,0,-1,1,1,-1,0,0,0;
      1,-1,-1,0,1,1,0,0,0,-1,1; 1,1,0,-1,0,1,1,1,1,0,0;
      1,0,1,1,-1,0,-1,1,1,0,-1; 0,-1,-1,-1,1,0,-1,-1,-1,0,0;
      1,-1,1,0,1,-1,1,1,1,0,0; 1,1,1,0,0,0,0,-1,1,1,1;
      -1,-1,-1,0,-1,1,-1,1,-1,1,0];
```

```
-1,-1,-1,0,-1,1,-1,1,-1,1,0];
tic, det1(A), toc, tic, det2(A), toc
```

It can be shown that the number of computational operations for the determinant of an $n \times n$ matrix is $(n-1)(n+1)! + n$. If $n = 25$, the number of floating-point operations (flops) for the computation is $9.679 \times 10^{27}$, which is equivalent to 204 years of workload on the fastest supercomputer in the world (Chinese Sunway TaihuLight 2017), with 93 petaflops. The computation of the determinant of large-scale matrices cannot be performed with such an algorithm. More efficient algorithms are needed to solve the problem faster.

### 3.1.3 MATLAB solution of determinant problems

Before the computation of a matrix determinant, and example is presented. Then a computation method is introduced and a general purpose MATLAB solver is formulated.

**Example 3.5.** Compute the determinant of the following matrix:

$$A = \begin{bmatrix} 0 & 1 & 0 & -1 \\ 0 & 0 & -1 & 0 \\ -1 & 2 & 2 & 0 \\ 2 & -1 & 2 & 1 \end{bmatrix}.$$

**Solutions.** Generally speaking, the algebraic cofactor method of the determinant of a $4 \times 4$ matrix can be converted to a problem of four determinants of $3 \times 3$ matrices. The conversion of large-scale matrices to small-scale matrices may be time demanding. For this specific example, the problem of computing the determinant of the $4 \times 4$ matrix can be converted into that for one $3 \times 3$ matrix, therefore, the computation load can be significantly reduced.

In real applications, certain algorithms can be considered to convert ordinary matrices to special forms, for instance, triangular factorization, also known as LU factorization, which will be fully discussed later, can be used to convert a matrix into the product of an upper-triangular matrix $U$, and a lower-triangular matrix $L$, i. e., $A = LU$. Matrix $L$ has the determinant of one. Meanwhile, since $U$ is an upper-triangular matrix, whose determinant equals to the product of diagonal terms, we get that the the determinant of matrix $A$ equals to $\det(A) = \prod_{i=1}^{n} u_{ii}$.

A built-in function det() is provided in MATLAB to compute the determinant with $d$=det(A). The determinant of a matrix $A$ can be computed directly. The function also works for a symbolic matrix $A$.

**Example 3.6.** Compute the determinant of the matrix $A$ in Example 3.2.

**Solutions.** It can be seen that the analytically computed value of the matrix determinant is zero, while the numerical solution is $5.1337 \times 10^{-13}$, meaning that errors exist. The algebraic cofactor method, since only simple matrix multiplications are involved, normally is unlikely to generate errors. The efficient det() in MATLAB, although it is very fast, may lead to small errors.

```
>> A=[16 2 3 13; 5 11 10 8; 9 7 6 12; 4 14 15 1];
   det(A), det(sym(A))
```

**Example 3.7.** Compute the determinant again for the matrix in Example 3.4 with det() function.

**Solutions.** With the numerical method, the result of function det() is 509.9999999999999, with 0.0121 seconds needed, while with the symbolic computation method, the elapsed time is 0.0809 seconds, and an analytical solution can be obtained.

```
>> A=[1,-1,0,0,1,1,0,-1,0,1,-1; 1,-1,1,0,-1,-1,-1,-1,1,1,0;
      0,0,-1,0,-1,1,-1,0,1,1,0; 0,0,-1,0,-1,1,1,-1,0,0,0;
      1,-1,-1,0,1,1,0,0,0,-1,1; 1,1,0,-1,0,1,1,1,1,0,0;
      1,0,1,1,-1,0,-1,1,1,0,-1; 0,-1,-1,-1,1,0,-1,-1,-1,0,0;
      1,-1,1,0,1,-1,1,1,1,0,0; 1,1,1,0,0,0,0,-1,1,1,1;
      -1,-1,-1,0,-1,1,-1,1,-1,1,0];
   tic, det(A), toc, tic, det(sym(A)), toc
```

**Example 3.8.** A large-scale Hilbert matrix approaches a singular matrix as its size increases infinitely. Compute the analytical value of the determinant of an 80×80 Hilbert matrix.

**Solutions.** Function hilb() can be used to generate an $80 \times 80$ Hilbert matrix, and it can be converted to a symbolic one. MATLAB function det() can be used to compute the analytical value of this matrix determinant.

```
>> tic, H=sym(hilb(80)); det(H), toc % construct a symbolic matrix
```

It can be seen that the analytical value of the determinant is

$$\det(\boldsymbol{H}) = \frac{1}{\underbrace{99\,030\,101\,466\,993\,477\,878\,867\,678\cdots000\,000\,000\,000}_{3\,790\text{ digits, many digits are omitted}}} \approx 1.00979 \times 10^{-3\,971}.$$

It can be seen that the analytical solution can be found within 1.34 seconds. A low-level triangular matrix decomposition rather than the algebraic cofactor method is used in MATLAB det() function.

**Example 3.9.** It seems that symbolic computation is perfect in finding the determinant. Can the symbolic computation approach be used in finding the determinant of any matrix?

**Solutions.** Compared with numerical computation, the accuracy of symbolic operation is much higher than that in numerical computation. A test is made here to compare the computation speed in determinant evaluation. To have a fair comparison, random matrices are used in the test, and loops can be used to test the relationship between the matrix size and elapsed time. The elapsed times are recorded in Table 3.1, the time used in numerical computation is similar, within milliseconds. It can be seen from the trend that, when the size of a matrix is increased by 10, the elapsed time almost doubles. Therefore, symbolic computations are not suitable for large-scale matrix computations.

```
>> for n=10:10:90, H=rand(n);
       tic, det(H); toc, tic, det(sym(H)); toc
   end
```

**Table 3.1:** Determinant evaluation and elapsed time measurement.

| matrix size | 10 | 20 | 30 | 40 | 50 | 60 | 70 | 80 | 90 |
|---|---|---|---|---|---|---|---|---|---|
| elapsed time (seconds) | 0.0128 | 0.124 | 0.602 | 1.764 | 4.451 | 9.41 | 17.85 | 32.73 | 51.75 |

**Example 3.10.** Derive the determinant formula for an arbitrary $4 \times 4$ matrix.

**Solutions.** An arbitrary $4 \times 4$ symbolic matrix can be generated first, then function det() can be executed directly to find the determinant. No special skills are needed since the function can be applied not only for double precision matrices, but also for symbolic matrices.

```
>> A=sym('a%d%d',4); d=det(A) % analytical solution for 4 × 4 matrix
```

The general formula obtained is

$$d = a_{11}a_{22}a_{33}a_{44} - a_{11}a_{22}a_{34}a_{43} - a_{11}a_{23}a_{32}a_{44} + a_{11}a_{23}a_{34}a_{42} + a_{11}a_{24}a_{32}a_{43}$$
$$- a_{11}a_{24}a_{33}a_{42} - a_{12}a_{21}a_{33}a_{44} + a_{12}a_{21}a_{34}a_{43} + a_{12}a_{23}a_{31}a_{44} - a_{12}a_{23}a_{34}a_{41}$$

$$- a_{12}a_{24}a_{31}a_{43} + a_{12}a_{24}a_{33}a_{41} + a_{13}a_{21}a_{32}a_{44} - a_{13}a_{21}a_{34}a_{42} - a_{13}a_{22}a_{31}a_{44}$$
$$+ a_{13}a_{22}a_{34}a_{41} + a_{13}a_{24}a_{31}a_{42} - a_{13}a_{24}a_{32}a_{41} - a_{14}a_{21}a_{32}a_{43} + a_{14}a_{21}a_{33}a_{42}$$
$$+ a_{14}a_{22}a_{31}a_{43} - a_{14}a_{22}a_{33}a_{41} - a_{14}a_{23}a_{31}a_{42} + a_{14}a_{23}a_{32}a_{41}.$$

It can be seen from the results that the first 6 terms in the determinant are related to the factor $a_{11}$, and the next 6 terms are related to the $a_{12}$ factor. Therefore, it is easier to express the determinant as $a_{11}A_{11} + a_{12}A_{12} + a_{13}A_{13} + a_{14}A_{14}$, where $A_{ij}$ is the algebraic cofactor of $a_{ij}$, which is the determinant of the submatrix, when the $i$th row and $j$th column are deleted. The result is then multiplied by $(-1)^{i+j}$.

To compute $A_{23}$, two methods can be adopted: one is direct computation with the following statements:

```
>> i=2; j=3; B=A; B(i,:)=[]; B(:,j)=[];
   A23=(-1)^(i+j)*det(B) % delete the current row and column
```

and the result is

$$A_{23} = -a_{11}a_{32}a_{44} + a_{11}a_{34}a_{42} + a_{12}a_{31}a_{44} - a_{12}a_{34}a_{41} - a_{14}a_{31}a_{42} + a_{14}a_{32}a_{41}.$$

The other method is to delete the terms without $a_{23}$ factor from $d$, then divide by $a_{23}$. The algebraic cofactor can be found easily and the result is exactly the same.

```
>> syms a23; A23_1=simplify((d-subs(d,a23,0))/a23) % method 2
```

**Example 3.11.** Compute the determinant of the complex matrix in Example 2.25.

**Solutions.** The matrix is first input into MATLAB, and then the following statements can be used to compute the determinant. The numerical solution is $-0.000000000000057 - 53.999999999999986$j while the symbolic solution is $-54$j. It can be seen from the example that a complex matrix can be handled in exactly the same way.

```
>> A=[1+9i,2+8i,3+7j; 4+6j 5+5i,6+4i; 7+3i,8+2j 1i];
   det(A), det(sym(A))
```

### 3.1.4 Determinants of special matrices of any sizes

MATLAB can only be used to handle matrices with given sizes and cannot handle problems with arbitrary size $n$. For some specific determinant problems, specific size $n$ can be given and the determinants can be found. The relationship between the size and the determinant can be summarized from the results.

**Example 3.12.** Compute the determinant of the following $n \times n$ matrix:

$$A = \begin{bmatrix} x-a & a & a & \cdots & a \\ a & x-a & a & \cdots & a \\ a & a & x-a & \cdots & a \\ \vdots & \vdots & \vdots & \ddots & \vdots \\ a & a & a & \cdots & x-a \end{bmatrix}.$$

**Solutions.** If $n$ is not given, the $n \times n$ matrix cannot be handled. A nominal number $n = 20$ can be selected. An $n \times n$ matrix can be generated, whose determinant is $d = -(18a + x)(2a - x)^{19}$.

```
>> n=20; syms a x; A=a*ones(n); A=(x-2*a)*eye(n)+A;
   d=simplify(det(A))
```

If $n$ is selected as 21, the determinant is $d = (19a + x)(2a - x)^{20}$. It can be seen that the determinant of the $n \times n$ matrix is $d = (-1)^{n+1}((n-2)a + x)(2a - x)^{n-1}$.

**Example 3.13.** Compute the determinant for the $n \times n$ tri-diagonal matrix

$$A = \begin{bmatrix} 2 & 1 & 0 & \cdots & 0 & 0 & 0 \\ 1 & 2 & 1 & \cdots & 0 & 0 & 0 \\ \vdots & \vdots & \vdots & \ddots & \vdots & \vdots & \vdots \\ 0 & 0 & 0 & \cdots & 1 & 2 & 1 \\ 0 & 0 & 0 & \cdots & 0 & 1 & 2 \end{bmatrix}.$$

**Solutions.** A loop structure can be used to compute the determinants of some given sizes $n = 2, 3, \ldots, 8$ as

```
>> for n=2:8,
       v=ones(n-1,1); A=2*eye(n)+diag(v,1)+diag(v,-1);
       det(sym(A))
   end
```

It can be seen that the results are respectively 3, 4, ..., 9, from which it can be concluded that the determinant of the $n \times n$ matrix is $n + 1$.

**Example 3.14.** Compute the determinant of the $(2n) \times (2n)$ matrix below, and find a general form of the determinant if

$$A = \begin{bmatrix} a & & & & & b \\ & \ddots & & & \iddots & \\ & & a & b & & \\ & & c & d & & \\ & \iddots & & & \ddots & \\ c & & & & & d \end{bmatrix}.$$

**Solutions.** It can be seen that the first $n$ forward diagonals are $a$'s, and the other $n$ elements are $d$'s. The first $n$ backward diagonal elements are $c$'s and the other $n$ are $b$'s. If one selects $n = 10$, the following statements can be used to generate the special matrix. Then, we can compute the determinant as

```
>> syms a b c d; n=10; v=ones(1,n);
   A=diag([a*v d*v])+rot90(diag([c*v b*v])); det(A)
```

and the result is $(ad - bc)^{10}$. If one selects $n = 11$, the determinant is $(ad - bc)^{11}$. It can be seen that the determinant of the $(2n) \times (2n)$ matrix is $(ad - bc)^n$.

### 3.1.5 Cramer's rule in linear equations

Determinants are directly applied to solve linear equations. The following equation with two unknowns can be considered, from which the relationship between the solution and determinants is demonstrated. The commonly used Cramer's rule is presented.

**Example 3.15.** Solve a simple system of linear equations with two unknowns:

$$\begin{cases} a_{11}x_1 + a_{12}x_2 = b_1 \\ a_{21}x_1 + a_{22}x_2 = b_2. \end{cases}$$

**Solutions.** The linear equations can be expressed in matrix form as $Ax = b$, then the `solve()` function in the Symbolic Math Toolbox can be used to solve directly the linear equation system

```
>> syms a11 a12 a21 a22 b1 b2 x1 x2
   [x1,x2]=solve(a11*x1+a12*x2==b1,a21*x1+a22*x2==b2)
```

and the solution of the equations is

$$x_1 = \frac{a_{22}b_1 - a_{12}b_2}{a_{11}a_{22} - a_{12}a_{21}}, \quad x_2 = \frac{a_{11}b_2 - a_{21}b_1}{a_{11}a_{22} - a_{12}a_{21}}.$$

It is obvious that the denominator is the determinant of the coefficient matrix $A$. Observing the numerators in the solutions, it is not difficult to find that

$$a_{22}b_1 - a_{12}b_2 = \begin{vmatrix} b_1 & a_{12} \\ b_2 & a_{22} \end{vmatrix}, \quad a_{11}b_2 - a_{21}b_1 = \begin{vmatrix} a_{11} & b_1 \\ a_{21} & b_2 \end{vmatrix},$$

where the numerator of $x_1$ is the determinant of the coefficient matrix, whose first column is substituted by vector $b$, whereas in the solution for $x_2$, the second column in matrix $A$ is replaced by vector $b$.

Swiss mathematician Gabriel Cramer (1704–1752) presented a determinant-based algorithm in 1750 to solve linear algebraic equations. The algorithm is also known as Cramer's rule.

**Theorem 3.10.** *If matrix **A** is a nonsingular square matrix, the unique solution of the equation **Ax** = **b** can be expressed as*

$$x_1 = \frac{\det(\boldsymbol{D}_1)}{\det(\boldsymbol{A})}, \quad x_2 = \frac{\det(\boldsymbol{D}_2)}{\det(\boldsymbol{A})}, \quad \ldots, \quad x_n = \frac{\det(\boldsymbol{D}_n)}{\det(\boldsymbol{A})}, \tag{3.1.6}$$

*where **D**$_j$ is the determinant of **A**, whose jth column is substituted by **b**.*

If **b** is a matrix rather than a column vector, each column can be processed individually to solve the equations.

Based on the Cramer's rule discussed earlier, a general purpose MATLAB solver for linear equations can be written. Equations with multiple column matrix **B** can be solved and also symbolic solutions can be obtained.

```
function x=cramer(A,B)
D=det(A); [n,m]=size(B);
if D==0, error('coefficient matrix is singular')
else
    for i=1:m, for j=1:n
        A1=A; A1(:,j)=B(:,i); x0(j)=det(A1)/D;
    end, x(:,i)=x0;
end, end
```

**Example 3.16.** Solve the following linear equation with Cramer's rule:

$$\begin{bmatrix} 17 & 24 & 1 & 8 & 15 \\ 23 & 5 & 7 & 14 & 16 \\ 4 & 6 & 13 & 20 & 22 \\ 10 & 12 & 19 & 21 & 3 \\ 11 & 18 & 25 & 2 & 9 \end{bmatrix} \boldsymbol{x} = \begin{bmatrix} -1 & -2 \\ 1 & -1 \\ -2 & -2 \\ 0 & 2 \\ 1 & 0 \end{bmatrix}.$$

**Solutions.** The two matrices should be entered into MATLAB first, and then the analytical and numerical solutions can be obtained directly.

```
>> A=[17,24,1,8,15; 23,5,7,14,16; 4,6,13,20,22;
      10,12,19,21,3; 11,18,25,2,9];
   B=[-1,-2; 1,-1; -2,-2; 0,2; 1,0];
   x=cramer(sym(A),B), x1=cramer(A,B)
```

The analytical and numerical solutions of the equation are

$$x = \begin{bmatrix} 127/975 & 31/975 \\ -68/975 & -14/975 \\ 62/975 & 56/975 \\ -68/975 & 61/975 \\ -68/975 & -179/975 \end{bmatrix}, \quad x_1 = \begin{bmatrix} 0.130256410256410 & 0.031794871794872 \\ -0.069743589743590 & -0.014358974358974 \\ 0.063589743589744 & 0.057435897435897 \\ -0.069743589743590 & 0.062564102564103 \\ -0.069743589743590 & -0.183589743589744 \end{bmatrix}.$$

It is worth mentioning that, due to the limitations in the evaluation of determinants, the linear equation solver provided here is not efficient. The solvers to be presented in Chapter 5 are recommended.

### 3.1.6 Positive and totally positive matrices

**Definition 3.4.** If all the elements in $A \in \mathcal{R}^{n \times m}$ are positive, it is referred to as a positive matrix, denoted as $A > 0$.

**Definition 3.5.** If all the elements in $A$ satisfy $a_{ij} \geq 0$, the matrix is referred to as a nonnegative matrix, denoted as $A \geq 0$.

**Definition 3.6.** If all the elements in matrices $A$ and $B$ satisfy $a_{ij} \geq b_{ij}$, this is denoted as $A \geq B$.

**Definition 3.7.** Assume that matrix $A$ can be written as

$$A = \begin{bmatrix} a_{11} & a_{12} & a_{13} & \cdots & a_{1n} \\ a_{21} & a_{22} & a_{23} & \cdots & a_{2n} \\ a_{31} & a_{32} & a_{33} & \cdots & a_{3n} \\ \vdots & \vdots & \vdots & \ddots & \vdots \\ a_{n1} & a_{n2} & a_{n3} & \cdots & a_{nn} \end{bmatrix}, \tag{3.1.7}$$

the determinants of the upper-left corner submatrices (minors) are referred to as main subdeterminants.

**Definition 3.8.** If all the determinants of the main minors are positive, the matrix is referred to as a totally positive matrix.

**Example 3.17.** Check whether the coefficient matrix $A$ in Example 3.16 is a positive or a totally positive matrix.

**Solutions.** Since all the elements in the matrix are positive, the matrix is positive. A loop structure can be used to evaluate the determinants of the upper-left submatrices, with $v = [17, -467, -5\,995, 56\,225, 5\,070\,000]$. Since there are negative entries, $A$ is not a totally positive matrix.

```
>> A=[17,24,1,8,15; 23,5,7,14,16; 4,6,13,20,22;
       10,12,19,21,3; 11,18,25,2,9]; A=sym(A);
   v=[]; for i=1:5, v=[v,det(A(1:i,1:i))]; end
```

**Example 3.18.** Judge whether $A$ is a totally positive matrix if

$$
A = \begin{bmatrix} 4 & 4 & 0 & -1 \\ -2 & 4 & 1 & 4 \\ 3 & -1 & 4 & 1 \\ 1 & 2 & 1 & 3 \end{bmatrix}.
$$

**Solutions.** Since there are negative elements in $A$, it is not a positive matrix. A loop structure can be used again to evaluate the determinants of the upper-left submatrices, with $v = [4, 24, 112, 217]$, which are all positive, meaning that $A$ is a totally positive matrix. It can be seen that a totally positive matrix may not be a positive matrix.

```
>> A=[4,4,0,-1; -2,4,1,4; 3,-1,4,1; 1,2,1,3];
   v=[]; for i=1:5, v=[v,det(A(1:i,1:i))]; end
```

## 3.2 Simple analysis of matrices

Matrices are very important objects in linear algebra. In this section, the properties of matrices are further analyzed, and the concepts of trace, rank, and linear independence are introduced; and the solutions to problems involving such concepts are given in MATLAB. Also, the definitions and computations of the norm measures are presented, and the concept of a vector space is addressed.

### 3.2.1 The traces

**Definition 3.9.** The trace of an $n \times n$ square matrix $A = \{a_{ij}\}$ is defined as the sum of all its diagonal elements,

$$
\mathrm{tr}(A) = \sum_{i=1}^{n} a_{ii}. \tag{3.2.1}
$$

There is no trace defined for nonsquare matrices. The properties of matrix trace are as follows:

**Theorem 3.11.** *For any matrices $A$, $B$ and constant $c$, the trace satisfies*

$$
\mathrm{tr}(A) = \mathrm{tr}(A^{\mathrm{T}}), \quad \mathrm{tr}(A + B) = \mathrm{tr}(A) + \mathrm{tr}(B), \quad \mathrm{tr}(cA) = c\,\mathrm{tr}(A). \tag{3.2.2}
$$

**Theorem 3.12.** *The trace of matrices satisfies* $\mathrm{tr}(AB) = \mathrm{tr}(BA)$.

**Theorem 3.13.** *For any constants n and m,* $\mathrm{tr}(m\boldsymbol{A} + n\boldsymbol{B}) = m\,\mathrm{tr}(\boldsymbol{A}) + n\,\mathrm{tr}(\boldsymbol{B})$.

**Theorem 3.14.** *For any column vector* $\boldsymbol{x}$, $\boldsymbol{x}^{\mathrm{T}}\boldsymbol{A}\boldsymbol{x} = \mathrm{tr}(\boldsymbol{x}\boldsymbol{x}^{\mathrm{T}}\boldsymbol{A}) = \mathrm{tr}(\boldsymbol{A}\boldsymbol{x}\boldsymbol{x}^{\mathrm{T}})$.

**Theorem 3.15.** *For any matrices* $\boldsymbol{A}$ *and* $\boldsymbol{B}$, $\mathrm{tr}(\boldsymbol{A} \otimes \boldsymbol{B}) = \mathrm{tr}(\boldsymbol{A})\,\mathrm{tr}(\boldsymbol{B})$.

It can be seen from algebra that the trace of a matrix equals to the sum of its eigenvalues. The trace of matrix $\boldsymbol{A}$ can be evaluated in MATLAB with `trace()` function, with $t$=`trace(A)`. If $\boldsymbol{A}$ is a rectangular (nonsquare) matrix, an error message is given. A low-level command $t$=`sum(diag(A))` can be used directly, even if the matrix is nonsquare.

**Example 3.19.** Compute the trace of the matrix in Example 3.2.

**Solutions.** It can be found with the following MATLAB commands that $\mathrm{tr}(\boldsymbol{A}) = 34$.

```
>> A=[16 2 3 13; 5 11 10 8; 9 7 6 12; 4 14 15 1];
   t=trace(A)
```

### 3.2.2 Linear independence and matrix rank

**Definition 3.10.** For a given set of column vectors $\boldsymbol{a}_1, \boldsymbol{a}_2, \ldots, \boldsymbol{a}_m$, if there exists a set of nonzero constants $k_1, k_2, \ldots, k_m$ such that

$$k_1\boldsymbol{a}_1 + k_2\boldsymbol{a}_2 + \cdots + k_m\boldsymbol{a}_m = \boldsymbol{0}, \tag{3.2.3}$$

then the set of such vectors is referred to as linearly dependent. If there does not exist such a set of constants, the vectors are linearly independent.

In other words, if one or more vectors can be expressed as a linear combination of other vectors in the set, they are linearly dependent; If none of them can be expressed as linear combinations of the others, they are linearly independent.

**Definition 3.11.** If $r_c$ column vectors in a matrix are linearly independent, the column rank of the matrix is $r_c$.

**Definition 3.12.** If $r_c = m$, $\boldsymbol{A}$ is referred to as a full column-rank matrix.

**Definition 3.13.** Similarly, if $r_r$ rows in $\boldsymbol{A}$ are linearly independent, the row rank is $r_r$. If $r_r = n$, $\boldsymbol{A}$ is referred to as a full row-rank matrix.

Some of the properties of rank and linearly independent sets of vectors are:

**Theorem 3.16.** *The row and column ranks are identical, and referred to as the rank of a matrix, denoted as* $\mathrm{rank}(\boldsymbol{A}) = r_c = r_r$.

**Theorem 3.17.** *If the column vectors* $\boldsymbol{a}_1, \boldsymbol{a}_2, \ldots, \boldsymbol{a}_m$ *are linearly dependent, no matter what column* $\boldsymbol{a}_{m+1}$ *is, the vectors* $\boldsymbol{a}_1, \boldsymbol{a}_2, \ldots, \boldsymbol{a}_m, \boldsymbol{a}_{m+1}$ *are linearly dependent.*

**Theorem 3.18.** *For m given $n \times 1$ vectors, if $m > n$, the m vectors must be linearly dependent.*

**Theorem 3.19.** *If a vector set A given by $a_1, a_2, \ldots, a_m$ is linearly independent, while set B formed by $a_1, a_2, \ldots, a_m, b$ is linearly dependent, then vector $b$ can be expressed as a linear combination of the vectors in set A, and the combination is unique.*

**Theorem 3.20.** *The following elementary row (column) operations do not affect the rank of the matrix:*
(1) *Any row (column) multiplication by a constant $\alpha$;*
(2) *Exchange of any two rows (columns);*
(3) *Any row (column) multiplication by $\alpha$ and addition to another row (column).*

**Theorem 3.21.** *The rank of any rectangular matrix satisfies*

$$\mathrm{rank}(A) = \mathrm{rank}(AA^T) = \mathrm{rank}(A^T A). \tag{3.2.4}$$

The rank of a matrix is the maximum size of a submatrix, the combination of any $k$ rows and $k$ columns, whose determinant is not zero.

There are many algorithms which can be used to evaluate the rank of a matrix, some of them are stable, while others may be unreliable. Singular value-based algorithm is implemented in MATLAB[4], that is, $n$ singular values $\sigma_i$, $i = 1, 2, \ldots, n$ of $A$ are obtained, and if $r$ of them is greater than the error tolerance $\varepsilon$, then $r$ is the numerical rank of matrix $A$. The concept of singular values will be presented later.

A built-in function rank() can be used to compute the rank, with

$r$=rank($A$), % default error tolerance

$r$=rank($A$,$\varepsilon$), % with given error tolerance $\varepsilon$

where $A$ is a given matrix, while $\varepsilon$ is the error tolerance. A rank() function is also provided in Symbolic Math Toolbox to compute the analytical value of the rank, and the syntax is rank($A$).

**Example 3.20.** Compute the rank of matrix $A$ in Example 3.2.

**Solutions.** With function rank($A$), the rank can be evaluated directly. The rank of the matrix is 3, less than the size of the matrix, indicating that $A$ is not a full-rank matrix, i. e., a singular matrix.

```
>> A=[16 2 3 13; 5 11 10 8; 9 7 6 12; 4 14 15 1];
   rank(A), det(A) % compute the rank and determinant
```

In fact, it is not a good way to check the singularity of a matrix with determinant and eigenvalues, since the numerical value of the determinant is $5.1337 \times 10^{-13}$, rather than zero. Misleading conclusions may be drawn, and one must be very careful in practical applications.

**Example 3.21.** Consider the $20 \times 20$ Hilbert matrix in Example 3.8. Compute the numerical and analytical ranks of this matrix, and validate the results.

**Solutions.** Consider the numerical method first. The following statements can be used, and it can be seen that the numerical rank is 12.

```
>> H=hilb(20); rank(H) % find the numerical rank, leading to wrong conclusions
```

A misleading conclusion may be obtained by this result: since the rank obtained is less than the size of the matrix, $H$ is not a full-rank matrix. If symbolic computation is adopted for the same example, it is found that the rank of the matrix is 20, meaning that the matrix is of full-rank. If a matrix is close to a singular one, the numerical ranks may not be correct.

```
>> H=sym(hilb(20));
   rank(H)    % analytical rank, meaning the matrix is nonsingular
```

### 3.2.3 Norms

Norms of a matrix are measures of the matrix. Before introducing the norms of a matrix, the norms of a vector are discussed. Then, matrix norms and their computation are presented.

**Definition 3.14.** For a vector $x$, if a function $\rho(x)$ satisfies the three conditions:
(1) $\rho(x) \geq 0$, and $\rho(x) = 0$ if and only if $x = 0$;
(2) $\rho(ax) = |a|\rho(x)$, where $a$ is any scalar;
(3) for vectors $x$ and $y$, $\rho(x + y) \leq \rho(x) + \rho(y)$,

then the function $\rho(x)$ is referred to as a norm of $x$.

There are various kinds of norms.

**Theorem 3.22.** *It can be shown that a family of functions satisfying all the three conditions presented above is*

$$\|x\|_p = \left( \sum_{i=1}^{n} |x_i|^p \right)^{1/p}, \quad p = 1, 2, \ldots, \quad and \quad \|x\|_\infty = \max_{1 \leq i \leq n} |x_i|. \tag{3.2.5}$$

*The notation of norm $\|x\|_p$ is used.*

**Definition 3.15.** The 2-norm of $x$ is $\|x\|_2 = \sqrt{x_1^2 + x_2^2 + \cdots + x_n^2}$, also known as the length of the vector $x$.

The definition of a matrix norm is slightly more complicated than that of a vector norm, with the mathematical definition given below.

**Definition 3.16.** The norm of matrix $A$ is defined as

$$\|A\| = \sup_{x \neq 0} \frac{\|Ax\|}{\|x\|}. \tag{3.2.6}$$

**Definition 3.17.** Similar to vector norms, there are some commonly used matrix norms, namely

$$\|A\|_1 = \max_{1 \leqslant j \leqslant n} \sum_{i=1}^{n} |a_{ij}|, \quad \|A\|_2 = \sqrt{s_{\max}(A^H A)}, \quad \|A\|_\infty = \max_{1 \leqslant i \leqslant n} \sum_{j=1}^{n} |a_{ij}|, \tag{3.2.7}$$

where $s(X)$ represents the eigenvalues of matrix $X$, and $s_{\max}(A^H A)$ is the maximum eigenvalue of $A^H A$. In fact, $\|A\|_2$ also equals to the maximum singular value of matrix $A$.

It can be seen from the definitions that the $\|A\|_1$ norm is the maximum value of the sum of the absolute values of the column elements, while the $\|A\|_\infty$ norm is the sum of the absolute values of the row elements. If $A$ is a symmetric matrix, $\|A\|_\infty = \|A\|_1$.

**Definition 3.18.** Another commonly used norm is the Frobenius norm, defined as $\|A\|_F = \sqrt{\text{tr}(A^H A)}$.

Function norm() in MATLAB can be used to evaluate the norms of a matrix, with $N$=norm($A$,options), where "options" are shown in Table 3.2. If no option is given, $\|A\|_2$ is computed. In new versions of MATLAB, this function can also be used to handle symbolic matrices.

**Example 3.22.** Compute various norms for the matrix in Example 3.2.

**Solutions.** Matrix $A$ can be input first, then the following MATLAB function calls can be written to directly compute the norms:

```
>> A=[16 2 3 13; 5 11 10 8; 9 7 6 12; 4 14 15 1]; n1=norm(A)
   n2=norm(A,2), n3=norm(A,1), n4=norm(A,Inf)
   n5=norm(A,'fro') % various norms
```

and the norms obtained are $\|A\|_1 = \|A\|_2 = \|A\|_\infty = 34$, $\|A\|_F = 38.6782$.

**Table 3.2:** Options in the norm() function.

| options | meaning and explanations |
|---|---|
| none | maximum singular value, i. e., $\|A\|_2$ |
| 2 | 2-norm of the matrix, i. e., $\|A\|_2$ |
| 1 | 1-norm of the matrix, i. e., $\|A\|_1$ |
| Inf or 'inf' | infinity norm $\|A\|_\infty$ |
| 'fro' | Frobenius norm, i. e., $\|A\|_F = \sqrt{\text{tr}(A^H A)}$ |
| number $p$ | any integer for vectors, while for matrices, only 1, 2, inf, and 'fro' |
| -inf | for vectors, the minimum absolute value $\|A\|_{-\infty} = \min(|a_1|, |a_2|, \ldots, |a_n|)$ |

It is worth mentioning that `norm(A)` and `norm(A,2)` yield the same results, namely $\|A\|_2$; also since for the magic matrix, all the sums of rows and columns are the same, $\|A\|_1 = \|A\|_\infty$. Generally speaking, $\|A\|_1 = \|A\|_\infty$ is not always satisfied.

Norm function `norm()` can be used to compute norms of a numeric matrix, but it cannot be used to measure norms containing variables. In the current versions of MAT-LAB, `norm()` function can also be used to handle symbolic matrices whose elements are numbers.

**Example 3.23.** Cramer's rule-based numerical solutions have been obtained for the equations in Example 3.16. Assess the precision of the results.

**Solutions.** The best way to assess the accuracy of equation solutions is to substitute the solutions back to the equations, and measure the error matrix. Since the error is a matrix, it is not good to assess individually each error element, norm measures should be adopted. The error matrix norm in this example is $1.8208 \times 10^{-15}$, and it can be seen that the error is relatively small under the double precision framework.

```
>> A=[17,24,1,8,15; 23,5,7,14,16; 4,6,13,20,22;
      10,12,19,21,3; 11,18,25,2,9];
   B=[-1,-2; 1,-1; -2,-2; 0,2; 1,0];
   x=cramer(A,B), norm(A*x-B)
```

Since there is no error in the analytical solution, the error matrix norm is zero.

```
>> x=cramer(sym(A),B), norm(A*x-B)
```

### 3.2.4 Vector space

**Definition 3.19.** Assume that $V$ is a set of $n$-dimensional column vectors, $a$ and $b$ are any two vectors in $V$, denoted as $a \in V$ and $b \in V$. If $a+b \in V$, and for any real number $k$, $ka \in V$, set $V$ is referred to as a linear space.

**Definition 3.20.** Assume that in the vector space $V$, there are $r$ column vectors $a_1, a_2, \ldots, a_r \in V$ such that
(1) $a_1, a_2, \ldots, a_r$ are linearly independent;
(2) Any column vector $V$ can be represented as a linear combination of vectors $a_1, a_2, \ldots, a_r$;

then the vector set $a_1, a_2, \ldots, a_r$ is referred to as a basis of vector space $V$, and $r$ is referred to as the dimension of the space; $V$ is also known as an $r$-dimensional vector space.

**Definition 3.21.** If the basis $a_1, a_2, \ldots, a_r$ is regarded as the coordinates of vector space $V$, any vector $v$ can be expressed as a linear combination of the basis

$$v = \text{span}(a_1, a_2, \ldots, a_r) = \sum_{j=1}^{n} x_j a_j, \tag{3.2.8}$$

then $(x_1, x_2, \ldots, x_r)$ can be regarded as the coordinates of the vector $v$.

**Definition 3.22.** Assume that $n$-dimensional vectors $e_1, e_2, \ldots, e_r$ form a basis of vector space $V$, $e_1, e_2, \ldots, e_r$ are orthogonal and are all unit vectors, then vectors $e_1, e_2, \ldots, e_r$ are said to be an orthonormal basis in $V$.

**Example 3.24.** Assume that the vector space is defined as $a_1 = [2, 2, -1]^T$, $a_2 = [2, -1, 2]^T$, $a_3 = [-1, 2, 2]^T$. Compute the new coordinates of vector $v = [1, 0, -4]^T$ in the vector space spanned by the vectors $a_1, a_2, a_3$.

**Solutions.** Assume that the new coordinates are $x_1, x_2$ and $x_3$. The following equation can be set up:

$$x_1 a_1 + x_2 a_2 + x_3 a_3 = v \implies [a_1, a_2, a_3]x = v.$$

The new coordinates can be found with the following statements, and the result is $x = [0, 4/3, 5/3]^T$.

```
>> a1=[1,2,-1]'; a2=[2,-1,1]'; a3=[-1,2,1]';
   v=[1,2,3]'; A=[a1 a2 a3]; x=cramer(sym(A),v)
```

## 3.3 Inverse and generalized inverse matrices

Inverse matrix is one of the most important concepts in linear algebra. The algebraic computations of matrices presented so far were additions, subtractions, and multiplications. Inverse matrices can be regarded as results of divisions of matrices. When solving linear equations and in many other fields, inverse matrices are inevitable. In this section, the definition and properties of inverse matrices are introduced. Different methods are described to compute the inverses, and a general MATLAB function for inverse computation is introduced. The concept of generalized inverse is presented.

### 3.3.1 Inverse matrices

**Definition 3.23.** For a given $n \times n$ square matrix $A$, if there exists a matrix $C$ of the same size, satisfying

$$AC = CA = I, \tag{3.3.1}$$

where $I$ is an identity matrix, then matrix $C$ is referred to as the inverse matrix of $A$, and it is denoted as $C = A^{-1}$.

It should be noted that rectangular (nonsquare) matrices do not have inverses. In traditional linear algebra courses, inverse matrices can be computed with adjoint matrices. In this section, adjoint matrices are also presented.

**Definition 3.24.** For a square matrix $A$, the transpose matrix of the algebraic cofactors $A_{ij}$ of the $i$th row and $j$th column is regarded as the an adjoint matrix of $A$, denoted as $A^*$,

$$A^* = \begin{bmatrix} A_{11} & A_{21} & \cdots & A_{n1} \\ A_{12} & A_{22} & \cdots & A_{n2} \\ \vdots & \vdots & \ddots & \vdots \\ A_{1n} & A_{2n} & \cdots & A_{nn} \end{bmatrix}. \tag{3.3.2}$$

Function adjoint() is provided in Symbolic Math Toolbox to compute the adjoint matrix, with the syntax $A_1$=adjoint($A$).

**Theorem 3.23.** *The inverse of matrix $A$ can be computed from*

$$A^{-1} = \frac{A^*}{\det(A)}. \tag{3.3.3}$$

**Example 3.25.** Compute the inverse if the complex matrix $B$ in Example 2.25 with adjoint matrix method, and validate the results.

**Solutions.** The complex matrix should be entered into MATLAB and converted to a symbolic matrix, since adjoint() function can only be used to process symbolic variables, then the adjoint and inverse matrices can be computed from the following statements:

```
>> B=[1+9i,2+8i,3+7j; 4+6j 5+5i,6+4i; 7+3i,8+2j 1i];
   B=sym(B); A=adjoint(B), iB=A/det(B), iB*B, B*iB
```

The adjoint and inverse matrices are obtained as given below. It can be validated such that $BB^{-1}$ and $B^{-1}B$ are both identity matrices, indicating the inverse matrix is correct:

$$B^* = \begin{bmatrix} -39j - 45 & 60j + 18 & 6j \\ 42j + 36 & -57j - 9 & -12j \\ 6j & -12j & 6j \end{bmatrix},$$

$$B^{-1} = \begin{bmatrix} 13/18 - 5j/6 & j/3 - 10/9 & -1/9 \\ 2j/3 - 7/9 & 19/18 - j/6 & 2/9 \\ -1/9 & 2/9 & -1/9 \end{bmatrix}.$$

As in the case of determinant methods, the efficiency of this method is not high, it is not suitable for finding inverses of large-scale matrices. High efficiency inverse matrix computing method should be considered.

**Theorem 3.24.** *If matrix **A** is invertible, the inverse of **A** is unique.*

**Theorem 3.25.** *If matrix **A** is invertible, the following equations hold:*

$$(\mathbf{A}^{\mathrm{T}})^{-1} = (\mathbf{A}^{-1})^{\mathrm{T}}, \quad (\mathbf{A}^{-1})^{-1} = \mathbf{A}, \quad (k\mathbf{A})^{-1} = \frac{1}{k}\mathbf{A}^{-1}, \quad k \neq 0. \tag{3.3.4}$$

**Theorem 3.26.** *If **A** and **B** matrices are invertible, $(\mathbf{AB})^{-1} = \mathbf{B}^{-1}\mathbf{A}^{-1}$.*

### 3.3.2 Derivatives of inverse matrix

Some theorems regarding the derivatives of inverse matrix-related functions are summarized, and the inverse matrix derivative method in MATLAB is presented. Direct methods for finding the high-order derivatives of any matrix functions are formulated.

**Theorem 3.27.** *For a matrix function **A**(t), the derivative of the inverse matrix satisfies*

$$\frac{\mathrm{d}\mathbf{A}^{-1}(t)}{\mathrm{d}t} = -\mathbf{A}^{-1}(t)\frac{\mathrm{d}\mathbf{A}(t)}{\mathrm{d}t}\mathbf{A}^{-1}(t). \tag{3.3.5}$$

**Theorem 3.28.** *For a matrix function **A**(t), the derivatives of matrix powers satisfy*

$$\frac{\mathrm{d}\mathbf{A}^n(t)}{\mathrm{d}t} = \sum_{i=1}^{n} \mathbf{A}^{i-1}(t)\frac{\mathrm{d}\mathbf{A}(t)}{\mathrm{d}t}\mathbf{A}^{n-i}(t) \tag{3.3.6}$$

$$\frac{\mathrm{d}\mathbf{A}^{-n}(t)}{\mathrm{d}t} = -\sum_{i=1}^{n} \mathbf{A}^{-i}(t)\frac{\mathrm{d}\mathbf{A}(t)}{\mathrm{d}t}\mathbf{A}^{-(n+1-i)}(t). \tag{3.3.7}$$

It can be seen that (3.3.5) is only a special case of (3.3.7). A general purpose MATLAB function can be written to implement the formulas in Theorem 3.28.

```
function A1=mpower_diff(A,n)
A1=zeros(size(A)); dA=diff(A); n0=abs(n); if n<0, iA=inv(A); end
for i=1:n0
    if n>=0, F=A^(i-1)*dA*A^(n-i); else, F=-iA^i*dA*iA^(n0+1-i); end
    A1=A1+F;
end
```

**Example 3.26.** Compute the inverse of matrix **A**(t) in Example 2.33, and find the second-order derivative of $\mathbf{A}^{-2}(t)$.

**Solutions.** Matrix **A**(t) should be entered into MATLAB environment as a symbolic function, then mpower_diff() can be used to find the first-order derivative of $\mathbf{A}^{-2}(t)$, and from the result, function diff() can be called again to find the second-order derivative of $\mathbf{A}^{-2}(t)$.

```
>> syms t;
   A(t)=[t^2/2+1,t,t^2/2; t,1,t; -t^2/2,-t,1-t^2/2]*exp(-t);
   A1=mpower_diff(A,-2); A2=simplify(diff(A1))
```

The second-order derivative can be obtained as

$$
A_2(t) = \begin{bmatrix} 8(t+1)^2 & -8-8t & 8t^2+16t+4 \\ -8-8t & 4 & -8-8t \\ -8t^2-16t-4 & 8t+8 & -8t(t+2) \end{bmatrix} e^{2t}.
$$

In fact, the following statements can be used directly with `diff()` function, and the same results are obtained:

```
>> A3=diff(A^(-2),2); simplify(A3)
```

**Example 3.27.** Compute the third-order derivative of the matrix function $B = e^{A(t)}A^{-1}(t)$.

**Solutions.** The following statements can be used to compute the third-order derivative directly. The results are too complicated to be listed here.

```
>> syms t;
   A(t)=[t^2/2+1,t,t^2/2; t,1,t; -t^2/2,-t,1-t^2/2]*exp(-t);
   B=exp(A)*inv(A); A1=simplify(diff(B,3))
```

In practical applications, there is no need to find derivatives according to Theorem 3.28, since the formulas involved are complicated. Function `diff()` can be used instead to compute directly derivative functions of complicated symbolic functions.

### 3.3.3 MATLAB based inverse matrix evaluation

Function $C=$`inv(A)` in MATLAB can be used to compute the inverse matrix $C$ for the given matrix $A$. This function can only be used to compute inverses for symbolic matrices. If $A$ is a symbolic matrix, analytical solution to the inverse can be found, otherwise, numerical solutions can be found, and the efficiency of the function is much higher than that of the existing functions.

**Example 3.28.** Compute the inverses of Hilbert matrices.

**Solutions.** Consider first a $4 \times 4$ Hilbert matrix. A direct call of `inv()` can be used to compute the inverse of it.

```
>> format long; H=hilb(4); H1=inv(H)
   norm(H*H1-eye(4)) % display more digits
```

The inverse thus obtained is shown below. The norm of the error matrix is $1.3931 \times 10^{-13}$:

$$H^{-1} = \begin{bmatrix} 15.999999999999 & -119.99999999999 & 239.99999999998 & -139.99999999999 \\ -119.99999999999 & 1\,199.9999999999 & -2\,699.9999999997 & 1\,679.9999999998 \\ 239.99999999998 & -2\,699.9999999997 & 6\,479.9999999994 & -4\,199.9999999996 \\ -139.99999999999 & 1\,679.9999999998 & -4\,199.9999999996 & 2\,799.9999999997 \end{bmatrix}.$$

It can be seen that the norm of the error matrix is very small, although it is larger than the usual $10^{-15} \sim 10^{-16}$ range. It can usually be accepted for this example.

Large-scale Hilbert matrices are very close to singular matrices. Normally, it is not recommended to call function inv() numerically, function invhilb() should be called instead, and the norm of the error is $5.684 \times 10^{-14}$.

```
>> H2=invhilb(4); norm(H*H2-eye(size(H))) % validate invhilb() function
```

It can be seen that for small-scale matrices, the accuracy of the result from function invhilb() is improved. Considering now a $10 \times 10$ Hilbert matrix, the errors of the two functions are respectively $n_1 = 1.4718 \times 10^{-4}$ and $n_2 = 1.6129 \times 10^{-5}$, which are relatively large in real applications.

```
>> H=hilb(10); H1=inv(H); n1=norm(H*H1-eye(size(H)))
   H2=invhilb(10); n2=norm(H*H2-eye(size(H))) % different algorithms
```

It can be seen that although the accuracy of invhilb() is higher than that of inv() function, if the size of the matrix is further increased, for instance, to 13, the two errors are $n_1 = 2.1315$ and $n_2 = 11.3549$, which are wrong, and it is necessary to say that the matrix approaches a singular one.

```
>> H=hilb(13); H1=inv(H); n1=norm(H*H1-eye(size(H)))
   H2=invhilb(13); n2=norm(H*H2-eye(size(H))) % large-scale matrix
```

Since inv() can also be used to find inverses of large-scale matrices, the above mentioned matrix can be analytically found. The inverse of the $6 \times 6$ Hilbert matrix can be found with

```
>> H=sym(hilb(6)); H1=inv(H) % symbolic computation
```

and the inverse matrix obtained is

$$H_1 = \begin{bmatrix} 36 & -630 & 3\,360 & -7\,560 & 7\,560 & -2\,772 \\ -630 & 14\,700 & -88\,200 & 211\,680 & -220\,500 & 83\,160 \\ 3\,360 & -88\,200 & 564\,480 & -1\,411\,200 & 1\,512\,000 & -582\,120 \\ -7\,560 & 211\,680 & -1\,411\,200 & 3\,628\,800 & -3\,969\,000 & 1\,552\,320 \\ 7\,560 & -220\,500 & 1\,512\,000 & -3\,969\,000 & 4\,410\,000 & -1\,746\,360 \\ -2\,772 & 83\,160 & -582\,120 & 1\,552\,320 & -1\,746\,360 & 698\,544 \end{bmatrix}.$$

In fact, Symbolic Math Toolbox can be used to compute inverses for large-scale Hilbert matrices. For instance, a $30 \times 30$ matrix can be used as a test example. It can be seen that the error is zero.

```
>> H=sym(hilb(30));
   norm(H*inv(H)-eye(size(H))) % the error is zero
```

**Example 3.29.** Find the inverse of the singular matrix in Example 3.2, and see what will happen in numerical computation.

**Solutions.** Input the model first, then the function inv() can be used to compute the inverse

```
>> A=[16 2 3 13; 5 11 10 8; 9 7 6 12; 4 14 15 1];
   B=inv(A), A*B % numerical computation
```

A warning message "Warning: Matrix is close to singular or badly scaled" is displayed, indicating that the matrix is close to being singular, and, in fact, it is singular. The inverse matrix thus obtained is not correct.

With the following commands, the "inverse" **B** can be found, and **AB** can also be computed as

$$
B = \begin{bmatrix}
-0.2649 & -0.7948 & 0.7948 & 0.2649 \\
-0.7948 & -2.384 & 2.384 & 0.7948 \\
0.7948 & 2.384 & -2.384 & -0.7948 \\
0.2649 & 0.7948 & -0.7948 & -0.2649
\end{bmatrix} \times 10^{15},
$$

$$
AB = \begin{bmatrix}
1.5 & 0 & 2 & 0.5 \\
-1 & -2 & 3 & 2.25 \\
-0.5 & -4 & 4 & 0.5 \\
-1.125 & -5.25 & 5.375 & 3.0313
\end{bmatrix}.
$$

Validating the above results, it can be seen that the errors are very large, and the inverse is incorrect.

**Example 3.30.** Solve the inverse problem of the matrix in Example 3.29 with symbolic methods.

**Solutions.** Direct solution commands can be used for the problem, and the result obtained is an error message "FAIL", indicating the original matrix is singular, meaning that the inverse does not exist.

```
>> A=[16 2 3 13; 5 11 10 8; 9 7 6 12; 4 14 15 1];
   A=sym(A); inv(A) % compute analytically the inverse
```

In fact, there are no inverses for singular matrices, satisfying the conditions in (3.3.1). For singular matrices, `inv()` cannot be used either. The concept of "inverse" should be extended, which will be discussed later.

**Example 3.31.** Compute the inverse of any $4 \times 4$ Hankel matrix.

**Solutions.** The inverse function in MATLAB applies to matrices with variables. The following statements can be used to compute the inverse of a $4 \times 4$ Hankel matrix, with `inv()` function.

```
>> a=sym('a',[1,4]); H=hankel(a);
   inv(H) % inverse of any 4 × 4 Hankel matrix
```

The following inverse can be obtained:

$$
H^{-1} = \begin{bmatrix}
0 & 0 & 0 & 1/a_4 \\
0 & 0 & 1/a_4 & -1/a_4^2 a_3 \\
0 & 1/a_4 & -1/a_4^2 a_3 & -1/a_4^3(a_2 a_4 - a_3^2) \\
1/a_4 & -1/a_4^2 a_3 & -1/a_4^3(a_2 a_4 - a_3^2) & -(a_1 a_4^2 - 2a_2 a_3 a_4 + a_3^3)/a_4^4
\end{bmatrix}.
$$

### 3.3.4 Reduced row echelon form

In classical linear algebra textbooks, elementary row transforms are usually used to compute the inverse of a matrix. Also, the inverse can be found in the reduced row echelon form. Function $H_1 = \text{rref}(H)$ can be used to find the reduced row echelon form for matrix $H$, and matrix $H_1$ is this reduced row echelon form. Matrix $H$ can be in numerical or symbolic form.

If an identity matrix is appended to the right of a square matrix $H$, then, performing elementary row transforms, the left-hand side can be transformed into an identity matrix, while the right-hand side becomes the inverse of the original matrix. If the left-hand side is not an identity matrix, there is no inverse matrix, since the original matrix is a singular one. Therefore, this method can be used to find the inverse of a nonsingular matrix. If the matrix is singular, a singular matrix is obtained on the left side, after the inversion process.

**Example 3.32.** Compute again the inverse of Example 3.31 with the elementary row transform method.

**Solutions.** The following MATLAB commands can be used to find the inverse matrix with elementary row transforms:

```
>> a=sym('a',[1,4]); H1=[hankel(a) eye(4)]; H2=rref(H1)
   H3=H2(:,5:8) % extract the last four columns
```

Matrix $H_3$ is identical to that obtained earlier. The intermediate matrix $H_2$ can also be obtained, where the left-hand side is an identity matrix, and the right-hand side is the needed inverse $H_3$:

$$H_2 = \begin{bmatrix} 1 & 0 & 0 & 0 & 0 & 0 & 0 & 1/a_4 \\ 0 & 1 & 0 & 0 & 0 & 0 & 1/a_4 & -1/a_4^2 a_3 \\ 0 & 0 & 1 & 0 & 0 & 1/a_4 & -1/a_4^2 a_3 & -1/a_4^3(a_2 a_4 - a_3^2) \\ 0 & 0 & 0 & 1 & 1/a_4 & -1/a_4^2 a_3 & -1/a_4^3(a_2 a_4 - a_3^2) & -(a_1 a_4^2 - 2a_2 a_3 a_4 + a_3^3)/a_4^4 \end{bmatrix}.$$

**Example 3.33.** Compute the inverse using elementary row transforms for the matrix in Example 3.29.

**Solutions.** An identity matrix can be appended to the original matrix, and then elementary row transforms are performed to get the results.

```
>> A=[16 2 3 13; 5 11 10 8; 9 7 6 12; 4 14 15 1];
   A=sym(A); A1=rref([A,eye(4)])
```

The reduced row echelon form is obtained below. Since the original matrix is singular, the left-hand side is not an identity matrix. Therefore, there is no inverse for the considered singular matrix:

$$A_1 = \begin{bmatrix} 1 & 0 & 0 & 1 & 0 & -21/136 & 25/136 & 1/34 \\ 0 & 1 & 0 & 3 & 0 & 111/136 & -35/136 & -15/34 \\ 0 & 0 & 1 & -3 & 0 & -49/68 & 13/68 & 8/17 \\ 0 & 0 & 0 & 0 & 1 & 3 & -3 & -1 \end{bmatrix}.$$

**Example 3.34.** Find the values of $a$ and $b$, such that the rank of the following matrix is 2:

$$A = \begin{bmatrix} a & 2 & 1 & 2 \\ 3 & b & 2 & 3 \\ 1 & 3 & 1 & 1 \end{bmatrix}.$$

**Solutions.** This problem is not an easy one to solve with MATLAB. Elementary row transforms can be tried. In order to simplify the problem, the columns with variables can be shifted to the right. The following statements can be used:

```
>> syms a b, A=[a,2,1,2; 3,b,2,3; 1,3,1,1];
   A1=A(:,[3 4 1 2]); H=rref(A1)
```

and the reduced row echelon form can be obtained as

$$H = \begin{bmatrix} 1 & 0 & 0 & 9-b \\ 0 & 1 & 0 & -(6a+b-ab-7)/(a-2) \\ 0 & 0 & 1 & -(b-5)/(a-2) \end{bmatrix}.$$

It can be seen from the results that $H$ looks like a full-rank matrix. In the reduced row echelon form formulation process, it is assumed automatically in MATLAB that $a \neq 2$, otherwise the matrix is not of full-rank. Obviously, to make the rank of $A$ equal to 2, one must set $a = 2$. Besides, the numerator $6a + b - ab - 7$ should also be zero, from which it can be seen that $b = 5$. Substituting the variables back to the original matrix, it can be seen that the rank is indeed 2.

```
>> A1=subs(A,{a,b},{2,5}); rank(A1)
```

### 3.3.5 Generalized inverse

It has been demonstrated that even though Symbolic Math Toolbox is introduced to handle the inverse problems for singular matrices, it is not possible to find the inverse of such a matrix, because singular matrices do not have inverses. Also, inverse matrices are not available for rectangular (nonsquare) matrices. The concept of "inverse" should be extended. This extension is also known as the generalized inverse.

**Definition 3.25.** For matrix $A$, if there exists a matrix $N$ such that

$$ANA = A, \tag{3.3.8}$$

then the matrix $N$ is referred to as a generalized inverse of $A$. Generalized inverse matrix is denoted as $N = A^-$.

If matrix $A$ is an $n \times m$ rectangular matrix, $N$ is an $m \times n$ matrix. There are infinitely many matrices satisfying the conditions of the generalized inverse.

**Theorem 3.29.** *If a norm-minimization criterion is defined as*

$$\min_{M} \|AM - I\|, \tag{3.3.9}$$

*then, for a given matrix $A$, there exists a unique matrix $M$, satisfying all the three conditions:*
*(1)* $AMA = A$;
*(2)* $MAM = M$;
*(3)* $AM$ *and* $MA$ *are Hermitian symmetric matrices.*

*Such a matrix $M$ is referred to as the Moore–Penrose generalized inverse of matrix $A$, also known as the pseudoinverse matrix, denoted as $M = A^+$.*

It can be seen that the first condition is the same as for the generalized inverse, and the second and third ensure that $M$ is unique.

The concept of Moore–Penrose generalized inverse was proposed independently by American mathematician Eliakim Hastings Moore (1862–1932) and British mathematical physicist Sir Roger Penrose (1931–).

**Theorem 3.30.** *Moore–Penrose generalized inverse has the following properties:*

$$(A^+)^+ = A, \quad A^+ = (A^H A)^+ A^H, \quad A^+ = A^H (AA^H)^+. \tag{3.3.10}$$

Function `pinv()` in MATLAB can be used to compute Moore–Penrose generalized inverse for any matrix, with the syntax

$M$=pinv$(A,\epsilon)$, % with the error tolerance $\epsilon$

where $\epsilon$ is the error tolerance. If it is omitted, default constant eps is adopted. The returned argument $M$ is the Moore–Penrose generalized inverse of $A$. If $A$ is a nonsingular matrix, the result is the inverse of $A$. The speed of the function is lower than that of function `inv()`. It is not recommended to use such a function to deal with nonsingular matrices.

**Example 3.35.** Compute the Moore–Penrose generalized inverse of the singular matrix in Example 3.29.

**Solutions.** It has been shown in Example 3.29 that the analytical inverse does not exist, and cannot be obtained with symbolic `inv()` function. Moore–Penrose generalized inverse can be tried with

```
>> A=[16 2 3 13; 5 11 10 8; 9 7 6 12; 4 14 15 1];
   B=pinv(A), A*B % pseudoinverse
```

where matrices $B$ and $AB$ can be obtained respectively as

$$B = \begin{bmatrix} 0.1011 & -0.0739 & -0.0614 & 0.0636 \\ -0.0364 & 0.0386 & 0.0261 & 0.0011 \\ 0.0136 & -0.0114 & -0.0239 & 0.0511 \\ -0.0489 & 0.0761 & 0.0886 & -0.0864 \end{bmatrix},$$

$$AB = \begin{bmatrix} 0.95 & -0.15 & 0.15 & 0.05 \\ -0.15 & 0.55 & 0.45 & 0.15 \\ 0.15 & 0.45 & 0.55 & -0.15 \\ 0.05 & 0.15 & -0.15 & 0.95 \end{bmatrix}.$$

It can be seen that $AB$ is no longer an identity matrix. It is the matrix minimizing the norm defined in Theorem (3.3.9). In other words, it is the matrix such that $AB$ is the closest to an identity matrix. The obtained Moore–Penrose generalized inverse is validated and the errors in the three conditions are all at the $10^{-14}$-level, indicating the matrix is the Moore–Penrose generalized inverse of $A$.

```
>> norm(A*B*A-A), norm(B*A*B-B)
   norm(A*B-(A*B)'), norm(B*A-(B*A)') % validating the conditions
```

Taking Moore–Penrose generalized inverse once again for the matrix $B$, it can be seen that the original matrix can be restored, i. e., $(A^+)^+ = A$.

```
>> pinv(B), norm(ans-A) % compute pseudoinverse to restore the matrix
```

**Example 3.36.** Compute the pseudoinverse in Example 3.35 with the symbolic method and observe the precision.

**Solutions.** The original matrix should be entered first into MATLAB in symbolic format, and then the solution can be found directly with `pinv()`

```
>> A=sym(magic(4)); B=pinv(A), A*B, B*A,
   norm(A*B*A-A) % symbolic pseudoinverse computation
```

It can be seen that the error matrix is zero, and the Moore–Penrose generalized inverse found is

$$B_1 = \begin{bmatrix} 55/544 & -201/2720 & -167/2720 & 173/2720 \\ -99/2720 & 21/544 & 71/2720 & 3/2720 \\ 37/2720 & -31/2720 & -13/544 & 139/2720 \\ -133/2720 & 207/2720 & 241/2720 & -47/544 \end{bmatrix}.$$

Substituting the result back to the common inverse formula shows that $BA$ and $AB$ are identical matrices:

$$AB = BA = \begin{bmatrix} 19/20 & -3/20 & 3/20 & 1/20 \\ -3/20 & 11/20 & 9/20 & 3/20 \\ 3/20 & 9/20 & 11/20 & -3/20 \\ 1/20 & 3/20 & -3/20 & 19/20 \end{bmatrix}.$$

**Example 3.37.** Compute the pseudoinverse of the rectangular matrix $A$,

$$A = \begin{bmatrix} 6 & 1 & 4 & 2 & 1 \\ 3 & 0 & 1 & 4 & 2 \\ -3 & -2 & -5 & 8 & 4 \end{bmatrix}.$$

**Solutions.** The following MATLAB commands can be used to analyze the behavior of the matrix, and it is found that the matrix is not of full-rank.

```
>> A=[6,1,4,2,1; 3,0,1,4,2; -3,-2,-5,8,4]; rank(A) % compute rank
```

Since matrix $A$ is singular, function `pinv()` can be used to compute Moore–Penrose generalized inverse. The following statements can also used to validate all the conditions and show that it is indeed the Moore–Penrose generalized inverse expected.

```
>> pinv(sym(A)), iA=pinv(A) % pseudoinverse computation
   norm(A*iA*A-A), norm(iA*A-A'*iA')
   norm(iA*A-A'*iA'), norm(A*iA-iA'*A')
```

The pseudoinverse obtained is

$$
A^+ = \begin{bmatrix} 183/2506 & 207/5012 & -111/5012 \\ 27/2506 & 5/2506 & -39/2506 \\ 115/2506 & 89/5012 & -193/5012 \\ 41/1253 & 54/1253 & 80/1253 \\ 41/2506 & 27/1253 & 40/1253 \end{bmatrix} \approx \begin{bmatrix} 0.073 & 0.0413 & -0.0221 \\ 0.0108 & 0.002 & -0.0156 \\ 0.0459 & 0.0178 & -0.0385 \\ 0.0327 & 0.0431 & 0.0638 \\ 0.0164 & 0.0215 & 0.0319 \end{bmatrix},
$$

and

$$
\begin{cases}
\|A^+AA^+ - A^+\| = 1.0263 \times 10^{-16} \\
\|AA^+A - A\| = 8.1145 \times 10^{-15} \\
\|A^+A - A^H(A^+)^H\| = 3.9098 \times 10^{-16} \\
\|AA^+ - (A^+)^HA^H\| = 1.6653 \times 10^{-16}.
\end{cases}
$$

## 3.4 Characteristic polynomials and eigenvalues

Eigenvalue problems are important in linear algebra and matrix analysis. Their applications can be found in all research areas. Characteristic polynomials and equations of matrices are discussed in this section. Then, the definitions of eigenvalues and eigenvectors are discussed, and the concepts of generalized eigenvalues are presented.

### 3.4.1 Characteristic polynomials

**Definition 3.26.** Introducing a scalar operator $s$, a matrix $sI - A$ can be constructed, whose determinant is a polynomial of $s$:

$$
C(s) = \det(sI - A) = s^n + c_1 s^{n-1} + \cdots + c_{n-1}s + c_n. \tag{3.4.1}
$$

Such a polynomial $C(s)$ is referred to as the characteristic polynomial of matrix $A$, where $c_i$, $i = 1, 2, \ldots, n$ are referred to as the characteristic polynomial coefficients.

**Theorem 3.31** (Cayley–Hamilton Theorem). *If the characteristic polynomial of matrix $A$ is*

$$
f(s) = \det(sI - A) = a_1 s^n + a_2 s^{n-1} + \cdots + a_n s + a_{n+1}, \tag{3.4.2}
$$

*then $f(A) = 0$, i. e.,*

$$
a_1 A^n + a_2 A^{n-1} + \cdots + a_n A + a_{n+1}I = 0. \tag{3.4.3}
$$

Function poly() is provided in MATLAB to find the characteristic polynomial coefficients with c=poly($A$), and the returned $c$ is a row vector whose components are the coefficients of the characteristic polynomial, in the descending order of $s$. Alternatively, if $A$ is a vector, the elements of which can be regarded as the roots of the characteristic polynomial, then characteristic polynomial coefficients can be obtained. If vector $A$ contains Inf or NaN, they are removed automatically, and the characteristic polynomial can be constructed again.

It is worth mentioning that, if $A$ is a symbolic matrix, poly() function in the new version is no longer applicable. Instead, an alternative charpoly() function can be called:

$p$=charpoly($A$), % characteristic polynomial coefficients $p$

$p$=charpoly($A$,$s$), % characteristic polynomial expression $p$

**Example 3.38.** Find the characteristic polynomial of matrix $A$ in Example 3.2.

**Solutions.** The characteristic polynomial can be obtained with poly() function, with the coefficient vector of $p \approx [1, -34, -80, 2720, 0]$. The error norm can be obtained as $2.6636 \times 10^{-12}$.

```
>> A=[16 2 3 13; 5 11 10 8; 9 7 6 12; 4 14 15 1]; p=poly(A)
   norm(p-[1,-34,-80,2720,0]) % find the error in characteristic polynomial
```

The function charpoly(sym($A$)) in Symbolic Math Toolbox can be used to find the analytical expression of the polynomial. Exact solution can be found directly.

Other numerical algorithms for finding the characteristic polynomials can also be adopted. For instance, the following recursive Faddeev–Le Verrier algorithm can be used to compute the characteristic polynomial coefficients, with the MATLAB implementation given below.

**Theorem 3.32** (Faddeev–Le Verrier recursive algorithm).

$$c_{k+1} = -\frac{1}{k} \operatorname{tr}(AR_k), \quad R_{k+1} = AR_k + c_{k+1}I, \quad k = 1, \ldots, n, \tag{3.4.4}$$

*where the initial values are* $R_1 = I$, $c_1 = 1$.

An identity matrix $I$ should be generated first and assigned to $R_1$. Then, for each $k$ the polynomial coefficient $c_k$ can be obtained, and matrix $R_k$ can be updated, such that all the polynomial coefficients can eventually be found. The MATLAB implementation of the algorithm is given in function poly1().

```
function c=poly1(A)
[nr,nc]=size(A); I=eye(nc); R=I; c=[1 zeros(1,nc)]; % initial setting
for k=1:nc, c(k+1)=-1/k*trace(A*R); R=A*R+c(k+1)*I; end % recursive
```

If function `poly1(A)` is used instead in Example 3.38, accurate coefficients can be obtained, with no errors, since in the code, only simple multiplications are involved.

**Example 3.39.** Generate a vector $B = [a_1, a_2, a_3, a_4, a_5]$, then construct the corresponding Hankel matrix. Compute the characteristic polynomial.

**Solutions.** The Hankel matrix $A$ can be constructed first, then `charpoly(A)` function can be used to compute the characteristic polynomial, and function `collect()` can be used to collect the like-terms.

```
>> syms x; a=sym('a%d',[1,5]); A=hankel(a);
   collect(charpoly(A,x),x)
```

The mathematical expression of the characteristic polynomial is

$$\det(xI - A) = x^5 + (-a_3 - a_5 - a_1)x^4 + (a_5 a_1 + a_3 a_1 + a_5 a_3 - 2a_4^2 - 2a_2^2 - a_2^2 - a_3^2)x^3$$
$$+ (-a_1 a_3 a_5 + 2a_5^3 - 2a_2 a_4 a_3 + a_2^2 a_5 + a_1 a_4^2 + a_3^3$$
$$+ a_1 a_5^2 + a_3 a_5^2 + a_5 a_4^2 + a_4^2 a_3 - 2a_2 a_5 a_4)x^2$$
$$+ (2a_2 a_5^2 a_4 + a_4^4 + a_5^4 + a_3^2 a_5^2 + a_5^2 a_4^2 - 3a_3 a_5 a_4^2 - a_1 a_5^3 - a_3 a_5^3)x - a_5^5.$$

### 3.4.2 Finding the roots of polynomial equations

**Definition 3.27.** The mathematical form of a polynomial $f(x)$ is

$$f(x) = a_1 x^n + a_2 x^{n-1} + a_3 x^{n-2} + \cdots + a_n x + a_{n+1}. \tag{3.4.5}$$

Two methods are provided in MATLAB for expressing the polynomial $f(x)$: one is by storing the coefficients in the descending order of $s$, i. e., $f = [a_1, a_2, \ldots, a_n, a_{n+1}]$, and the other is to use a symbolic expression.

If the polynomial is expressed by the coefficient vector $f$, the roots can be computed from $r=$`roots`$(f)$, where $f$ can be a double precision vector or a symbolic vector, with the numerical solution found in the former case, and the analytical solution found in the latter.

**Example 3.40.** Compute all the roots of

$$f(x) = x^{11} - 3x^{10} - 15x^9 + 53x^8 + 30x^7 - 218x^6 + 190x^5 - 202x^4 + 225x^3 + 477x^2 + 81x + 405.$$

**Solutions.** The polynomial coefficients can be entered into MATLAB first, and then polynomial equation can be solved with

```
>> f=[1,-3,-15,53,30,-218,190,-202,225,477,81,405];
   r1=roots(f), f=sym(f); r2=roots(f)
```

The numerical and analytical solutions are:

$$
r_1 = \begin{bmatrix}
-2.999999999999996 + 0.000000000000000j \\
-2.999999999999996 + 0.000000000000000j \\
3.000000060673409 + 0.000000000000000j \\
2.999999939326572 + 0.000000000000000j \\
2.000000000000002 + 1.000000000000002j \\
2.000000000000002 - 1.000000000000002j \\
-0.999999999999999 + 0.000000000000000j \\
0.000000002580916 + 1.000000014179606j \\
0.000000002580916 - 1.000000014179606j \\
-0.000000002580915 + 0.999999985820393j \\
-0.000000002580915 - 0.999999985820393j
\end{bmatrix}, \quad
r_2 = \begin{bmatrix}
-3 \\
-3 \\
-1 \\
3 \\
3 \\
-j \\
-j \\
j \\
j \\
2-j \\
2+j
\end{bmatrix}.
$$

It can be seen that there exist multiple complex roots, the errors are relatively large in the numerical results and they may lead to fatal errors. For instance, for the pair of multiple roots at j, the numerical solutions are different. It will be discussed later that the eigenvector matrix may lead to singularity, and wrong results may be obtained.

### 3.4.3 Eigenvalues and eigenvectors

**Definition 3.28.** For a given matrix $A$, if there exists a nonzero vector $x$ and a scalar $\lambda$, satisfying

$$Ax = \lambda x, \tag{3.4.6}$$

then $\lambda$ is referred to as an eigenvalue of $A$ and $x$ is referred to an eigenvector corresponding to $\lambda$.

Strictly speaking, $x$ is referred to as a right-eigenvector of $A$. Similarly, left-eigenvectors can also be defined. However, right-eigenvectors are more popular, so the left-eigenvectors are not discussed any more in the book.

**Theorem 3.33.** *The product of all the eigenvalues of a matrix equals to the determinant of the matrix, i. e., $\lambda_1\lambda_2\cdots\lambda_n = \det(A)$.*

**Theorem 3.34.** *The sum of all the eigenvalues of a matrix equals to the trace of the matrix, i. e., $\lambda_1 + \lambda_2 + \cdots + \lambda_n = \mathrm{tr}(A)$.*

Various algorithms can be used in computing the eigenvalues of a matrix. The most popular ones are the Jacobian algorithm for finding real symmetric matrices, origin shift QR decomposition algorithm, and two-step QR algorithm. Powerful eigenvalue and eigenvector solvers are provided in standard software packages such as EISPACK[8, 22]. Function eig() can be used for finding eigenvalues and eigenvectors of any complex matrices. If there exist multiple eigenvalues, the eigenvector matrix may approach singularity. Therefore one must be very careful in dealing with such a matrix.

If there are no repeated eigenvalues of a complex matrix $A$, all the columns of the eigenvector matrix are linearly independent. Therefore, a nonsingular matrix can be composed with these eigenvectors, with it similarity transform can be taken, and the original matrix can be transformed into a diagonal matrix whose diagonal elements are the eigenvalues of the matrix. Eigenvalues and eigenvector matrix can easily be obtained with function eig(), with the syntaxes

$d$=eig($A$), % computing eigenvalues

[$V$,$D$]=eig($A$), % computing eigenvalues and eigenvectors

where $d$ is a vector containing all the eigenvalues, $D$ is a diagonal matrix with eigen-values, and the corresponding eigenvectors are returned in each column of matrix $V$. If the eigenvector matrix is of full-rank, one has $AV = VD$, and the sum of squared eigenvectors, i. e., 2-norm, equals to one. If there is only one returned argument in the function call, the eigenvalues are returned alone, in a column vector. Even though $A$ is a complex matrix, function eig() can still be used to compute the eigenvalues and eigenvector matrix.

The eigenvalues and the roots of the characteristic polynomials are equivalent. If the polynomial coefficients are exactly known, the eigenvalues can also be evaluated with roots().

**Example 3.41.** Compute the eigenvalues and eigenvectors of $A$ in Example 3.2.

**Solutions.** Function eig() can be used to evaluate directly the numerical eigenvalues of $A$, and the results are 34, $\pm$8.9443, $-2.2348 \times 10^{-15}$.

```
>> A=[16 2 3 13; 5 11 10 8; 9 7 6 12; 4 14 15 1];
   eig(A), [v,d]=eig(A), norm(A*v-v*d)
```

The eigenvalue and eigenvector matrix are as follows, and the norm of the error matrix is $1.2284 \times 10^{-14}$:

$$
v = \begin{bmatrix}
-0.5 & -0.8236 & 0.3764 & -0.2236 \\
-0.5 & 0.4236 & 0.0236 & -0.6708 \\
-0.5 & 0.0236 & 0.4236 & 0.6708 \\
-0.5 & 0.3764 & -0.8236 & 0.2236
\end{bmatrix},
$$

$$
d = \begin{bmatrix}
34 & 0 & 0 & 0 \\
0 & 8.9443 & 0 & 0 \\
0 & 0 & -8.9443 & 0 \\
0 & 0 & 0 & -2.2348 \times 10^{-15}
\end{bmatrix}.
$$

An overload function `eig()` in Symbolic Math Toolbox is provided for finding exact eigenvalues and eigenvector matrices. Theoretically speaking, any large-scale matrices can be handled, and high-precision solutions can be obtained. For the given **A** matrix, the exact solutions of the eigenvalues are $0$, $34$, $\pm 4\sqrt{5}$.

```
>> eig(sym(A)), vpa(ans,70)
   [v,d]=eig(sym(A)) % analytical computation
```

The corresponding matrices are

$$
v = \begin{bmatrix}
-1 & 1 & 12\sqrt{5}/31 - 41/31 & -12\sqrt{5}/31 - 41/31 \\
-3 & 1 & 17/31 - 8\sqrt{5}/31 & 8\sqrt{5}/31 + 17/31 \\
3 & 1 & -4\sqrt{5}/31 - 7/31 & 4\sqrt{5}/31 - 7/31 \\
1 & 1 & 1 & 1
\end{bmatrix},
$$

$$
d = \begin{bmatrix}
0 & 0 & 0 & 0 \\
0 & 34 & 0 & 0 \\
0 & 0 & -4\sqrt{5} & 0 \\
0 & 0 & 0 & 4\sqrt{5}
\end{bmatrix}.
$$

It can be seen that the `eig()` function is called twice, and since the numbers of output arguments are different, the first statement returns only the eigenvalues, while the latter returns the eigenvalue and eigenvector matrices. When **A** is a symbolic matrix, the orders of the eigenvalues are different, and the eigenvector matrix is no longer normalized.

In the function call, the numerical eigenvalues are sorted in descending order, while symbolic eigenvalues are sorted in ascending order.

**Example 3.42.** Compute the eigenvalues and eigenvectors of the following matrix:

$$
A = \begin{bmatrix}
1 & 6 & -2 & -2 \\
-1 & -5 & 0 & 1 \\
1 & 2 & -3 & -1 \\
-2 & -5 & 1 & 0
\end{bmatrix}.
$$

**Solutions.** The following MATLAB commands can be directly used to compute the eigenvalues and eigenvectors of the matrix:

```
>> A=[1,6,-2,-2; -1,-5,0,1; 1,2,-3,-1; -2,-5,1,0];
   [v,d]=eig(A), norm(A*v-v*d)
```

The eigenvalue and eigenvector matrices can be obtained as follows, and the norm of the error matrix is $2.9437 \times 10^{-15}$:

$$d = \begin{bmatrix} -1 + 6.6021 \times 10^{-8}j & 0 & 0 & 0 \\ 0 & -1 - 6.6021 \times 10^{-8}j & 0 & 0 \\ 0 & 0 & -3 & 0 \\ 0 & 0 & 0 & -2 \end{bmatrix},$$

$$v = \begin{bmatrix} 0.6325 & 0.6325 & -0.5774 & 0.7559 \\ -0.3162 + 1.044 \times 10^{-8}j & -0.3162 - 1.044 \times 10^{-8}j & 0.57735 & -0.3780 \\ 0.3162 - 1.044 \times 10^{-8}j & 0.3162 + 1.044 \times 10^{-8}j & 3.343 \times 10^{-16} & 0.3780 \\ -0.6325 + 2.088 \times 10^{-8}j & -0.6325 - 2.088 \times 10^{-8}j & 0.57735 & -0.3780 \end{bmatrix}.$$

With the symbolic algorithm

```
>> [v,d]=eig(sym(A))
```

the following results are obtained:

$$d = \begin{bmatrix} -3 & 0 & 0 & 0 \\ 0 & -2 & 0 & 0 \\ 0 & 0 & -1 & 0 \\ 0 & 0 & 0 & -1 \end{bmatrix}, \quad v = \begin{bmatrix} -1 & -2 & -1 \\ 1 & 1 & 1/2 \\ 0 & -1 & -1/2 \\ 1 & 1 & 1 \end{bmatrix}.$$

It can be seen that there is a double eigenvalue of −1, therefore, the eigenvector matrix only has three columns. Due to the limitations in the double precision algorithm, small errors are introduced, and misjudgement is made, i. e., the two eigenvalues are considered as different, such that the eigenvector matrix is close to a singular one.

If there exist repeated eigenvalues, theoretically speaking, $V$ is close to a singular matrix. In numerical computation, inaccurate eigenvalues may be obtained, such that singular $V$ matrix may be returned and wrong results may be found. Symbolic computation is recommended for dealing with such problems.

**Example 3.43.** Construct a matrix from the characteristic polynomial in Example 3.40. Then compute the eigenvalues.

**Solutions.** It can be seen that there are infinitely many matrices having the characteristic polynomial in Example 3.40, and the simplest way is to construct a companion matrix. Eigenvalues can be found with the following commands:

```
>> f=[1,-3,-15,53,30,-218,190,-202,225,477,81,405];
   f=sym(f); A=compan(f), eig(A)
```

The constructed matrix is as follows, and the eigenvalues obtained are exactly the same as those in Example 3.40:

$$
A = \begin{bmatrix}
3 & 15 & -53 & -30 & 218 & -190 & 202 & -225 & -477 & -81 & -405 \\
1 & 0 & 0 & 0 & 0 & 0 & 0 & 0 & 0 & 0 & 0 \\
0 & 1 & 0 & 0 & 0 & 0 & 0 & 0 & 0 & 0 & 0 \\
0 & 0 & 1 & 0 & 0 & 0 & 0 & 0 & 0 & 0 & 0 \\
0 & 0 & 0 & 1 & 0 & 0 & 0 & 0 & 0 & 0 & 0 \\
0 & 0 & 0 & 0 & 1 & 0 & 0 & 0 & 0 & 0 & 0 \\
0 & 0 & 0 & 0 & 0 & 1 & 0 & 0 & 0 & 0 & 0 \\
0 & 0 & 0 & 0 & 0 & 0 & 1 & 0 & 0 & 0 & 0 \\
0 & 0 & 0 & 0 & 0 & 0 & 0 & 1 & 0 & 0 & 0 \\
0 & 0 & 0 & 0 & 0 & 0 & 0 & 0 & 1 & 0 & 0 \\
0 & 0 & 0 & 0 & 0 & 0 & 0 & 0 & 0 & 1 & 0
\end{bmatrix}.
$$

### 3.4.4 Generalized eigenvalues and eigenvectors

**Definition 3.29.** If there exists a scalar $\lambda$ and a nonzero vector $x$ such that

$$
Ax = \lambda Bx, \tag{3.4.7}
$$

where matrix $B$ is a symmetric positive definite matrix, then $\lambda$ is referred to as a generalized eigenvalue, and vector $x$ is referred to as a generalized eigenvector.

**Definition 3.30.** Equation $\det(\lambda B - A) = 0$ is referred to as the generalized characteristic equation, the roots of the equation are also the generalized eigenvalues.

Generalized eigenvalues can also be obtained with MATLAB. In fact, ordinary eigenvalue problem can be regarded as a special case of the generalized eigenvalue problem, where $B = I$. The generalized eigenvalue problem in (3.4.7) can be converted to the ordinary eigenvalue problem.

If the $B$ matrix is nonsingular, the generalized eigenvalue problem can also be converted to the eigenvalue problem of $B^{-1}A$

$$
B^{-1}Ax = \lambda x, \tag{3.4.8}
$$

where $\lambda$ and $x$ are respectively the eigenvalue and eigenvector of matrix $B^{-1}A$. The QZ algorithm in [20] is used to directly solve generalized eigenvalue problems. The function eig() can be used to handle generalized eigenvalue problems with the syntaxes

$d$=eig($A$,$B$), % generalized eigenvalue problem

[$V$,$D$]=eig($A$,$B$), % generalized eigenvalues and eigenvectors

The generalized eigenvalues can be returned in a column vector $d$ in the first syntax, while in the second, generalized eigenvector matrix is returned in $V$, and generalized eigenvalues are returned in a diagonal matrix $D$, such that $AV = BVD$. It is worth mentioning that generalized eigenvalue problem solutions for singular matrices $B$ can be obtained, but the function cannot be used to handle symbolic matrix problems.

**Example 3.44.** For the given matrices $A$ and $B$, compute the generalized eigenvalues and eigenvectors

$$A = \begin{bmatrix} -4 & -6 & -4 & -1 \\ 1 & 0 & 0 & 0 \\ 0 & 1 & 0 & 0 \\ 0 & 0 & 1 & 0 \end{bmatrix}, \quad B = \begin{bmatrix} 2 & 6 & -1 & -2 \\ 5 & -1 & 2 & 3 \\ -3 & -4 & 1 & 10 \\ 5 & -2 & -3 & 8 \end{bmatrix}.$$

**Solutions.** There exist multiple eigenvalues, namely $-1$, of matrix $A$, thus inaccurate solutions may be found with numerical methods. With the following statements, the generalized eigenvalues and eigenvectors in $(A, B)$ can be obtained:

```
>> B=[2,6,-1,-2; 5,-1,2,3; -3,-4,1,10; 5,-2,-3,8];
   A=[-4,-6,-4,-1; 1,0,0,0; 0,1,0,0; 0,0,1,0];
   [V,D]=eig(A,B), norm(A*V-B*V*D)
```

The generalized eigenvalues and eigenvectors can be obtained as provided below, and the error obtained is $6.3931 \times 10^{-15}$:

$$V = \begin{bmatrix} 0.0268 & -1 & -0.2413 & 0.0269 \\ -1 & 0.7697 & -0.6931 & 0.0997 \\ -0.1252 & -0.3666 & 1 & 0.0796 \\ -0.3015 & 0.4428 & -0.1635 & -1 \end{bmatrix},$$

$$D = \begin{bmatrix} -1.2830 & & & \\ & 0.1933 & & \\ & & -0.2422 & \\ & & & -0.0096 \end{bmatrix}.$$

Ordinary and generalized characteristic polynomials can be obtained with the following statements:

```
>> syms lam, p1=charpoly(sym(A),lam), p2=det(lam*B-A)
```

The results are

$$p_1(\lambda) = \lambda^4 + 4\lambda^3 + 6\lambda^2 + 4\lambda + 1, \quad p_2(\lambda) = -1736\lambda^4 - 2329\lambda^3 - 50\lambda^2 + 104\lambda + 1.$$

**Example 3.45.** If matrix $B$ in Example 3.44 is changed to a singular magic matrix, compute the generalized eigenvalues and eigenvectors.

**Solutions.** The following statements can be used and it can be seen that there are infinite eigenvalues, and $AV - BVD$ is not a zero matrix, meaning the function cannot be used to handle problems with singular $B$ matrices.

```
>> A=[-4,-6,-4,-1; 1,0,0,0; 0,1,0,0; 0,0,1,0];
   B=magic(4); [V,D]=eig(A,B), A*V-B*V*D
```

The following statements can be used to compute generalized characteristic polynomials, with the result $p(\lambda) = -1\,632\lambda^3 - 560\lambda^2 + 94\lambda + 1$.

```
>> syms lam; p=det(lam*B-A)
```

### 3.4.5 Gershgorin discs and diagonally dominant matrices

Gershgorin disc theorem was proposed in 1931 by the Soviet Union mathematician Semyon Aranovich Gershgorin (1901–1933), where the positions of complex eigenvalues are estimated. This idea can be used to assess whether a matrix is diagonally dominant or not.

**Definition 3.31.** Let $A \in \mathscr{C}^{n \times n}$ be a complex square matrix, and $R_i = \sum_{j \neq i} |a_{ij}|$ be the sum of absolute values of the off-diagonal elements in the $i$th row. Some discs can be constructed as $\mathscr{D}(a_{ii}, R_i)$, $i = 1, 2, \ldots, n$, meaning the disc centered at $a_{ii}$ with radius $R_i$. The discs are referred to as row Gershgorin discs. Similarly, column Gershgorin discs can also be defined.

**Theorem 3.35** (Gershgorin disc theorem). *All the eigenvalues of matrix $A$ are located inside at least one of the Gershgorin discs $\mathscr{D}(a_{ii}, R_i)$, i. e.,*

$$G(A) = \bigcup_{i=1}^{n} \{z \in \mathscr{C} : |z - a_{ii}| \leqslant R_i\}. \tag{3.4.9}$$

**Definition 3.32.** For a given square matrix $A \in \mathscr{C}^{n \times n}$, if the diagonal element is larger than the sum of the absolute values of the off-diagonal elements,

$$|a_{ii}| > \sum_{j=1, j \neq i}^{n} |a_{ij}|, \quad i = 1, 2, \ldots, n, \tag{3.4.10}$$

matrix $A$ is referred to as row diagonally dominant matrix. Similarly, column diagonally dominant matrix can be defined.

**Theorem 3.36.** *If the origin $(0, 0)$ is not covered by any of the Gershgorin discs, matrix $A$ is a diagonally dominant matrix.*

**Example 3.46.** For the given matrix $A$, validate Gershgorin disc theorem, and see whether the matrix is diagonally dominant or not

$$A = \begin{bmatrix} 12 - 11j & -1 & 2 & 2 & 2 - j \\ 1 + j & -10 + 17j & 2 + 2j & 2 + 3j & 2j \\ 3 + 4j & 3j & 13 + 13j & 1 + 4j & 4 + 4j \\ 2 + 4j & 1 & 3 + 2j & -10 - 13j & 2 + j \\ -1 + 3j & -1 + j & 2 + j & 2 + j & -15 + 2j \end{bmatrix}.$$

**Solutions.** The matrix should be entered into MATLAB first, then the following statements can be used to compute the radius of each Gershgorin disc, and then draw it, as shown in Figure 3.1. In the figure, circles represent the eigenvalues, and × indicate the centers of the Gershgorin discs.

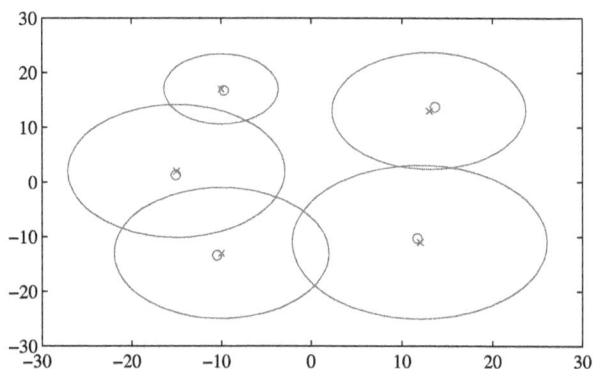

**Figure 3.1:** Gershgorin discs and eigenvalues.

```
>> A=[12-11i,-1,2,2,2-1i; 1+1i,-10+17i,2+2i,2+3i,0+2i;
      3+4i,0+3i,13+13i,1+4i,4+4i; 2+4i,1,3+2i,-10-13i,2+1i;
      -1+3i,-1+1i,2+1i,2+1i,-15+2i];
   t=linspace(0,2*pi,100); c=cos(t); s=sin(t);
   A1=A-diag(diag(A)); R=sum(abs(A1));   % compute radius
   for i=1:size(A,1)
       plot(real(A(i,i)),imag(A(i,i)),'x'), hold on
       plot(real(A(i,i))+R(i)*c,imag(A(i,i))+R(i)*s)
   end
   d=eig(A); plot(real(d),imag(d),'o')
```

In this example, the eigenvalues are very close to the centers. Besides, the Gershgorin discs do not cover the origin, thus matrix **A** is diagonally dominant.

## 3.5 Matrix polynomials

Extending the regular univariate polynomials, matrix polynomials can be constructed. In this section, the concept of matrix polynomials is presented, and MATLAB representation of regular polynomials is introduced.

### 3.5.1 Solutions of matrix polynomial problems

Matrix polynomial is an extension of a univariate polynomial in Definition 3.27, i. e., the independent variable can be replaced by the matrix, and the zeroth power of the independent variable can be replaced by an identity matrix.

**Definition 3.33.** The mathematical form of a matrix polynomial is

$$B = f(A) = a_1 A^n + a_2 A^{n-1} + \cdots + a_n A + a_{n+1} I, \tag{3.5.1}$$

where $A$ is a square matrix, $I$ is the identity matrix of the same size. The returned matrix $B$ is the matrix polynomial.

The matrix polynomial can be evaluated in MATLAB environment with function polyvalm(), with the syntax $B$=polyvalm($a$,$A$), where $a$ is the coefficient vector in the descending orders of $s$, with $a=[a_1, a_2, \ldots, a_n, a_{n+1}]$.

Function polyvalm() can be used to compute numerical matrix polynomials, and as an extension, symbolic matrices are supported by the following new MATLAB function:

```
function B=polyvalmsym(p,A)   %polyvalm() for symbolic matrices
E=eye(size(A)); B=zeros(size(A)); n=length(A);
for i=n+1:-1:1, B=B+p(i)*E; E=E*A; end
```

**Definition 3.34.** Dot operation-based polynomial function is defined as

$$C = a_1 x.\hat{\ } n + a_2 x.\hat{\ }(n-1) + \cdots + a_{n+1}. \tag{3.5.2}$$

Matrix $C$ can be evaluated directly with $C$=polyval($a$,$x$).

If Symbolic Math Toolbox is used, and the polynomial is expressed as a symbolic expression $p$, function subs() can be used to compute the dot operation of the polynomial, with the syntax $C$=subs($p$,$s$,$x$).

**Example 3.47.** Assuming that $A$ is a Vandermonde matrix, validate that the matrix satisfies Cayley–Hamilton Theorem:

$$A = \begin{bmatrix} 1 & 1 & 1 & 1 & 1 & 1 & 1 \\ 64 & 32 & 16 & 8 & 4 & 2 & 1 \\ 729 & 243 & 81 & 27 & 9 & 3 & 1 \\ 4096 & 1024 & 256 & 64 & 16 & 4 & 1 \\ 15625 & 3125 & 625 & 125 & 25 & 5 & 1 \\ 46656 & 7776 & 1296 & 216 & 36 & 6 & 1 \\ 117649 & 16807 & 2401 & 343 & 49 & 7 & 1 \end{bmatrix}.$$

**Solutions.** The following MATLAB statements can be used to validate Cayley–Hamilton Theorem.

```
>> A=vander([1 2 3 4 5 6 7]), a=poly(A);
   B=polyvalm(a,A); e=norm(B) % direct validation
```

Since the MATLAB function `poly()` may lead to errors, huge errors may be obtained in matrix polynomials, such as $e = 3.5654 \times 10^5$. An erroneous conclusion may be reached. It can be seen that the errors of `poly()` sometimes may lead to nonnegligible errors. If the function `poly()` is replaced by `poly1()`, the obtained **B** matrix is a zero matrix, which validates the Cayley–Hamilton Theorem.

```
>> a1=poly1(A); B1=polyvalm(a1,A);
   norm(B1) % correct result with poly1() function
```

**Example 3.48.** Show that for an arbitrary $5 \times 5$ matrix, Cayley–Hamilton Theorem is satisfied.

**Solutions.** An arbitrary $5 \times 5$ matrix **A** can be generated first, then `charpoly()` can be used to get the characteristic polynomial coefficient vector. Since `polyvalm()` does not support symbolic variables, function `polyvalmsym()` can be used instead to evaluate the matrix polynomial. It can be found that after simplification, the resulted matrix is zero, which validates Cayley–Hamilton Theorem. The elapsed time is about 1.32 seconds. If an arbitrary $6 \times 6$ matrix is tested, the elapsed time is 13.17 seconds.

```
>> A=sym('a%d%d',5); tic, p=charpoly(A);
   E=polyvalmsym(p,A); simplify(E), toc % theorem validation
```

### 3.5.2 Minimal polynomials of matrices

Characteristic polynomials of matrices were discussed earlier, and `charpoly()` function was used to compute them. For certain special matrices, the concept of a minimal polynomial is introduced.

**Definition 3.35.** The so-called minimal polynomial is the least-order polynomial satisfying $f(\mathbf{A}) = \mathbf{0}$.

Function `minpoly()` in MATLAB is used to compute the minimal polynomial. It is known from Cayley–Hamilton Theorem that the characteristic polynomial satisfies $f(\mathbf{A}) = \mathbf{0}$, so the minimal polynomial has the lowest order among polynomials with this property. In fact, only a very small category of characteristic polynomials have minimal polynomials. An example is given below.

**Example 3.49.** Compute the minimal polynomial and characteristic polynomial for the following matrix:

$$A = \begin{bmatrix} 1 & 1 & 0 \\ 0 & 1 & 0 \\ 0 & 0 & 1 \end{bmatrix}.$$

**Solutions.** It can be found that the characteristic polynomial of the matrix is $p_1 = s^3 - 3s^2 + 3s - 1$, and the minimal polynomial is $p_2 = s^2 - 2s + 1$. Substituting $A$ matrix into polynomial $p_2$, it can be seen that the matrix polynomial is zero.

```
>> A=[1,1,0; 0,1,0; 0,0,1]; A=sym(A); syms s
   p1=charpoly(A,s), p2=minpoly(A,s), A^2-2*A+eye(3)
```

### 3.5.3 Conversions between symbolic and numerical polynomials

Two types of description methods of polynomials are supported in MATLAB. The first method is a numerical one, where the coefficients can be extracted in the descending order of $s$ to form a row vector. The other is the symbolic expression of the polynomial.

There are limitations in processing numerical representation of polynomials. Symbolic expressions can further be processed with functions. For instance, function `collect()` can be used to collect like-terms in a polynomial, and `expand()` can be used to expand a polynomial.

If a numerical vector is described as $p = [a_1, a_2, \dots, a_{n+1}]$, the command $f$=`poly2sym`$(p)$ or $f$=`poly2sym`$(p,x)$ in the Symbolic Math Toolbox can be used to convert it into symbolic form. If the symbolic expression of a polynomial is known, the command $p$=`sym2poly`$(f)$ can be used to convert it into a coefficient vector, under the double precision format.

**Example 3.50.** For a given polynomial $f = s^5 + 2s^4 + 3s^3 + 4s^2 + 5s + 6$, express it in MATLAB in different formats.

**Solutions.** The polynomial can be expressed in two formats. For instance, the polynomial can be expressed in the numerical format, then using a corresponding method converted into the symbolic form, and afterwards changed back into a vector.

```
>> syms v; P=[1 2 3 4 5 6]; f=poly2sym(P,v)
   P1=sym2poly(f) % exchange of the two formats
```

MATLAB function $C$=`coeffs`$(P,x)$ can be used to extract polynomial coefficients of $x$ in the descending order of $x$. If $x$ is the only symbolic variable in a polynomial expression $P$, it can be omitted from the function call.

**Example 3.51.** Extract the coefficients of the polynomial of $x$ from the symbolic expression $x(x^2 + 2y)^8$.

**Solutions.** It is obvious that the original expression can be a polynomial of $x$, also a polynomial of $y$. The following commands can be used to extract the coefficients of $x$ in the ascending order of $x$, and the zeroth order coefficient is missing.

```
>> syms x y; P=x*(x^2+2*y)^8;
   p=coeffs(P,x) % extract the coefficients of x in P
   p1=p(end:-1:1); % sort in descending order of x
```

The result is

$$p = [256y^8, 1\,024y^7, 1\,792y^6, 1\,792y^5, 1\,120y^4, 448y^3, 112y^2, 16y, 1].$$

The vector $p_1$ here is sorted in the descending order of $x$.

From the information presentation point of view, this presentation method is inappropriate. In real applications, since necessary information may be missing, such a coefficient vector cannot be used to reconstruct the original polynomial, while function sym2poly() cannot be used to handle polynomials with other parameters, since the even-order terms of $x$ are missing. It is necessary to develop a new algorithm and MATLAB function to extract the coefficients of a polynomial.

**Theorem 3.37.** *Assume that a polynomial model is given by*

$$p(x) = a_1 x^n + a_2 x^{n-1} + \cdots + a_n x + a_{n+1}, \tag{3.5.3}$$

*where the coefficients $a_i$ are independent of $x$. The polynomial coefficients can be evaluated recursively from*

$$a_{n+1} = p(0), \quad and \quad a_i = \frac{1}{(n-i+1)!} \left.\frac{d^{n-i+1}p(x)}{dx^{n-i+1}}\right|_{t=0}, \quad i = 1, 2, \ldots, n. \tag{3.5.4}$$

Based on the recursive algorithm, a loop structure can be used to effectively extract the polynomial coefficients, and a MATLAB function can be written:

```
function c=polycoef(p,x)
c=[]; n=0; p1=p; n1=1; nn=1; if nargin==1, x=symvar(p); end
while (1),
   c=[c subs(p1,x,0)]; p1=diff(p1,x); n=n+1; n1=n1*n; nn=[nn,n1];
   if p1==0, c=c./nn(1:end-1); c=c(end:-1:1); break;
end, end
```

**Example 3.52.** Consider the polynomial problem in Example 3.51 again. Compute the coefficients of $x$ and fill in the zero coefficients in the vector.

**Solutions.** The new function can be called to compute the coefficients of the polynomial in the descending order of $x$

```
>> syms x y; P=x*(x^2+2*y)^8; p=polycoef(P,x)
```

The coefficient vector is obtained below. Note that all the coefficients are extracted, including the zero ones. Meanwhile, the coefficients are arranged in the descending order of $x$:

$$p = [1, 0, 16y, 0, 112y^2, 0, 448y^3, 0, 1120y^4, 0, 1792y^5, 0, 1792y^6, 0, 1024y^7, 0, 256y^8, 0].$$

The polynomial can be restored with the following statements:

```
>> P1=poly2sym(p,x); simplify(P1)
```

There are nine elements in the polynomial vector $p$ in Example 3.51. This polynomial may be misunderstood as a polynomial of the eighth order. In fact, since the zero coefficients are not extracted, there exists ambiguity in the result. With the new function polycoef(), all the coefficients, including those which are zero, can be extracted. The result can be used to restore polynomial $P(x)$ easily.

## 3.6 Problems

3.1    Compute the determinants

$$(1) \begin{vmatrix} \sin\alpha & \cos\alpha & \sin(\alpha+\delta) \\ \sin\beta & \cos\beta & \sin(\beta+\delta) \\ \sin\gamma & \cos\gamma & \sin(\gamma+\delta) \end{vmatrix}, \quad (2) \begin{vmatrix} (a^x+a^{-x})^2 & (a^x-a^{-x})^2 & 1 \\ (b^y+b^{-y})^2 & (b^y-b^{-y})^2 & 1 \\ (c^z+c^{-z})^2 & (c^z-c^{-z})^2 & 1 \end{vmatrix}.$$

3.2    Compute the determinant of a Vandermonde matrix and find its simplest form

$$A = \begin{bmatrix} a^4 & a^3 & a^2 & a & 1 \\ b^4 & b^3 & b^2 & b & 1 \\ c^4 & c^3 & c^2 & c & 1 \\ d^4 & d^3 & d^2 & d & 1 \\ e^4 & e^3 & e^2 & e & 1 \end{bmatrix}.$$

3.3    Given an $n \times n$ matrix, compute its determinant

$$\begin{vmatrix} n & -1 & 0 & 0 & \cdots & 0 & 0 \\ n-1 & x & -1 & 0 & \cdots & 0 & 0 \\ n-2 & 0 & x & -1 & \cdots & 0 & 0 \\ \vdots & \vdots & \vdots & \vdots & \ddots & \vdots & \vdots \\ 2 & 0 & 0 & 0 & \cdots & x & -1 \\ 1 & 0 & 0 & 0 & \cdots & 0 & x \end{vmatrix}.$$

3.4    Validate that the 100th or any lower even-order magic matrices are all singular.

3.5 Select some orders $n$ and generate symbolic forms of random matrices. Then compute the inverse matrices and record elapsed time. Find the relationship between order and elapsed time.

3.6 Analyze the following two matrices with MATLAB. Check whether they are singular or not. Compute their ranks, determinants, traces, and inverses. Validate the inverses obtained if

$$A = \begin{bmatrix} 2 & 7 & 5 & 7 & 7 \\ 7 & 4 & 9 & 3 & 3 \\ 3 & 9 & 8 & 3 & 8 \\ 5 & 9 & 6 & 3 & 6 \\ 2 & 6 & 8 & 5 & 4 \end{bmatrix}, \quad B = \begin{bmatrix} 703 & 795 & 980 & 137 & 661 \\ 547 & 957 & 271 & 12 & 284 \\ 445 & 523 & 252 & 894 & 469 \\ 695 & 880 & 876 & 199 & 65 \\ 621 & 173 & 737 & 299 & 988 \end{bmatrix}.$$

3.7 Compute with MATLAB the trace and characteristic polynomial of the $n \times n$ matrix in Problem 3.3.

3.8 For a given set of column vectors, find from them the maximal group of vectors which are linearly independent if

$$v_1 = \begin{bmatrix} 2 \\ 2 \\ 2 \end{bmatrix}, \quad v_2 = \begin{bmatrix} 1 \\ 0 \\ 1 \end{bmatrix}, \quad v_3 = \begin{bmatrix} 2 \\ 1 \\ 2 \end{bmatrix}, \quad v_4 = \begin{bmatrix} 1 \\ 1 \\ 1 \end{bmatrix}, \quad v_5 = \begin{bmatrix} 2 \\ 0 \\ 2 \end{bmatrix}, \quad v_6 = \begin{bmatrix} 0 \\ 1 \\ 3 \end{bmatrix}.$$

3.9 For the given matrices $A$ and $B$, compute their characteristic polynomials, eigenvalues, and eigenvector matrices. Validate Cayley–Hamilton Theorem, and explain how errors can be eliminated if

$$A = \begin{bmatrix} 5 & 7 & 6 & 5 & 1 & 6 & 5 \\ 2 & 3 & 1 & 0 & 0 & 1 & 4 \\ 6 & 4 & 2 & 0 & 6 & 4 & 4 \\ 3 & 9 & 6 & 3 & 6 & 6 & 2 \\ 10 & 7 & 6 & 0 & 0 & 7 & 7 \\ 7 & 2 & 4 & 4 & 0 & 7 & 7 \\ 4 & 8 & 6 & 7 & 2 & 1 & 7 \end{bmatrix}, \quad B = \begin{bmatrix} 3 & 5 & 5 & 0 & 1 & 2 & 3 \\ 3 & 2 & 5 & 4 & 6 & 2 & 5 \\ 1 & 2 & 1 & 1 & 3 & 4 & 6 \\ 3 & 5 & 1 & 5 & 2 & 1 & 2 \\ 4 & 1 & 0 & 1 & 2 & 0 & 1 \\ -3 & -4 & -7 & 3 & 7 & 8 & 12 \\ 1 & -10 & 7 & -6 & 8 & 1 & 5 \end{bmatrix}.$$

3.10 Compute the inverse matrix for the problem in Example 3.9 with reduced row echelon form, and compare the result of inv() function.

3.11 For the matrices $A_1, A_2, A_3$ below, validate Cayley–Hamilton Theorem:

$$A_1 = \begin{bmatrix} a_{11} & a_{12} & a_{13} \\ a_{21} & a_{22} & a_{23} \\ a_{31} & a_{32} & a_{33} \end{bmatrix}, \quad A_2 = \begin{bmatrix} a_{11} & a_{12} & a_{13} & a_{14} \\ a_{21} & a_{22} & a_{23} & a_{24} \\ a_{31} & a_{32} & a_{33} & a_{34} \\ a_{41} & a_{42} & a_{43} & a_{44} \end{bmatrix},$$

$$A_3 = \begin{bmatrix} a_{11} & a_{12} & a_{13} & a_{14} & a_{15} \\ a_{21} & a_{22} & a_{23} & a_{24} & a_{25} \\ a_{31} & a_{32} & a_{33} & a_{34} & a_{35} \\ a_{41} & a_{42} & a_{43} & a_{44} & a_{45} \\ a_{51} & a_{52} & a_{53} & a_{54} & a_{55} \end{bmatrix}.$$

3.12 Compute the eigenvalues and eigenvectors of the matrices in Problem 3.6.

3.13 Try finite order $n$, for instance, $n = 50$, and validate that the characteristic polynomial is $s^n - a_1 a_2 \cdots a_n$:

$$A = \begin{bmatrix} 0 & a_1 & 0 & \cdots & 0 \\ 0 & 0 & a_2 & \cdots & 0 \\ \vdots & \vdots & \vdots & \ddots & \vdots \\ 0 & 0 & 0 & \cdots & a_{n-1} \\ a_n & 0 & 0 & \cdots & 0 \end{bmatrix}.$$

3.14 Find all the roots of the polynomial equation $f(s) = 0$. Compute the numerical and analytical solutions of polynomial equation, and assess the precision of the results:

(1) $f(s) = s^5 + 8s + 1$;

(2) $f(s) = s^9 + 11s^8 + 51s^7 + 139s^6 + 261s^5 + 353s^4 + 373s^3 + 333s^2 + 162s + 108$.

3.15 For the matrix $A$ given below, compute the characteristic polynomial $f(s)$. Compute also the matrix polynomial $f(A)$. Besides, compute the characteristic polynomial $g(s)$ of $A$, and compute $g(A)$ if

$$A = \begin{bmatrix} 3 & 1 & 2 & 3 \\ 2 & 2 & 0 & 1 \\ 1 & 3 & 1 & 3 \\ 1 & 0 & 0 & 1 \end{bmatrix}.$$

3.16 Compute the characteristic polynomial and minimal polynomial of the following matrices:

$$A = \begin{bmatrix} 3 & -2 & 2 & 2 \\ 1 & 0 & 1 & 1 \\ -2 & 2 & -1 & -2 \\ 1 & -1 & 1 & 2 \end{bmatrix}, \quad B = \begin{bmatrix} -8 & 7 & -7 & -1 \\ -12 & 11 & -13 & -1 \\ -3 & 3 & -5 & 0 \\ -9 & 10 & -10 & -3 \end{bmatrix}.$$

3.17 Extract the coefficient vector from the following mathematical expression:

$$f(x) = (ax^2 + b)^4 (x^2 \sin c + x \cos d + 3)^2.$$

# 4 Fundamental transformation and decomposition of matrices

Matrix transforms and decomposition are important issues in matrix analysis. Normally a certain transformation can be introduced to convert a matrix into other easy-to-manipulate formats. Similarity transform is one of the most commonly used transformation methods. In Section 4.1, the concept and properties of the matrix similarity transform are introduced. The concept of an orthogonal matrix is presented. In Section 4.2, elementary row transforms are presented, and three commonly used elementary row transform rules are demonstrated. Based on the transformation methods, matrix inverse and pivot selection algorithms are studied. In Section 4.3, the well-established Gaussian elimination method is introduced for linear equations, and triangular factorization methods are presented for ordinary matrices. Section 4.4 introduces Cholesky factorization for symmetric matrices, and the concepts of positive definite and regular matrices are studied. In Section 4.5, methods are demonstrated to transform ordinary matrices into their companions, as well as diagonal and Jordan forms. In Section 4.6, the singular value decomposition technique is discussed, and the concept of a condition number is addressed. In Section 4.7, Givens and Householder transforms are presented.

## 4.1 Similarity transform and orthogonal matrices

The similarity transform is a kind of commonly used matrix transformation methods. By selecting appropriate transformation matrices, the matrices of interest can be converted into any other prespecified format, without affecting their important properties.

In this chapter, the similarity transform of matrices is presented. Orthogonal matrices and orthonormal bases are introduced, followed by the concepts of positive and totally positive matrices.

### 4.1.1 Similarity transform

**Definition 4.1.** For a given square matrix $A$ and a nonsingular matrix $T$, the following transform can be made on the original matrix $A$:

$$X = T^{-1}AT, \tag{4.1.1}$$

to transform it into another form. This kind of transform is also known as the similarity transformation, and $T$ is the similarity transform matrix.

**Theorem 4.1.** *The ranks, traces, determinants, and eigenvalues of matrices are not affected by the similarity transform.*

https://doi.org/10.1515/9783110666991-004

Through appropriate selection of the transformation matrix $T$, the format of an arbitrary matrix $A$ can be deliberately changed into other meaningful one, without affecting important properties of the original matrix.

**Example 4.1.** For the following $A$, if the transformation matrix $T$ is selected below, find the similarity transform:

$$A = \begin{bmatrix} 2 & 3 & 5 & 9 \\ 5 & 9 & 8 & 3 \\ 0 & 3 & 2 & 4 \\ 3 & 4 & 5 & 8 \end{bmatrix}, \quad T = \begin{bmatrix} 0 & 9 & 119 & 1974 \\ 0 & 3 & 128 & 2508 \\ 0 & 4 & 49 & 974 \\ 1 & 8 & 123 & 2098 \end{bmatrix}.$$

**Solutions.** The two matrices should be entered into MATLAB workspace, and then the similarity transform can be carried out

```
>> A=[2,3,5,9; 5,9,8,3; 0,3,2,4; 3,4,5,8]; A=sym(A);
   T=[0,9,119,1974; 0,3,128,2508; 0,4,49,974; 1,8,123,2098];
   A1=inv(T)*A*T
```

The transformed matrix is given below. It can be seen that the format of the matrix can be changed by a certain similarity transform:

$$A_1 = \begin{bmatrix} 0 & 0 & 0 & -120 \\ 1 & 0 & 0 & -84 \\ 0 & 1 & 0 & -46 \\ 0 & 0 & 1 & 21 \end{bmatrix}.$$

### 4.1.2 Orthogonal matrices and orthonormal bases

**Definition 4.2.** Given a set of column vectors $v_1, v_2, \ldots, v_m$, if

$$v_i^T v_j = \begin{cases} 0, & \text{if } i \neq j, \\ \delta_{ij}, & \text{if } i = j, \end{cases} \tag{4.1.2}$$

then such vectors are known as orthogonal.

**Definition 4.3.** For a given matrix, if $H$ satisfies $H^H = H^{-1}$, it is referred to as a Hermitian matrix. Hermitian matrices are orthogonal.

Hermitian matrices are named after French mathematician Charles Hermite (1822–1901).

**Theorem 4.2.** *Orthogonal matrix $Q$ satisfies the following conditions:*

$$Q^H Q = I, \quad \text{and} \quad QQ^H = I \quad \text{where } I \text{ is an } n \times n \text{ identity matrix.} \tag{4.1.3}$$

A MATLAB function $Q$=orth($A$) is provided for finding an orthonormal basis $Q$ from the columns of matrix $A$. If $A$ is a nonsingular matrix, this orthonormal basis $Q$ satisfies the conditions in (4.1.3).

**Theorem 4.3.** *If $A$ is a singular matrix, the number of columns in $Q$ is the rank of matrix $A$, and $Q^H Q = I$, but not $QQ^H = I$.*

**Example 4.2.** Compute an orthonormal basis from the columns of matrix $A$.

**Solutions.** Such an orthonormal basis can be obtained directly with function orth(). It can also be validated with the following statements:

```
>> A=[2,3,5,9; 5,9,8,3; 0,3,2,4; 3,4,5,8];
   Q=orth(A), norm(Q'*Q-eye(4))
   norm(Q*Q'-eye(4)) % computing and validating orthonormal basis
```

The orthogonal matrix and error matrix norms are

$$\|Q^H Q - I\| = 6.7409 \times 10^{-16}, \quad \|QQ^H - I\| = 6.9630 \times 10^{-16},$$

$$Q = \begin{bmatrix} -0.5198 & 0.5298 & 0.1563 & -0.6517 \\ -0.6197 & -0.7738 & 0.0262 & -0.1286 \\ -0.2548 & 0.1551 & -0.9490 & 0.1017 \\ -0.5300 & 0.3106 & 0.2725 & 0.7406 \end{bmatrix}.$$

Symbolic computation can also be used in finding an orthonormal basis

```
>> Q=simplify(orth(sym(A))), norm(Q'*Q-eye(4))
```

and the analytical solution obtained is

$$Q = \begin{bmatrix} \sqrt{38}/19 & -6\sqrt{15\,238}/7\,619 & 696\sqrt{678\,091}/678\,091 & -17\sqrt{1\,691}/1\,691 \\ 5\sqrt{38}/38 & 27\sqrt{15\,238}/15\,238 & -363\sqrt{678\,091}/678\,091 & -13\sqrt{1\,691}/1\,691 \\ 0 & 3\sqrt{15\,238}/401 & 205\sqrt{678\,091}/678\,091 & 12\sqrt{1\,691}/1\,691 \\ 3\sqrt{38}/38 & -37\sqrt{15\,238}/15\,238 & 141\sqrt{678\,091}/678\,091 & 33\sqrt{1\,691}/1\,691 \end{bmatrix}.$$

**Example 4.3.** Consider again the singular matrix in Example 3.2. Compute its orthonormal basis and validate the orthogonality of the matrices.

**Solutions.** The following statements can be used to compute and validate the orthonormal basis matrices. Note that, since $A$ is a singular matrix, the obtained $Q$ matrix is a $4 \times 3$ rectangular one.

```
>> A=[16,2,3,13; 5,11,10,8; 9,7,6,12; 4,14,15,1];
   Q=orth(A), norm(Q'*Q-eye(3)), Q1=simplify(orth(sym(A)))
```

The numerical and analytical solutions of the orthonormal basis matrices are obtained below, and the error matrix norm is $\|Q^H Q - I\| = 1.0140 \times 10^{-15}$ in numerical computation:

$$
Q = \begin{bmatrix}
-0.5 & 0.6708 & 0.5 \\
-0.5 & -0.2236 & -0.5 \\
-0.5 & 0.2236 & -0.5 \\
-0.5 & -0.6708 & 0.5
\end{bmatrix},
$$

$$
Q_1 = \begin{bmatrix}
8\sqrt{42}/63 & -635\sqrt{255\,738}/767\,214 & 109\sqrt{30\,445}/60\,890 \\
5\sqrt{42}/126 & 391\sqrt{255\,738}/383\,607 & -163\sqrt{30\,445}/60\,890 \\
\sqrt{42}/14 & 11\sqrt{255\,738}/42\,623 & -197\sqrt{30\,445}/60\,890 \\
2\sqrt{42}/63 & 1\,117\sqrt{255\,738}/767\,214 & 211\sqrt{30\,445}/60\,890
\end{bmatrix}.
$$

## 4.2 Elementary row transforms

The algebraic cofactor method for computing determinants was presented earlier. The computation load is rather heavy for large-scale matrices, and it is difficult to implement symbolically. Special treatment needs to be introduced to convert certain elements to zeros, while not affecting the values in the determinants. A successful way of doing this is to introduce a class of elementary row transform techniques. Three types of elementary row transform patterns are introduced. Also, pivot base elementary row transforms can be adopted.

### 4.2.1 Three types of elementary row transforms

There are three types of elementary row transforms. The transforms are demonstrated in this section through examples, and MATLAB implementation is also presented.

**Definition 4.4.** The following three types of transforms are known as elementary row transforms:
(1) Multiply all the elements in one row by $k$;
(2) Multiply one row by $k$ and add it to another row;
(3) Exchange any two rows.

**Definition 4.5.** If the "row" in the above definition is changed to "column", elementary column transforms are defined.

The three types of elementary row transforms are presented below, and demonstrated with examples.
(1) Multiplying all the elements in a row by $k$. The target here is to find a matrix $E$ such that $EA$ multiplies all the elements in a given row of $A$ by $k$, while the other elements in the matrix are unchanged.

**Theorem 4.4.** *Take an identity matrix $E$ and let $E(i,i) = k$, then $EA$ multiplies the ith row in $A$ by $k$, while the other elements are unchanged. Also, one has $\det(EA) = k \det(A)$. Matrix $F = E^{-1}$ is also started from an identity matrix, and then we set $F(i,i) = 1/k$. To compute $AE$, the ith column in $A$ is multiplied by $k$, while the other elements are unchanged.*

**Example 4.4.** Consider the matrix in Example 4.1, Design a left matrix $E$ such that $EA$ multiplies all the elements in the second row of $A$ by 1/5.

**Solutions.** To configure the transformation matrix, one should set $E$ to an identity matrix first. Then, since one wants to multiply all the elements in the second row of $A$ by 1/5, one must set $E(2,2) = 1/5$. After elementary row transform

```
>> A=[2,3,5,9; 5,9,8,3; 0,3,2,4; 3,4,5,8];
   E=eye(4); E(2,2)=1/5, A1=E*A, inv(E)
```

the transformed matrices are obtained as

$$
E = \begin{bmatrix} 1 & 0 & 0 & 0 \\ 0 & 1/5 & 0 & 0 \\ 0 & 0 & 1 & 0 \\ 0 & 0 & 0 & 1 \end{bmatrix}, \quad A_1 = \begin{bmatrix} 2 & 3 & 5 & 9 \\ 1 & 9/5 & 8/5 & 3/5 \\ 0 & 3 & 2 & 4 \\ 3 & 4 & 5 & 8 \end{bmatrix}, \quad E^{-1} = \begin{bmatrix} 1 & 0 & 0 & 0 \\ 0 & 5 & 0 & 0 \\ 0 & 0 & 1 & 0 \\ 0 & 0 & 0 & 1 \end{bmatrix}.
$$

If command $A*E$ is used, all the elements in the second column of $A$ are multiplied by 1/5, with the other elements left unchanged.

(2) Multiply a row by a constant and add to another row. It is known from Theorem 3.8 that if one row is multiplied by constant $k$ and added to another row, the determinant is unchanged. Here, such an elementary transform is introduced.

**Theorem 4.5.** *If the ith row is to be multiplied by $k$ and then added to the jth row, an identity matrix $E$ should be taken first. Then setting $E(i,j) = k$, the determinant $\det(EA) = \det(A)$. The initial value of $F = E^{-1}$ is also an identity matrix, and then we set $F(i,j) = -k$. If $AE$ is computed, the ith column of $A$ will be multiplied by $k$, and added to the jth column.*

**Example 4.5.** Consider matrix $A_1$ in Example 4.4. If we want to multiply its second row by $-3$ and add to the fourth row, how can we select $E$?

**Solutions.** Consider the converted matrix in Example 4.4. An initial identity matrix $E$ is selected, and then we set $E(4,2) = -3$. For this task the following statements can be used:

```
>> A=[2,3,5,9; 5,9,8,3; 0,3,2,4; 3,4,5,8];
   E=eye(4); E(2,2)=1/5; A1=E*A; E=eye(4); E(4,2)=-3
   A2=E*A1, inv(E)
```

The results are

$$
E = \begin{bmatrix} 1 & 0 & 0 & 0 \\ 0 & 1 & 0 & 0 \\ 0 & 0 & 1 & 0 \\ 0 & -3 & 0 & 1 \end{bmatrix}, \quad
A_2 = \begin{bmatrix} 2 & 3 & 5 & 9 \\ 1 & 9/5 & 8/5 & 3/5 \\ 0 & 3 & 2 & 4 \\ 0 & -7/5 & 1/5 & 31/5 \end{bmatrix}, \quad
E^{-1} = \begin{bmatrix} 1 & 0 & 0 & 0 \\ 0 & 1 & 0 & 0 \\ 0 & 0 & 1 & 0 \\ 0 & 3 & 0 & 1 \end{bmatrix}.
$$

(3) Exchange two rows. The target is to construct a matrix $E$ such that $EA$ exchanges any given two rows of $A$, without affecting the other rows.

**Theorem 4.6.** *To exchange the ith and jth rows in matrix $A$, an identity matrix $E$ should be created first, then set $E(i, i) = E(j, j) = 0$ and $E(i, j) = E(j, i) = 1$. Therefore, $EA$ is the expected matrix, and $\det(EA) = -\det(A)$. The inverse matrix $E$ satisfies $E^{-1} = E$. If $AE$ is computed, the ith and jth columns of $A$ are exchanged.*

**Example 4.6.** Consider the matrix $A$ in Example 4.5. Design an $E$ matrix such that the first and second rows of $A_2$ are exchanged. What will happen if $A_2E$ is computed.

**Solutions.** The matrix $A_2$ in Example 4.5 can be created again. An identity matrix $E$ can be constructed, then setting $E(1, 1) = E(2, 2) = 0$, $E(1, 2) = E(2, 1) = 1$, the transformed matrix $A_3$ can be obtained with $EA_2$.

```
>> A=[2,3,5,9; 5,9,8,3; 0,3,2,4; 3,4,5,8];
   E1=eye(4); E1(2,2)=1/5, E2=eye(4); E2(4,2)=-3; A2=E2*E1*A;
   E=eye(4); E([1,2],[1,2])=[0 1; 1 0]; A3=E*A2, inv(E), A2*E
```

The related matrices are given below. The transformation matrix $E$ is the same as its inverse. In $A_2E$, the first and second columns of $A_2$ are exchanged.

$$
E = E^{-1} = \begin{bmatrix} 0 & 1 & 0 & 0 \\ 1 & 0 & 0 & 0 \\ 0 & 0 & 1 & 0 \\ 0 & 0 & 0 & 1 \end{bmatrix}, \quad
A_3 = \begin{bmatrix} 1 & 9/5 & 8/5 & 3/5 \\ 2 & 3 & 5 & 9 \\ 0 & 3 & 2 & 4 \\ 0 & -7/5 & 1/5 & 31/5 \end{bmatrix}.
$$

### 4.2.2 Inverse through elementary row transforms

In ordinary linear algebra course, an elementary row transform based inverse matrix computation method is introduced. An augmented matrix can be created, and a series of elementary row transforms is carried out to find the inverse matrix. This method is efficient and can be adopted to deal with inverses of large-scale matrices. With the three elementary row transform methods discussed earlier, if one wants to eliminate a certain element in a given row, a transform matrix should be established first. The processing above is rather complicated, and an illustrating example is given to demonstrate better algorithms.

**Example 4.7.** Consider the matrix in Example 4.1. If one wants to have the upper-left element changed to 1, while the other elements in the first column are changed to zero, how can matrix $E$ be selected?

**Solutions.** To complete the two steps, the first row should be multiplied by 1/2, and the second, third, and fourth rows should be respectively changed by adding –5, 0, –3 times the first row. Therefore, the following statements can be used:

```
>> A=[2,3,5,9; 5,9,8,3; 0,3,2,4; 3,4,5,8];
   E1=eye(4); E1(1,1)=1/2, E2=eye(4);
   E2([2,3,4],1)=[-5; 0; -3], E=E2*E1, A1=E*A
```

The resulting matrices are

$$E_2 = \begin{bmatrix} 1 & 0 & 0 & 0 \\ -5 & 1 & 0 & 0 \\ 0 & 0 & 1 & 0 \\ -3 & 0 & 0 & 1 \end{bmatrix}, \quad E = \begin{bmatrix} 1/2 & 0 & 0 & 0 \\ -5/2 & 1 & 0 & 0 \\ 0 & 0 & 1 & 0 \\ -3/2 & 0 & 0 & 1 \end{bmatrix},$$

$$A_1 = \begin{bmatrix} 1 & 3/2 & 5/2 & 9/2 \\ 0 & 3/2 & -9/2 & -39/2 \\ 0 & 3 & 2 & 4 \\ 0 & -1/2 & -5/2 & -11/2 \end{bmatrix}.$$

More generally, the two steps can be combined into one, by selecting factor matrix $E$ as follows:

```
>> E=eye(4); ii=1:4; j=1; ii=ii(ii~=j);
   E(ii,1)=-A(ii,1); E(:,1)=E(:,1)/A(1,1); E*A
```

With the elementary row transform methods, the following ideas can be employed in finding the inverse of $A$. The original matrix $A$ can be augmented as $A_1 = [A, I]$. For each column, a transformation should be carried out inside a loop structure to find the inverse. After transformation, the left side of $A_1$ will be converted to an identity matrix, while the right half of $A_1$ will be the inverse of $A$.

```
function [A2,E0]=new_inv1(A)
[n,m]=size(A); E=eye(n); A1=[A E]; aa=[]; E0=E;
for i=1:n
    ij=1:n; ij=ij(ij~=i); E1=E; a=A1(i,i);
    E1(i,i)=1/a; E1(ij,i)=-A1(ij,i)/a; A1=E1*A1; E0=E1*E0;
end
A2=A1(:,n+1:end);
```

**Example 4.8.** Compute the inverse matrix for matrix $A$ in Example 4.1.

**Solutions.** The following statements can be used to find the numerical and analytical solutions of the inverse matrix, and the errors can be found ˙

```
>> A=[2,3,5,9; 5,9,8,3; 0,3,2,4; 3,4,5,8];
   A1=new_inv1(A), [A2,E0]=new_inv1(sym(A)),
   e1=norm(A*A1-eye(4)), e2=norm(inv(A)*A-eye(4)), A*A2
```

The numerical and analytical solutions obtained are as follows. The errors in numerical solutions are $e_1 = 2.2206 \times 10^{-15}$ and $e_2 = 2.1696 \times 10^{-15}$. It can be seen that the accuracy of the algorithm is similar to that of function $\mathtt{inv()}$. The analytical inverse can be found, and the product of the inverse matrix and the original one is exactly an identity matrix:

$$A_1 = \begin{bmatrix} -0.65833 & -0.091667 & -0.3 & 0.925 \\ -0.48333 & -0.016667 & 0.4 & 0.35 \\ 1.0083 & 0.24167 & -0.3 & -1.075 \\ -0.14167 & -0.10833 & 0.1 & 0.275 \end{bmatrix},$$

$$A_2 = \begin{bmatrix} -79/120 & -11/120 & -3/10 & 37/40 \\ -29/60 & -1/60 & 2/5 & 7/20 \\ 121/120 & 29/120 & -3/10 & -43/40 \\ -17/120 & -13/120 & 1/10 & 11/40 \end{bmatrix},$$

$$E_0 = \begin{bmatrix} -79/120 & -11/120 & -3/10 & 37/40 \\ -29/60 & -1/60 & 2/5 & 7/20 \\ 121/120 & 29/120 & -3/10 & -43/40 \\ -17/120 & -13/120 & 1/10 & 11/40 \end{bmatrix}.$$

### 4.2.3 Computing inverses with pivot methods

There may be a fatal error in the above algorithm. That is, in the transformation process, if $A_1(i, i)$ is zero, the algorithm cannot be carried out any longer. This phenomenon must be avoided. It is also known from the numerical computation viewpoint that the larger the value of the divider, the smaller the overall error. Therefore, in each step $i$ in the loop, the maximum absolute value among $A_1(i, i)$ and the elements below in the column is found, and the row containing that maximum is exchanged with the $i$th row. The maximum value here is referred to as a pivot. The algorithm considering the pivots is also referred to as a pivot-based algorithm.

Extending slightly the function $\mathtt{new\_inv1()}$, pivots can be selected, followed the conventional transformation process. The inverse matrix can be obtained, as given in the following MATLAB function. In each step in the loop structure, row exchange is carried out to have the pivot row moved the $i$th row. A new function $\mathtt{new\_inv2()}$ can be written as follows:

```
function [A2,E0]=new_inv2(A)
[n,m]=size(A); E=eye(n); A1=[A E]; aa=[];
E0=E; if strcmp(class(A),'sym'), E0=sym(E); end
for i=1:n
    [a,i0]=max(abs(A1(i:end,i))); i0=i0(1)+i-1;
    if i~=i0
        E1=E; E1([i i0],[i i0])=[0 1; 1 0]; A1=E1*A1; E0=E1*E0;
    end
    ij=1:n; ij=ij(ij~=i); E1=E; a=A1(i,i);
    E1(i,i)=1/a; E1(ij,i)=-A1(ij,i)/a; A1=E1*A1; E0=E1*E0;
end
A2=A1(:,n+1:end);
```

**Example 4.9.** Compute again the inverse matrix in Example 4.1 with the pivot transform method.

**Solutions.** The problem can be solved directly with the following statements, the error
obtained is $e_1 = 1.5846 \times 10^{-15}$, slightly lower than that obtained with function `inv()`.
It can be seen that the efficiency of function `new_inv2()` is lower than of the algorithm
in `inv()`.

```
>> A=[2,3,5,9; 5,9,8,3; 0,3,2,4; 3,4,5,8];
   A1=new_inv2(A), tic, A2=new_inv2(sym(A)); toc
   e1=norm(A*A1-eye(4)), A*A2, tic, inv(sym(A)); toc
```

Employing the reduced row echelon form with function `rref()`, the target is the
same as that studied here. The two methods append the right-hand side of the matrix
with an identity matrix. After transformation, the left-hand side is converted into an
identity matrix, and the right-hand side becomes the inverse matrix. Readers can add
display statements inside the function, and witness the intermediate steps.

## 4.3 Triangular decomposition of matrices

Triangular decomposition is originated from the solution of linear algebra equations.
Gaussian elimination and its MATLAB implementation are introduced. Triangular decomposition of matrices follows.

### 4.3.1 Gaussian elimination of linear equations

Various algorithms are available for solving linear algebraic equations. A direct algorithm – Gaussian elimination method – is presented in this section. Gaussian elim-

ination method is named after German mathematician Johann Carl Friedrich Gauss (1777–1855). This method is similar to the elementary row transform method shown earlier.

**Theorem 4.7.** *Linear equations* $AX = B$ *can be solved directly with Gaussian elimination method. An augmented matrix* $C = [A, B]$ *can be constructed, and* $C$ *can be transformed into the form*

$$
C^{(n)} = \begin{bmatrix}
a_{11}^{(1)} & a_{12}^{(1)} & a_{13}^{(1)} & \cdots & a_{1n}^{(1)} & b_1^{(1)} \\
0 & a_{22}^{(2)} & a_{13}^{(2)} & \cdots & a_{2n}^{(2)} & b_2^{(2)} \\
0 & 0 & a_{33}^{(3)} & \cdots & a_{3n}^{(3)} & b_3^{(3)} \\
\vdots & \vdots & \vdots & \ddots & \vdots & \vdots \\
0 & 0 & 0 & \cdots & a_{nn}^{(n)} & b_n^{(n)}
\end{bmatrix}, \tag{4.3.1}
$$

*where for* $k = 1, 2, \ldots, n - 1$, *the elimination equations are obtained as*

$$
d_{ik} = a_{ik}^{(k)} / a_{nn}^{(k)}, \quad a_{ik}^{(k+1)} = 0, \tag{4.3.2}
$$

$$
a_{ij}^{(k+1)} = a_{ij}^{(k)} - d_{ik} a_{kj}^{(k)}, \quad b_i^{(k+1)} = b_i^{(k)} - d_{ik} b_k^{(k)}, \tag{4.3.3}
$$

*with* $i = k + 1, k + 2, \ldots, n$ *and* $j = k + 1, k + 2, \ldots, n$.

*The elimination equations can be solved, with the back substitution method as*

$$
x_n = b_n^{(n)} / a_{nn}^{(n)}, \quad x_k = \frac{b_k^{(k)} - \sum\limits_{j=k+1}^{n} a_{kj}^{(k)} x_j}{a_{kk}^{(k)}}, \quad k = n - 1, n - 2, \ldots, 1. \tag{4.3.4}
$$

The recursive algorithm in Theorem 4.7 can be implemented in the following MATLAB solver. The function can also be used in solving the linear equation, where the right-hand side $B$ is a matrix.

```
function x=gauss_eq(A,B)
n=length(A);
for k=1:n, i=k+1:n; j=k+1:n; d=A(i,k)/A(k,k);
    if length(d)>0,
        A(i,j)=A(i,j)-d*A(k,j); B(i,:)=B(i,:)-d*B(k,:); A(i,k)=0;
end, end
x(n,:)=B(n,:)/A(n,n);
for k=n-1:-1:1, x(k,:)=(B(k,:)-A(k,k+1:n)*x(k+1:n,:))/A(k,k); end
```

**Example 4.10.** Compute the inverse matrix of $A$ in Example 4.8 with Gaussian elimination method.

**Solutions.** Inverse matrix computation is a special case of solving linear equations. If matrix $B$ is selected as an identity matrix, then the inverse is the solution of equation

$AX = B$. The analytical solution $X$ is the inverse of the matrix, which is exactly the same as that in Example 4.8.

```
>> A=[2,3,5,9; 5,9,8,3; 0,3,2,4; 3,4,5,8];
   X=gauss_eq(A,sym(eye(4)))
```

The algorithms and implementation have bugs. That is, if a certain $A_i^{(i)}(i,i)$ equals to zero, problems may occur. Pivot algorithm can be introduced to correct the solution procedure.

### 4.3.2 Triangular factorization and implementation

The triangular factorization method was proposed by Polish mathematician Tadeusz Banachiewicz (1882–1954), and it can be used to solve matrix equations.

**Theorem 4.8.** *Triangular factorization of a matrix is also known as LU factorization. The purpose of the factorization is that a matrix can be factorized as the product of a lower triangular matrix $L$ and an upper-triangular $U$, i. e., $A = LU$, where $L$ and $U$ matrices can be written as*

$$L = \begin{bmatrix} 1 & & & \\ l_{21} & 1 & & \\ \vdots & \vdots & \ddots & \\ l_{n1} & l_{n2} & \cdots & 1 \end{bmatrix}, \quad U = \begin{bmatrix} u_{11} & u_{12} & \cdots & u_{1n} \\ & u_{22} & \cdots & u_{2n} \\ & & \ddots & \vdots \\ & & & u_{nn} \end{bmatrix}. \tag{4.3.5}$$

*The matrices can be obtained with matrix $A$ with entries*

$$\begin{aligned}
&a_{11} = u_{11}, &&a_{12} = u_{12}, &&\cdots &&a_{1n} = u_{1n}, \\
&a_{21} = l_{21}u_{11}, &&a_{22} = l_{21}u_{12} + u_{22}, &&\cdots &&u_{2n} = l_{21}u_{1n} + u_{2n}, \\
&\vdots &&\vdots &&\ddots &&\vdots \\
&a_{n1} = l_{n1}u_{11}, &&a_{n2} = l_{n1}u_{12} + l_{n2}u_{22}, &&\cdots &&a_{nn} = \sum_{k=1}^{n-1} l_{nk}u_{kn} + u_{nn}.
\end{aligned} \tag{4.3.6}$$

*From (4.3.6), the entries $l_{ij}$ and $u_{ij}$ can be obtained recursively from*

$$l_{ij} = \frac{a_{ij} - \sum_{k=1}^{j-1} l_{ik}u_{kj}}{u_{jj}} \quad (j < i), \quad u_{ij} = a_{ij} - \sum_{k=1}^{i-1} l_{ik}u_{kj} \quad (j \geq i). \tag{4.3.7}$$

*The initial values of the variables are*

$$u_{1i} = a_{1i}, \quad i = 1, 2, \ldots, n. \tag{4.3.8}$$

Note that pivot selection is not used in the above algorithm, and sometimes the divisor may be zero, which may lead to a failure of the LU algorithm. Based on the algorithm, the following MATLAB function can be written. If it fails, the function `lu()` in MATLAB should be used instead.

```
function [L,U]=lusym(A)
n=length(A); U=sym(zeros(size(A))); L=sym(eye(size(A)));
U(1,:)=A(1,:); L(:,1)=A(:,1)/U(1,1);
for i=2:n,
    for j=2:i-1, L(i,j)=(A(i,j)-L(i,1:j-1)*U(1:j-1,j))/U(j,j); end
    for j=i:n, U(i,j)=A(i,j)-L(i,1:i-1)*U(1:i-1,j); end
end
```

### 4.3.3 MATLAB triangular factorization solver

It has been commented that there may be problems in the traditional LU factorization algorithm, specifically, when $a_{ii}$ equals to zero, the factorization algorithm may fail. Pivot-based LU factorization algorithm should be used instead. The `lu()` function is provided in MATLAB, with the syntaxes

$[L,U]$=lu$(A)$, % LU factorization $A = LU$

$[L,U,P]$=lu$(A)$, % $P$ is the permutation matrix, $A = P^{-1}LU$

where $L$ and $U$ are respectively lower- and upper-triangular matrices. Pivot selection is implemented in `lu()` function, and reliable results can be obtained. When a pivot is considered, the returned lower-triangular matrix $L$ may not be a genuine lower-triangular one. If one really wants to have a genuine lower-triangular matrix $L$, permutation in (3.1.1) is accompanied such that $A$ matrix can be factorized as $A = P^{-1}LU$. In the new versions of MATLAB, $A$ can also be a symbolic matrix.

**Example 4.11.** Reconsider the LU factorization problem in Example 3.2. The two LU solvers can be used to perform the factorization.

**Solutions.** Inputting $A$ matrix and performing LU factorization

```
>> A=[16 2 3 13; 5 11 10 8; 9 7 6 12; 4 14 15 1]; [L1,U1]=lu(A)
```

the results are

$$
L_1 = \begin{bmatrix} 1 & 0 & 0 & 0 \\ 0.3125 & 0.7685 & 1 & 1 \\ 0.5625 & 0.4352 & 1 & 0 \\ 0.25 & 1 & 0 & 0 \end{bmatrix}, \quad U_1 = \begin{bmatrix} 16 & 2 & 3 & 13 \\ 0 & 13.5 & 14.25 & -2.25 \\ 0 & 0 & -1.8889 & 5.6667 \\ 0 & 0 & 0 & 3.55 \times 10^{-15} \end{bmatrix}.
$$

It can be seen that $L_1$ is not a lower-triangular matrix, since pivot algorithm is used. Consider now an alternative lu() syntax

```
>> [L,U,P]=lu(A), inv(P)*L % pivot based LU factorization
```

The factorization matrices are

$$
L = \begin{bmatrix} 1 & 0 & 0 & 0 \\ 0.25 & 1 & 0 & 0 \\ 0.3125 & 0.7685 & 1 & 0 \\ 0.5625 & 0.4352 & 1 & 1 \end{bmatrix}, \quad
U = \begin{bmatrix} 16 & 2 & 3 & 13 \\ 0 & 13.5 & 14.25 & -2.25 \\ 0 & 0 & -1.8889 & 5.6667 \\ 0 & 0 & 0 & 3.55 \times 10^{-15} \end{bmatrix},
$$

$$
P = \begin{bmatrix} 1 & 0 & 0 & 0 \\ 0 & 0 & 0 & 1 \\ 0 & 1 & 0 & 0 \\ 0 & 0 & 1 & 0 \end{bmatrix}, \quad
P^{-1}L = \begin{bmatrix} 1 & 0 & 0 & 0 \\ 0.3125 & 0.7685 & 1 & 0 \\ 0.5625 & 0.4352 & 1 & 1 \\ 0.25 & 1 & 0 & 0 \end{bmatrix}.
$$

Note that matrix $P$ is not an identity matrix, in fact, it is a permutation of an identity matrix. Considering the obtained matrix $L_1$, it can be seen that $P$ has $p_{2,4} = 1$, indicating the fourth row is moved to the second row in $L_1$, while $p_{3,2} = p_{4,3} = 1$ similarly indicate that the second row is moved to the third and the third is moved to the fourth in $L_1$. In this case, a genuine lower-triangular $L$ can be constructed. Substituting $P$, $L$ and $U$ back into the formula, the relationship can be validated, and matrix $A$ can be restored with command inv(P)*L*U.

With analytical function, the original matrix can be processed with triangular factorization solver

```
>> [L2,U2]=lu(sym(A)) % symbolic LU factorization
```

and the analytical solution obtained is

$$
L_2 = \begin{bmatrix} 1 & 0 & 0 & 0 \\ 5/16 & 1 & 0 & 0 \\ 9/16 & 47/83 & 1 & 0 \\ 1/4 & 108/83 & -3 & 1 \end{bmatrix}, \quad
U_2 = \begin{bmatrix} 16 & 2 & 3 & 13 \\ 0 & 83/8 & 145/16 & 63/16 \\ 0 & 0 & -68/83 & 204/83 \\ 0 & 0 & 0 & 0 \end{bmatrix}.
$$

**Example 4.12.** Perform triangular factorization to any $3 \times 3$ matrix.

**Solutions.** The following statements can be used to perform triangular factorization to given arbitrary matrices:

```
>> A=sym('a%d%d',3); [L U]=lu(A) % LU factorization for any matrix
```

and the results are

$$
L = \begin{bmatrix}
1 & 0 & 0 \\
a_{21}/a_{11} & 1 & 0 \\
a_{31}/a_{11} & (a_{32} - a_{12}a_{31}/a_{11})(a_{22} - a_{12}a_{21}/a_{11}) & 1
\end{bmatrix},
$$

$$
U = \begin{bmatrix}
a_{11} & a_{12} & a_{13} \\
0 & a_{22} - a_{12}a_{21}/a_{11} & a_{23} - a_{13}a_{21}/a_{11} \\
0 & 0 & a_{33} - \dfrac{(a_{23} - a_{13}a_{21}/a_{11})(a_{32} - a_{12}a_{31}/a_{11})}{a_{22} - a_{12}a_{21}/a_{11}} - \dfrac{a_{13}a_{31}}{a_{11}}
\end{bmatrix}.
$$

Since $a_{11}$ and $a_{22} - a_{12}a_{21}/a_{11}$ appear on the denominators of the results, when $a_{11} = 0$ and $a_{11}a_{22} - a_{12}a_{21} = 0$, triangular factorization cannot be used. Function lu() can be called instead, or one can modify the factorization function by considering pivot selections.

## 4.4 Cholesky factorization

Symmetric matrices are the object in this section, where Cholesky factorization is mainly presented. The factorization is named after French mathematician André-Louis Cholesky (1875–1918). The concept and judgement of positive-definite and regular matrices are presented.

### 4.4.1 Cholesky factorization of symmetric matrices

**Theorem 4.9.** *If $A$ is a symmetric matrix, it can be factorized in a similar way as in LU factorization, such that the original matrix $A$ can be factorized as*

$$
A = LL^{T} = \begin{bmatrix}
l_{11} & & & \\
l_{21} & l_{22} & & \\
\vdots & \vdots & \ddots & \\
l_{n1} & l_{n2} & \cdots & l_{nn}
\end{bmatrix}
\begin{bmatrix}
l_{11} & l_{21} & \cdots & l_{n1} \\
 & l_{22} & \cdots & l_{n2} \\
 & & \ddots & \vdots \\
 & & & l_{nn}
\end{bmatrix}. \tag{4.4.1}
$$

*Using the properties of symmetric matrices, the factorization matrix can be recursively computed from*

$$
l_{ii} = \sqrt{a_{ii} - \sum_{k=1}^{i-1} l_{ik}^2}, \quad l_{ji} = \frac{1}{l_{jj}}\left(a_{ij} - \sum_{k=1}^{j-1} l_{ik}l_{jk}\right), \quad j < i, \tag{4.4.2}
$$

*with the initial conditions $l_{11} = \sqrt{a_{11}}$ and $l_{j1} = a_{j1}/l_{11}$. This algorithm is also known as Cholesky factorization algorithm.*

Function `chol()` is provided in MATLAB to perform Cholesky factorization to compute matrix $D$, with $D$=`chol`$(A)$, where $D$ is an upper-triangular matrix, $D = L^T$. In the new versions of MATLAB, $A$ can also be a symbolic matrix.

**Example 4.13.** Compute Cholesky factorization for a $4 \times 4$ symmetrical matrix $A$ given by

$$A = \begin{bmatrix} 9 & 3 & 4 & 2 \\ 3 & 6 & 0 & 7 \\ 4 & 0 & 6 & 0 \\ 2 & 7 & 0 & 9 \end{bmatrix}.$$

**Solutions.** The following commands can be used to perform Cholesky factorization

```
>> A=[9,3,4,2; 3,6,0,7; 4,0,6,0; 2,7,0,9];
   D=chol(A), D1=chol(sym(A))
```

and the analytical and numerical results are respectively

$$D = \begin{bmatrix} 3 & 1 & 1.3333 & 0.6667 \\ 0 & 2.2361 & -0.5963 & 2.8324 \\ 0 & 0 & 1.9664 & 0.4068 \\ 0 & 0 & 0 & 0.6065 \end{bmatrix},$$

$$D_1 = \begin{bmatrix} 3 & 1 & 4/3 & 2/3 \\ 0 & \sqrt{5} & -4\sqrt{5}/15 & 19\sqrt{5}/15 \\ 0 & 0 & \sqrt{15}\sqrt{58}/15 & 2\sqrt{15}\sqrt{58}/145 \\ 0 & 0 & 0 & 4\sqrt{2}\sqrt{87}/87 \end{bmatrix}.$$

### 4.4.2 Quadratic forms of symmetric matrices

**Definition 4.6.** If a symmetric matrix $A$ is given by

$$A = \begin{bmatrix} a_{11} & a_{12} & \cdots & a_{1n} \\ a_{12} & a_{22} & \cdots & a_{2n} \\ \vdots & \vdots & \ddots & \vdots \\ a_{1n} & a_{2n} & \cdots & a_{nn} \end{bmatrix}, \tag{4.4.3}$$

the quadratic form of the matrix is

$$f(x) = x^T A x = a_{11}x_1^2 + \cdots + a_{nn}x_n^2 + 2a_{12}x_1x_2 + \cdots + 2a_{(n-1)n}x_{n-1}x_n, \tag{4.4.4}$$

where $x = [x_1, x_2, \ldots, x_n]^T$.

Symbolic computation can be used to find the quadratic form of a given matrix, and this will be demonstrated through examples.

**Example 4.14.** Get the quadratic form of the symmetric matrix in Example 4.17.

**Solutions.** Quadratic form can easily be obtained with the following statements:

```
>> A=[7,5,5,8; 5,6,9,7; 5,9,9,0; 8,7,0,1];
   x=sym('x',[4,1]); F=expand(x.'*A*x)
```

and the quadratic form obtained is

$$F(x) = 7x_1^2 + 6x_2^2 + 9x_3^2 + x_4^2 + 10x_1x_2 + 10x_1x_3 + 16x_1x_4 + 18x_2x_3 + 14x_2x_4.$$

It should be noted that an asymmetric matrix $A$ also has its quadratic form. It is recommended to use symmetric forms, however, and it is better to use $B = (A + A^T)/2$, whose quadratic form is equivalent, i. e., $x^T A x = z^T B z$.

**Example 4.15.** Write down the quadratic form of the asymmetric matrix below. Also find a symmetric matrix which has the same quadratic form if

$$A = \begin{bmatrix} 16 & 2 & 3 & 0 \\ 5 & 11 & 10 & 8 \\ 9 & 7 & 6 & 12 \\ 0 & 14 & 15 & 1 \end{bmatrix}.$$

**Solutions.** The following MATLAB statements can be used to find the quadratic form of the given matrix:

```
>> A=[16,2,3,0; 5,11,10,8; 9,7,6,12; 0,14,15,1];
   x=sym('x',[4,1]); assume(x,'real')
   f=expand(x'*A*x), B=(A+A')/2
```

The quadratic form is

$$f(x) = 16x_1^2 + 6x_3^2 + x_4^2 + 7x_1x_2 + 12x_1x_3 + 11x_2^2 + 17x_2x_3 + 22x_2x_4 + 27x_3x_4.$$

A symmetric matrix with the same quadratic form can be written as

$$B = \begin{bmatrix} 16 & 3.5 & 6 & 0 \\ 3.5 & 11 & 8.5 & 11 \\ 6 & 8.5 & 6 & 13.5 \\ 0 & 11 & 13.5 & 1 \end{bmatrix}.$$

### 4.4.3 Positive definite and regular matrices

Positive definite and regular matrices are all defined using symmetric matrices. The concepts and judgements of these matrices are presented in this section.

**Definition 4.7.** For a given $n \times n$ Hermitian matrix $A$, if for any vector $x$, the quadratic form $x^H A x$ is always positive, the matrix $A$ is referred to as a positive-definite matrix.

Similarly, the concepts of semipositive definiteness and negative definiteness of symmetric matrices are introduced.

**Definition 4.8.** If for any $x$, the quadratic form $x^H A x \geqslant 0$, the matrix is called a semidefinite matrix; if for any $x$, $x^H A x < 0$, the matrix is a negative definite matrix.

MATLAB function `chol()` can be used to judge whether a matrix is positive definite or not. An alternative syntax is $[D,p]$=`chol(A)`, where, for a positive definite matrix $A$, $p = 0$. For matrix $A$, if a positive $p$ is returned, the matrix is not positive definite, and $D$ is a $(p - 1) \times (p - 1)$ square matrix.

**Example 4.16.** Judge the positive definiteness of the quadratic form $x_1^2 + 3x_2^2 + 5x_3^2 + 2x_1x_2 - 4x_1x_3$.

**Solutions.** To test whether a quadratic form is positive definite or not, from the quadratic form $x^T Q x$, a symmetric matrix $Q$ should be created. It is not difficult to write down a symmetric $Q$ matrix

$$Q = \begin{bmatrix} 1 & 1 & -2 \\ 1 & 3 & 0 \\ -2 & 0 & 5 \end{bmatrix}.$$

For the given matrix $Q$, Cholesky factorization is performed, and it can be seen that $p = 2 \neq 0$, meaning the quadratic form is not positive definite.

```
>> Q=[1 1 -2; 1 3 0; -2 0 5]; [a p]=chol(Q)
```

**Definition 4.9.** If a complex square matrix satisfies

$$A^H A = A A^H, \tag{4.4.5}$$

where $A^H$ is the Hermitian transpose of matrix $A$, the matrix is a regular matrix.

Regularity can be tested with the command `norm(A'*A-A*A')`<$\epsilon$. If the result is 1, then $A$ is a regular matrix.

**Example 4.17.** Judge whether symmetric matrix $A$ is positive definite or not. Also find the Cholesky factorization of the matrix if

$$A = \begin{bmatrix} 7 & 5 & 5 & 8 \\ 5 & 6 & 9 & 7 \\ 5 & 9 & 9 & 0 \\ 8 & 7 & 0 & 1 \end{bmatrix}.$$

**Solutions.** With the following statements, Cholesky factorization can be performed for matrix $A$, and the size of matrix $D$ is 2, meaning the size of positive definiteness is 2. It can be seen that the matrix is not a positive definite matrix, since $p \neq 0$.

```
>> A=[7,5,5,8; 5,6,9,7; 5,9,9,0; 8,7,0,1];
   [D,p]=chol(A)  % test whether matrix is positive definite or not
```

It can be seen that $p = 3 \neq 0$, and the $2 \times 2$ matrix $D$ is given below as

$$D = \begin{bmatrix} 2.6458 & 1.8898 \\ 0 & 1.5584 \end{bmatrix}.$$

Note that, for an asymmetric matrix $A$, function chol() can still be used, however, the result is not correct. It forced the asymmetric matrix into a symmetric one. A matrix which is not positive definite does not have a real Cholesky factorization.

### 4.4.4 Cholesky factorization of nonpositive definite matrices

It has been indicated that function chol() can only be used for the Cholesky factorization of positive definite matrices. In fact, based on Theorem 4.9, low-level Cholesky factorization function can be written, where $A$ is a symbolic matrix.

```
function D=cholsym(A)
n=length(A); D(1,1)=sqrt(A(1,1)); D(1,2:n)=A(2:n,1)/D(1,1);
for i=2:n, k=1:i-1; D(i,i)=sqrt(A(i,i)-sum(D(k,i).^2));
    for j=i+1:n, D(i,j)=(A(j,i)-sum(D(k,j).*D(k,i)))/D(i,i);
end, end
```

**Example 4.18.** Perform Cholesky factorization to matrix $A$ in Example 4.17.

**Solutions.** It has been shown that the matrix in Example 4.17 is not positive definite. Function chol() cannot be used to compute its factorization. If function cholsym() is used for the symbolic matrix, the following statements can be employed:

```
>> A=[7,5,5,8; 5,6,9,7; 5,9,9,0; 8,7,0,1];
   D=cholsym(sym(A))
```

With the above statements, the upper-triangular matrix can be obtained, and it can be seen that there are complex terms in the matrix, hence the matrix is not positive definite:

$$D = \begin{bmatrix} \sqrt{7} & 5\sqrt{7}/7 & 5\sqrt{7}/7 & 8\sqrt{7}/7 \\ 0 & \sqrt{7}\sqrt{17}/7 & 38\sqrt{7}\sqrt{17}/119 & 9\sqrt{7}\sqrt{17}/119 \\ 0 & 0 & \sqrt{17}\sqrt{114}\,j/17 & 73\sqrt{17}\sqrt{114}\,j/969 \\ 0 & 0 & 0 & 2\sqrt{31}\sqrt{57}/57 \end{bmatrix}.$$

## 4.5 Companion and Jordan transforms

As indicated earlier, similarity transform can be used to convert matrices from one format into another. One may deliberately choose transformation matrices. In this section, transformation to a companion form is presented first, then we discuss how to convert matrices into diagonal or Jordan form. Also the matrices with repeated complex eigenvalues are discussed.

### 4.5.1 Companion form transformation

**Theorem 4.10.** *For a given matrix $A$, if there exists a column vector $x$ such that matrix $T = [x, Ax, \dots, A^{n-1}x]$ is nonsingular, matrix $T$ can be used to transform the matrix into its companion.*

There are infinitely many transformation matrices which can be used to convert the matrix into the same companion form. The companion matrix is different from the standard form in (2.2.6). Left–right flipping of an identity matrix can be selected as the transformation matrix. This phenomenon is illustrated in the example below.

**Example 4.19.** Convert $A$ in Example 3.44 into a standard companion matrix.

**Solutions.** A random vector $x$ can be selected such that $T$ is a nonsingular matrix. Similarity transform can be used to transform the matrix into a companion form.

```
>> A=[5,7,6,5; 7,10,8,7; 6,8,10,9; 5,7,9,10];  % input the matrix
   while(1),
       x=randi([0,1],[4,1]); T=sym([x A*x A^2*x A^3*x]);
       if rank(T)==4, break;    % find a full-rank integer matrix
   end, end, T, A1=inv(T)*A*T  % similarity transform
```

The transformation and the transformed matrices are

$$T = \begin{bmatrix} 1 & 11 & 326 & 9\,853 \\ 0 & 15 & 453 & 13\,696 \\ 1 & 16 & 472 & 14\,296 \\ 0 & 14 & 444 & 13\,489 \end{bmatrix}, \quad A_1 = \begin{bmatrix} 0 & 0 & 0 & -1 \\ 1 & 0 & 0 & 100 \\ 0 & 1 & 0 & -146 \\ 0 & 0 & 1 & 35 \end{bmatrix}.$$

It can be seen that matrix $T$ is not unique. Matrix $A_1$ obtained is similar to the companion matrix in (2.2.6). The following statements can be used to convert the matrix into a standard companion form:

```
>> T1=inv(T*fliplr(eye(4)))'
   A2=inv(T1)*A*T1 % standard companion form
```

The final transformation matrix and its companion form are:

$$T = \frac{1}{14\,053} \begin{bmatrix} -318 & 10\,591 & -29\,493 & 19\,064 \\ -176 & 5\,243 & 3\,298 & -11\,368 \\ 318 & -10\,591 & 29\,493 & -5\,011 \\ 75 & -1\,835 & -13\,063 & 2\,928 \end{bmatrix},$$

$$A_2 = \begin{bmatrix} 35 & -146 & 100 & -1 \\ 1 & 0 & 0 & 0 \\ 0 & 1 & 0 & 0 \\ 0 & 0 & 1 & 0 \end{bmatrix}.$$

### 4.5.2 Matrix diagonalization

**Theorem 4.11.** *If all the eigenvalues of matrix $A$ are distinct, the eigenvector matrix $T$ is nonsingular. It can be selected as a transformation matrix to transform the original matrix into a diagonal one.*

Since MATLAB can be used to handle complex matrices in the same manner, matrices having complex eigenvalues can also be transformed into diagonal matrices containing complex diagonal entities. The transformation matrices are also complex.

**Example 4.20.** Convert the given matrix $A$ into a diagonal one and find its transformation matrix if

$$A = \begin{bmatrix} 3 & 2 & 2 & 2 \\ 1 & 2 & -2 & -2 \\ -1 & -2 & 0 & -2 \\ 0 & 1 & 3 & 5 \end{bmatrix}.$$

**Solutions.** It can be seen that the eigenvalues of the matrix are 1, 2, 3, 4, and they are all distinct, so the transform matrices can be constructed directly with the following statements:

```
>> A=[3,2,2,2; 1,2,-2,-2; -1,-2,0,-2; 0,1,3,5];
   [v,d]=eig(sym(A)); A1=inv(v)*A*v
```

the transformation and diagonal matrices are:

$$V = \begin{bmatrix} 1 & 0 & -1 & 0 \\ -1 & 0 & 1 & -1 \\ -1 & -1 & 1 & 0 \\ 1 & 1 & -2 & 1 \end{bmatrix}, \quad A_1 = \begin{bmatrix} 1 & 0 & 0 & 0 \\ 0 & 2 & 0 & 0 \\ 0 & 0 & 3 & 0 \\ 0 & 0 & 0 & 4 \end{bmatrix}.$$

If there exist multiple eigenvalues in $A$, the matrix cannot be converted into a diagonal matrix, in general, since the eigenvector matrix may be singular or close to singular.

**Example 4.21.** Perform diagonal transformation to the matrix $A$ with complex eigenvalues if

$$A = \begin{bmatrix} 1 & 0 & 4 & 0 \\ 0 & -3 & 0 & 0 \\ -2 & 2 & -3 & 0 \\ 0 & 0 & 0 & -2 \end{bmatrix}.$$

**Solutions.** Eigenvector matrices for matrices with complex eigenvalues can also be found with eig() function directly. The following statements can be used to compute the eigenvalues, eigenvector matrix, and the diagonal transformation.

```
>> A=[1,0,4,0; 0,-3,0,0; -2,2,-3,0; 0,0,0,-2];
   [V,D]=eig(sym(A)), A1=inv(V)*A*V % eigenvalues and eigenvectors
```

The matrices obtained are

$$V = \begin{bmatrix} -1 & 0 & -1+j & -1-j \\ -1 & 0 & 0 & 0 \\ 1 & 0 & 1 & 1 \\ 0 & 1 & 0 & 0 \end{bmatrix}, \quad D = A_1 = \begin{bmatrix} -3 & 0 & 0 & 0 \\ 0 & -2 & 0 & 0 \\ 0 & 0 & -1-2j & 0 \\ 0 & 0 & 0 & -1+2j \end{bmatrix}.$$

### 4.5.3 Jordan transform

If there exist multiple eigenvalues in matrix $A$, it cannot be transformed into diagonal matrix, in general, since the eigenvector matrix $V$ may be singular, or have fewer columns.

**Example 4.22.** Compute the eigenvalues and eigenvector matrix of matrix $A$ with numerical and analytical methods when

$$A = \begin{bmatrix} -71 & -65 & -81 & -46 \\ 75 & 89 & 117 & 50 \\ 0 & 4 & 8 & 4 \\ -67 & -121 & -173 & -58 \end{bmatrix}.$$

**Solutions.** The following statements can be used to compute numerically and analyt-ically the eigenvalues of the given matrix:

```
>> A=[-71,-65,-81,-46; 75,89,117,50; 0,4,8,4; -67,-121,-173,-58];
   D=eig(A), [v,d]=eig(sym(A)) % numerical and analytical solutions
```

The numerical and analytical solutions are

$$D = \begin{bmatrix} -8.0039 + 0.0039j \\ -8.0039 - 0.0039j \\ -7.9961 + 0.0039j \\ -7.9961 - 0.0039j \end{bmatrix}, \quad v = \begin{bmatrix} -17/19 \\ 13/19 \\ -8/19 \\ 1 \end{bmatrix}, \quad d = \begin{bmatrix} -8 & 0 & 0 & 0 \\ 0 & -8 & 0 & 0 \\ 0 & 0 & -8 & 0 \\ 0 & 0 & 0 & -8 \end{bmatrix}.$$

There exists a quadruple eigenvalue of −8 in the matrix, and with numerical method, there seems to be problems. With the analytical method, accurate results can be obtained, while the four columns of the eigenvector matrix are exactly the same, thus only one column is returned. Diagonal transform is not possible.

In numerical methods, the eigenvalues are considered erroneously as four differ-ent values. If the numerical eigenvector matrix is used in the similarity transform

```
>> [V D]=eig(A), A1=inv(V)*A*V
```

the result is not really a diagonal matrix

$$A_1 = \begin{bmatrix} -8.0028 + 0.0035j & 0.0011 - 0.0005j & 0.0011 - 0.0005j & 0.0011 - 0.0005j \\ 0.0010 + 0.0008j & -8.003 - 0.0037j & 0.0010 + 0.0008j & 0.0010 + 0.0008j \\ -0.0011 - 0.0008j & -0.0011 - 0.0008j & -7.9971 + 0.0032j & -0.0011 - 0.0008j \\ -0.0011 + 0.0011j & -0.0011 + 0.0011j & -0.0011 + 0.0011j & -7.9972 - 0.0028j \end{bmatrix}.$$

**Definition 4.10.** Matrices with repeated eigenvalues can be converted into Jordanian block diagonal matrix, with the $i$th Jordanian block expressed as

$$J_i = \begin{bmatrix} \lambda_i & 1 & & & \\ & \lambda_i & 1 & & \\ & & \lambda_i & \ddots & \\ & & & \ddots & 1 \\ & & & & \lambda_i \end{bmatrix} = \lambda_i I + H, \tag{4.5.1}$$

where $I$ is an identity matrix while $H$ is a nilpotent matrix whose first superdiagonal elements are 1's, and the other elements are zeros.

The concept of Jordan matrix was proposed by French mathematician Marie En-nemond Camille Jordan (1838–1922). For a given matrix, Symbolic Math Toolbox func-tion jordan() can be used to find the Jordan canonical form, and establish nonsingu-lar generalized eigenvector matrix. The syntaxes of the function are

$J$=jordan($A$), % returns only Jordan matrix $J$

[$V,J$]=jordan($A$), % returns $J$ and eigenvector matrix $V$

With the generalized matrix $V$, Jordan canonical form can be obtained with $J = V^{-1}AV$.

**Example 4.23.** Perform Jordan transformation for the matrix in Example 4.22.

**Solutions.** Jordan transformation of a symbolic matrix can be obtained directly with function jordan()

```
>> A=[-71,-65,-81,-46; 75,89,117,50; 0,4,8,4; -67,-121,-173,-58];
   [V,J]=jordan(sym(A)) % analytical Jordan transform
```

The matrices obtained are

$$V = \begin{bmatrix} -18\,496 & 2176 & -63 & 1 \\ 14\,144 & -800 & 75 & 0 \\ -8\,704 & 32 & 0 & 0 \\ 20\,672 & -1504 & -67 & 0 \end{bmatrix}, \quad J = \begin{bmatrix} -8 & 1 & 0 & 0 \\ 0 & -8 & 1 & 0 \\ 0 & 0 & -8 & 1 \\ 0 & 0 & 0 & -8 \end{bmatrix},$$

and $V$ is a full-rank matrix. Taking its inverse, the expected Jordan transformation can obtained. If numerical algorithms are used, matrices with repeated eigenvalues may cause problems. This problem will be demonstrated in later examples.

### 4.5.4 Real Jordan transforms for matrices with repeated complex eigenvalues

In practical numerical computation, Jordan transforms for matrices with complex eigenvalues may cause problems. Especially, when matrices have repeated complex eigenvalues, serious problems may appear. Also, if a matrix has complex eigenvalues, the Jordan canonical form and eigenvector matrix may contain complex entities. Some methods are introduced to convert the original complex matrices to equivalent real ones. The limitation of the MATLAB function here is that it may deal with matrices having at most double complex eigenvalues.

```
function [V,J]=jordan_real(A)
[V,J]=jordan(A); n=length(V); i=0;
vr=real(V); vi=imag(V); n1=n; k=[];
while(i<n1),
    i=i+1; V(:,i)=vr(:,i); v=vi(:,i); % extract real and imaginary parts
    if any(v~=0), k=[k,i+1];
        for j=i+1:n, if all(vi(:,j)+v==0), V(:,j)=v; n1=n1-1;
end, end, end, end
```

```
E=eye(size(V)); E(:,k)=E(:,k(end:-1:1));
V=V*E; J=inv(V)*A*V; % compute Jordan transform
```

**Example 4.24.** Consider the matrix with repeated complex eigenvalues in Example 4.21. Compute the real eigenvalue and eigenvector matrices.

**Solutions.** The diagonal matrices after conversion may contain complex diagonal elements. If the eigenvectors containing complex conjugate entries are replaced by their real and imaginary parts, real matrices can be obtained. The following statements can be entered to reconstruct the relevant matrices:

```
>> A=[1,0,4,0; 0,-3,0,0; -2,2,-3,0; 0,0,0,-2];      % input matrix
   [V,D]=eig(sym(A)); [V1,A1]=jordan_real(sym(A)) % build Jordan form
```

The new conversion matrix and real Jordan matrix are

$$
V_1 = \begin{bmatrix} -1 & 0 & -1 & 1 \\ -1 & 0 & 0 & 0 \\ 1 & 0 & 1 & 0 \\ 0 & 1 & 0 & 0 \end{bmatrix}, \quad A_1 = \left[ \begin{array}{cc:cc} -3 & 0 & 0 & 0 \\ 0 & -2 & 0 & 0 \\ \hdashline 0 & 0 & -1 & -2 \\ 0 & 0 & 2 & -1 \end{array} \right].
$$

It can be seen that the obtained matrix $A_1$ is no longer a diagonal matrix. The upper left-submatrix in $A_1$ contains a real Jordan block. The real matrix is a variation of standard Jordan matrix.

**Example 4.25.** Compute Jordan forms and transformation matrices for

$$
A = \begin{bmatrix} 0 & -1 & 0 & 0 & -1 & 1 \\ 0.5 & 0 & -0.5 & 0 & -1 & 0.5 \\ -0.5 & 0 & -0.5 & 0 & 0 & 0.5 \\ 468.5 & 452 & 304.5 & 577 & 225 & 360.5 \\ -468 & -450 & -303 & -576 & -223 & -361 \\ -467.5 & -451 & -303.5 & -576 & -223 & -361.5 \end{bmatrix}.
$$

**Solutions.** The following $A$ matrix can be entered into MATLAB workspace, with the eigenvalues directly obtained

```
>> A=[0,-1,0,0,-1,1; 0.5,0,-0.5,0,-1,0.5; -0.5,0,-0.5,0,0,0.5;
      468.5,452,304.5,577,225,360.5; -468,-450,-303,-576,-223,-361;
      -467.5,-451,-303.5,-576,-223,-361.5]; % input matrix
   A=sym(A); eig(A), [v,J]=jordan(A) % eigenvalues and Jordanian matrix
```

The eigenvalues of the matrix are $-2$, $-2$, $-1 \pm j2$, $-1 \pm j2$, i.e., it has a double real eigenvalue of $-2$, and double complex eigenvalues of $-1 \pm j2$. The conventional Jordan matrix can be found as

$$J = \begin{bmatrix} -2 & 1 & 0 & 0 & 0 & 0 \\ 0 & -2 & 0 & 0 & 0 & 0 \\ 0 & 0 & -1+j2 & 1 & 0 & 0 \\ 0 & 0 & 0 & -1+j2 & 0 & 0 \\ 0 & 0 & 0 & 0 & -1-j2 & 1 \\ 0 & 0 & 0 & 0 & 0 & -1-j2 \end{bmatrix}.$$

It can be seen that the matrices have complex elements. If one wants to have matrices with real elements only, the following statements can be used:

```
>> [V,J]=jordan_real(sym(A)) % convert the matrices to real ones
```

The new real transformation and Jordan block matrices are obtained as

$$J = \begin{bmatrix} -2 & 1 & 0 & 0 & 0 & 0 \\ 0 & -2 & 0 & 0 & 0 & 0 \\ 0 & 0 & -1 & -2 & 1 & 0 \\ 0 & 0 & 2 & -1 & 0 & 1 \\ 0 & 0 & 0 & 0 & -1 & -2 \\ 0 & 0 & 0 & 0 & 2 & -1 \end{bmatrix},$$

$$V = \begin{bmatrix} 423/25 & -543/125 & 851/100 & 757/100 & 334/125 & -9\,321/1\,000 \\ -423/25 & 7\,431/250 & 2\,459/100 & 663/100 & -7\,431/500 & -509/1\,000 \\ 423/5 & -471/10 & -757/40 & 851/40 & 471/20 & -1\,887/80 \\ 4\,371/25 & -70\,677/250 & -47\,327/400 & -9\,191/100 & 70\,677/500 & 247\,587/4\,000 \\ -4\,653/25 & 31\,353/125 & 16\,263/200 & 15\,991/200 & -31\,353/250 & -96\,843/2\,000 \\ -5\,922/25 & 76\,539/250 & 22\,507/200 & 12\,399/200 & -76\,539/500 & -74\,767/2\,000 \end{bmatrix}.$$

### 4.5.5 Simultaneous diagonalization

Simultaneous diagonalization methods of two positive-definite matrices are studied in this section [15].

**Theorem 4.12.** *Assuming that $A, B \in \mathcal{R}^{n \times n}$ are both positive definite matrices, there exists a nonsingular matrix $Q$ such that*

$$Q^{T}AQ = D, \quad Q^{T}BQ = I, \tag{4.5.2}$$

*where $D$ is a diagonal matrix, whose eigenvalues are the diagonal elements in the matrix $AB^{-1}$.*

From the given positive-definite matrix $B$, Cholesky decomposition is carried out so as to find matrix $L$, such that $B = LL^{T}$. Let $C = L^{-1}AL^{-T}$, there exists an orthogonal matrix $P$, such that $P^T C P = D$, where, $D$ is a diagonal matrix. Therefore, the transformation matrix $Q = L^{-T} P$ can be used to diagonalize the two matrices simultaneously. An example is given below to demonstrate simultaneous diagonalization of two matrices.

**Example 4.26.** For two positive-definite matrices $A$, $B$ given below, find matrix $Q$ to simultaneously diagonalize them.

$$A = \begin{bmatrix} 4 & 1 & 2 & 0 \\ 1 & 2 & -1 & 0 \\ 2 & -1 & 4 & 2 \\ 0 & 0 & 2 & 4 \end{bmatrix}, \quad B = \begin{bmatrix} 2 & 0 & 0 & -1 \\ 0 & 4 & -1 & -1 \\ 0 & -1 & 4 & 2 \\ -1 & -1 & 2 & 4 \end{bmatrix}$$

**Solutions.** The two matrices $A$ and $B$ can be entered into MATLAB workspace. Then based on the procedures presented above, transformation matrix $Q$ can be constructed. Validations are made to the simultaneous diagonalize matrices.

```
>> A=[4,1,2,0; 1,2,-1,0; 2,-1,4,2; 0,0,2,4];
   B=[2,0,0,-1; 0,4,-1,-1; 0,-1,4,2; -1,-1,2,4];
   L=chol(B); L=L'; C=inv(L)*A*inv(L');
   [P,D]=eig(C); Q=inv(L')*P, Q'*A*Q, Q'*B*Q
```

The transformation matrix is given below. It can be seen that the two matrices $A$ and $B$ are simultaneously diagonalized.

$$Q = \begin{bmatrix} 0.6864 & 0.2967 & 0.2047 & 0.0396 \\ 0.1774 & -0.4138 & 0.1774 & -0.1996 \\ 0.0842 & -0.3635 & -0.2937 & 0.3615 \\ 0.2902 & 0.1855 & -0.1371 & -0.5230 \end{bmatrix}$$

## 4.6 Singular value decomposition

The concept of singular values is introduced in numerical linear algebra research. They can be regarded as a measure of a matrix. Before introducing singular values of matrices, let us consider a simple example first. Then a formal description to the concepts of the singular value decomposition technique and condition number are presented.

**Example 4.27.** In the past, we usually computed the rank of a rectangular matrix. Alternative ways in Theorem 3.21 can be used to convert the problem into rank problems for square matrices $AA^T$ and $A^T A$, choosing that which is smaller in size. Assume that matrix $A$ is given below, where $\mu = 5\text{eps}$. Compute the rank of matrix $A^{[14]}$ if

$$A = \begin{bmatrix} 1 & 1 \\ \mu & 0 \\ 0 & \mu \end{bmatrix}.$$

**Solutions.** Since $\mu$ is not ignorable under the double precision framework, it is immediately seen that the rank of $A$ is 2. The same results can also be obtained with the following statements:

```
>> A=[1 1; 5*eps,0; 0,5*eps]; rank(A) % direct rank computation
```

It can also be seen that $AA^T$ is a $3 \times 3$ matrix, while $A^T A$ is a $2 \times 2$ one, so $A^T A$ matrix can alternatively used in finding the rank of matrix $A$. Here

$$A^T A = \begin{bmatrix} 1+\mu^2 & 1 \\ 1 & 1+\mu^2 \end{bmatrix}.$$

Under the double precision framework, since the theoretic value of $\mu^2$ is around $10^{-30}$, when it is added to 1, it vanishes, and $1 + \mu^2 = 1$ is the result. Matrix $A^T A$ is thus a matrix of ones, whose rank is 1, from which it is concluded that the rank of $A$ is 1. This is, of course, a wrong conclusion. To avoid this kind of phenomenon, a new measurement quantity should be introduced, i. e., the concept of a singular value is needed.

### 4.6.1 Singular values and condition numbers

**Theorem 4.13.** *Assume that $A \in \mathscr{R}^{n \times m}$ can then be decomposed as*

$$A = LA_1 M^T, \tag{4.6.1}$$

*where $L$ and $M$ are respectively $n \times n$ and $m \times m$ orthogonal matrices, while $A_1 = \text{diag}(\sigma_1, \ldots, \sigma_p)$ is an $n \times m$ diagonal matrix, with the diagonal elements $\sigma_1 \geqslant \cdots \geqslant \sigma_p \geqslant 0$, where $p = \min(n, m)$. If $\sigma_p = 0$, the matrix $A$ is singular. The rank of $A$ equals to the number of nonzero diagonal elements in matrix $A_1$.*

**Theorem 4.14.** *If $A \in \mathscr{R}^{n \times m}$, then $\|A\|_2 = \sigma_1$, $\|A\|_F = \sqrt{\sigma_1^2 + \cdots + \sigma_{\min(n,m)}^2}$.*

**Theorem 4.15.** *Matrices $A^T A$ and $AA^T$ have identical nonnegative eigenvalues $\lambda_i$, and the square roots of them are referred to as the singular values of $A$, denoted as $\sigma_i(A) = \sqrt{\lambda_i(A^T A)}$.*

The concept of singular value decomposition was independently proposed by Italian mathematician Eugenio Beltrami (1835–1900) and French mathematician Marie

Ennemond Camille Jordan (1838–1922) in 1873 and 1874, respectively. British mathematician James Joseph Sylvester (1814–1897) independently proposed in 1889 singular value decomposition result for real square matrices. Later, various decomposition algorithms were proposed, and by 1970, the decomposition algorithm currently used today was proposed[9].

Singular value decomposition function svd() can be used with the syntaxes

$S$=svd($A$), % compute singular values

$[L,A_1,M]$=svd($A$), % compute singular values and transform matrices

where $A$ is the original matrix, the resulted $A_1$ is a diagonal matrix, while matrices $L$ and $M$ are orthogonal matrices, with $A = LA_1M^{\mathrm{T}}$.

The sizes of singular values usually determine the characteristics of matrices. If the variations in singular values are very large, a tiny change in one element in the matrix may significantly affect the behavior of the whole matrix. Such matrices are referred to as ill- or badly-conditioned. When there are zero singular values, a matrix is referred to as singular.

**Definition 4.11.** The ratio of maximum singular value $\sigma_{\max}$ and the minimum $\sigma_{\min}$ is referred to as the condition number, denoted as cond($A$), i. e., cond($A$) = $\sigma_{\max}/\sigma_{\min}$.

The larger the condition number, the more sensitive the matrix to variations of its elements. The maximum and minimum singular values are also denoted as $\bar{\sigma}(A)$ and $\underline{\sigma}(A)$.

In MATLAB, the function $c$=cond($A$) can also be used to compute the condition number of matrix $A$.

**Example 4.28.** Perform singular value decomposition to the matrix studied in Example 4.27.

**Solutions.** The following statements can be used to input the matrix and obtain singular values:

```
>> A=[1 1; 5*eps,0; 0,5*eps]; [u b v]=svd(A)
   d=cond(A)
```

The decomposed singular matrices can be found as follows. Also the condition number is found to be $d = 1.2738 \times 10^{15}$, while

$$
u = \begin{bmatrix} -1 & 0 & 0 \\ 0 & -0.7071 & 0.7071 \\ 0 & 0.7071 & 0.7071 \end{bmatrix}, \quad b = \begin{bmatrix} 1.4142 & 0 \\ 0 & 1.11 \times 10^{-15} \\ 0 & 0 \end{bmatrix}, \quad v = \begin{bmatrix} -0.7071 & -0.7071 \\ -0.7071 & 0.7071 \end{bmatrix}.
$$

**Example 4.29.** Perform singular value decomposition to the matrix in Example 3.2.

**Solutions.** Function svd() can be used to decompose the given matrix to find the matrices $L$, $A_1$ and $M$, and also the condition number of the matrix can be computed.

```
>> A=[16,2,3,13; 5,11,10,8; 9,7,6,12; 4,14,15,1];
   [L,A1,M]=svd(A), cond(A) % singular value decomposition
```

The decomposition matrices are

$$
L = \begin{bmatrix} -0.5 & 0.6708 & 0.5 & -0.2236 \\ -0.5 & -0.2236 & -0.5 & -0.6708 \\ -0.5 & 0.2236 & -0.5 & 0.6708 \\ -0.5 & -0.6708 & 0.5 & 0.2236 \end{bmatrix}, \quad A_1 = \begin{bmatrix} 34 & 0 & 0 & 0 \\ 0 & 17.8885 & 0 & 0 \\ 0 & 0 & 4.4721 & 0 \\ 0 & 0 & 0 & 0 \end{bmatrix},
$$

$$
M = \begin{bmatrix} -0.5 & 0.5 & 0.6708 & -0.2236 \\ -0.5 & -0.5 & -0.2236 & -0.6708 \\ -0.5 & -0.5 & 0.2236 & 0.6708 \\ -0.5 & 0.5 & -0.6708 & 0.2236 \end{bmatrix}.
$$

It can be seen that the matrix has zero singular values, therefore the original matrix is singular. The condition number of the matrix can be computed with function cond($A$) as $4.7133 \times 10^{17}$, very close to $\infty$. Under the double precision framework, errors may be introduced. If the matrix $A$ is converted to a symbolic one, function svd() can be employed for a more accurate solution.

### 4.6.2 Singular value decomposition of rectangular matrices

**Example 4.30.** For an $n \neq m$ matrix $A$, singular value decomposition can also be performed. For instance, singular value decomposition can be carried out for the following rectangular matrix, and we can validate the results:

$$
A = \begin{bmatrix} 1 & 3 & 5 & 7 \\ 2 & 4 & 6 & 8 \end{bmatrix}.
$$

**Solutions.** The following commands can be used to perform singular value decomposition and check the results:

```
>> A=[1,3,5,7; 2,4,6,8]; [L,A1,M]=svd(A), A2=L*A1*M'
   norm(A-A2) % singular value decomposition
```

The following decomposition is achieved and the matrices are as follows. It is also found that $\|LA_1M^T - A\| = 9.7277 \times 10^{-15}$, indicating that the original matrix $A$ can be reconstructed as $LA_1M^T$ where

$$
L = \begin{bmatrix} -0.6414 & -0.7672 \\ -0.7672 & 0.6414 \end{bmatrix}, \quad A_1 = \begin{bmatrix} 14.2691 & 0 & 0 & 0 \\ 0 & 0.6268 & 0 & 0 \end{bmatrix},
$$

$$M = \begin{bmatrix} -0.1525 & 0.8226 & -0.3945 & -0.38 \\ -0.3499 & 0.4214 & 0.2428 & 0.8007 \\ -0.5474 & 0.0201 & 0.6979 & -0.4614 \\ -0.7448 & -0.3812 & -0.5462 & 0.0407 \end{bmatrix}.$$

**Example 4.31.** Traditional singular value decomposition was defined for real matrices. Compute singular value decomposition for the complex matrix in Example 2.2, and validate the results if

$$A = \begin{bmatrix} -1+6j & 5+3j & 4+2j & 6-2j \\ j & 2-j & 4 & -2-2j \\ 4 & -j & -2+2j & 5-2j \end{bmatrix}.$$

**Solutions.** The function svd() can be used directly to handle complex matrices.

```
>> A=[-1+6i,5+3i,4+2i,6-2i; 1i,2-1i,4,-2-2i; 4,-1i,-2+2i,5-2i];
   [u,A1,v]=svd(A), A2=u*A1*v', norm(A-A2)
```

The obtained matrices $u$ and $v$ are complex. The real diagonal matrix $A_1$ is given below, and the original matrix $A$ is restored in $A_2$, with an error of $8.0234 \times 10^{-15}$. The restoration condition is $uA_1v^H$, rather than $uA_1v^T$.

$$A_1 = \begin{bmatrix} 12.3873 & 0 & 0 & 0 \\ 0 & 7.7087 & 0 & 0 \\ 0 & 0 & 1.4599 & 0 \end{bmatrix}.$$

### 4.6.3 SVD based simultaneous diagonalization

In Section 4.5.5, a simultaneous diagonalization method is introduced for positive definite matrices. An alternative method introduced in [15] can also be used.

**Theorem 4.16.** *If positive definite matrices A and B can be factorized with Cholesky factorization as $A = L_A L_A^T$, $B = L_B L_B^T$, singular value decomposition of matrix $L_B^{-1} L_A$ can be performed such that*

$$L_B^{-1} L_A = U \Sigma V. \tag{4.6.2}$$

*The transformation matrix $Q = L_B^{-T} U$ can be selected to implement simultaneous diagonalization*

$$Q^T A Q = \Sigma^2, \quad Q^T B Q = I. \tag{4.6.3}$$

**Example 4.32.** Solve again the simultaneous diagonalization problem in Example 4.26 with the SVD method.

**Solutions.** The matrices **A** and **B** should be entered into MATLAB first, then with the procedures discussed earlier we can construct the **Q** matrix. The diagonalization behavior can then be validated.

```
>> A=[4,1,2,0; 1,2,-1,0; 2,-1,4,2; 0,0,2,4];
   B=[2,0,0,-1; 0,4,-1,-1; 0,-1,4,2; -1,-1,2,4];
   La=chol(A)'; Lb=chol(B)'; C=inv(Lb)*La;
   [U,S,V]=svd(C); Q=inv(Lb')*U, Q'*A*Q, Q'*B*Q
```

The transformation matrix constructed is as follows. It can be validated that the matrix may simultaneously diagonalize matrices **A** and **B**. Compared with Example 4.26, there are slight differences in the two transformation matrices, while they both reached the same target

$$Q = \begin{bmatrix} -0.6864 & 0.0396 & 0.2047 & 0.2967 \\ -0.1774 & -0.1996 & 0.1774 & -0.4138 \\ -0.0842 & 0.3615 & -0.2937 & -0.3635 \\ -0.2902 & -0.5230 & -0.1371 & 0.1855 \end{bmatrix}.$$

## 4.7 Givens and Householder transforms

Two other transforms – Givens and Householder transforms – are introduced in this section. Givens transform is named after American mathematician and computer scientist James Wallace Givens, Jr. (1910–1993), usually used for coordinate rotation processing. Householder transform is named after American mathematician Alston Scott Householder (1903–1993), which can be used to apply reflective mapping through a vector transform. The transforms are very practical in numerical linear algebra.

### 4.7.1 Two-dimensional coordinate rotation transform

**Definition 4.12.** Assume that a two-dimensional coordinate $(x, y)$ is given. If one wants to rotate the point by $\theta$ radians in the counterclockwise direction around the origin, a rotation matrix is needed to get the transformed coordinate $(x_1, y_1)$:

$$\begin{bmatrix} x_1 \\ y_1 \end{bmatrix} = \begin{bmatrix} \cos\theta & \sin\theta \\ -\sin\theta & \cos\theta \end{bmatrix} \begin{bmatrix} x \\ y \end{bmatrix}. \tag{4.7.1}$$

This kind of transform is also known as Givens transform.

**Example 4.33.** The mathematical model of a fractal tree is given as follows. Select an initial point $(x_0, y_0)$ in a two-dimensional plane. Assume we have a set of uniformly distributed samples in the interval $[0,1]$, denoted as $y_i$. Based on these values, a new point $(x_1, y_1)$ can be obtained[30]:

$$(x_1, y_1) \Leftarrow \begin{cases} x_1 = 0, y_1 = y_0/2, & \text{when } y_i < 0.05, \\ x_1 = 0.42(x_0 - y_0), y_1 = 0.2 + 0.42(x_0 + y_0), & \text{when } 0.05 \leqslant y_i < 0.45, \\ x_1 = 0.42(x_0 + y_0), y_1 = 0.2 - 0.42(x_0 - y_0), & \text{when } 0.45 \leqslant y_i < 0.85, \\ x_1 = 0.1x_0, y_1 = 0.2 + 0.1y_0, & \text{otherwise.} \end{cases}$$

A set of 10 000 points can be generated, and a fractal tree can be drawn with dots. Rotate respectively by $45°$ and $200°$ in the counterclockwise direction with Givens transform, and draw the rotated fractal trees.

**Solutions.** The following statements can be used to generate a random vector, and compute the points in the fractal tree. The dot-to-dot plot can be used to draw the fractal tree, as shown in Figure 4.1.

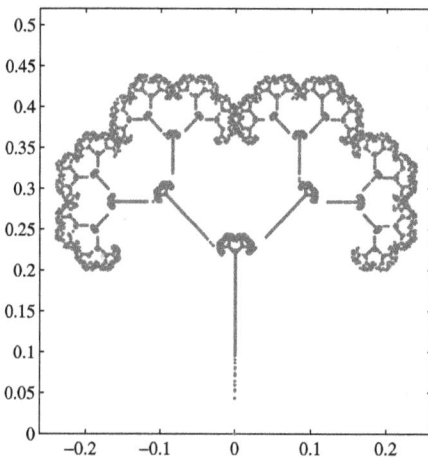

**Figure 4.1:** Fractal tree.

```
>> v=rand(10000,1); N=length(v); x=0; y=0;
   for k=2:N, gam=v(k);
       if gam<0.05, x(k)=0; y(k)=0.5*y(k-1);
       elseif gam<0.45,
           x(k)=0.42*(x(k-1)-y(k-1)); y(k)=0.2+0.42*(x(k-1)+y(k-1));
       elseif gam<0.85,
           x(k)=0.42*(x(k-1)+y(k-1)); y(k)=0.2-0.42*(x(k-1)-y(k-1));
       else, x(k)=0.1*x(k-1); y(k)=0.1*y(k-1)+0.2;
   end, end
   plot(x,y,'.')
```

If one wants to rotate it 45° in the counterclockwise rotation, the following statements can be used, Givens transformation matrix is computed, and the rotation is shown in Figure 4.2(a). It can be seen that through Givens transform, the original dots in the figure are rotated in the new coordinates.

```
>> theta=45;
    G1=[cosd(theta),sind(theta); -sind(theta) cosd(theta)];
    X=[x.' y.']; X1=X*G1; x1=X1(:,1); y1=X1(:,2); plot(x1,y1,'.')
```

Similarly, the 200° counterclockwise rotation can be obtained from the following statements, and the result is illustrated in Figure 4.2(b). It can be seen that the figure illustrates the expected results.

```
>> theta=200;
    G2=[cosd(theta),sind(theta); -sind(theta) cosd(theta)];
    X2=X*G2; x1=X2(:,1); y1=X2(:,2); plot(x1,y1,'.')
```

Note that the 200° rotation can be performed through –160° rotation, i. e., 160° rotation in the clockwise direction, and the results are exactly the same as those shown in Figure 4.2(b).

```
>> theta=-160;
    G3=[cosd(theta),sind(theta); -sind(theta) cosd(theta)];
    X=[x.' y.']; X1=X*G3; x1=X1(:,1); y1=X1(:,2); plot(x1,y1,'.')
```

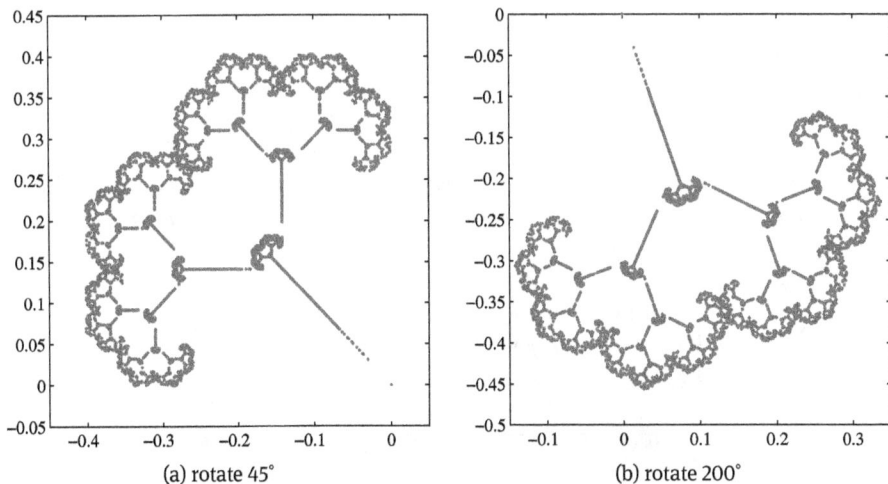

(a) rotate 45°      (b) rotate 200°

**Figure 4.2:** Fractal tree after rotation.

The transformation matrices are respectively

$$G_1 = \begin{bmatrix} 0.70711 & 0.70711 \\ -0.70711 & 0.70711 \end{bmatrix}, \quad G_2 = G_3 = \begin{bmatrix} -0.93969 & -0.34202 \\ 0.34202 & -0.93969 \end{bmatrix}.$$

### 4.7.2 Givens transform of ordinary matrices

The above mentioned Givens rotation transform can only be used to process two-dimensional problems. For multidimensional problems, the following rotation transform matrices are introduced.

**Definition 4.13.** If one wants to rotate the elements in the $(i,j)$th rows and $(i,j)$th columns in matrix $A$, Givens transform is introduced as

$$G(i,j,\theta) = \begin{bmatrix} 1 & \cdots & 0 & \cdots & 0 & \cdots & 0 \\ \vdots & \ddots & \vdots & & \vdots & & \vdots \\ 0 & \cdots & \cos\theta & \cdots & \sin\theta & \cdots & 0 \\ \vdots & & \vdots & \ddots & \vdots & & \vdots \\ 0 & \cdots & -\sin\theta & \cdots & \cos\theta & \cdots & 0 \\ \vdots & & \vdots & & \vdots & \ddots & \vdots \\ 0 & \cdots & 0 & \cdots & 0 & \cdots & 1 \end{bmatrix} \begin{matrix} \\ \\ \leftarrow ith\ row \\ \\ \leftarrow jth\ row \\ \\ \end{matrix} \qquad (4.7.2)$$

$$ith\ column \qquad jth\ column.$$

**Theorem 4.17.** *Givens transform matrix $G(i,j,\theta)$ is orthogonal, and $G^{-1} = G^{T}$.*

Based on the definition of the above Givens matrix, the following MATLAB generating function can be written

```
function G=givens_mat(n,i,j,theta)
G=sym(eye(n)); c=cosd(theta); s=sind(theta);
G(i,i)=c; G(j,j)=c; G(i,j)=s; G(j,i)=-s;
```

**Theorem 4.18.** *Assume that one of the subvectors in matrix $A$ is $v = [x_i, x_j]^{T}$. Through Givens transform, the transform result is $(y_i, y_j)$. Denoting $c = \cos\theta$ and $s = \sin\theta$, the general mapping formula of $y_k$ is*

$$y_k = \begin{cases} cx_i + sx_j, & k = i, \\ -sx_i + cx_j, & k = j, \\ x_k, & k \neq i,j;\ the\ other\ rows\ are\ not\ affected. \end{cases} \qquad (4.7.3)$$

*If $y_j$ is rotated to zero, the parameters in the rotation matrix are*

$$c = \frac{x_i}{\sqrt{x_i^2 + x_j^2}}, \quad s = \frac{x_j}{\sqrt{x_i^2 + x_j^2}}. \qquad (4.7.4)$$

**Theorem 4.19.** *Through the selected Givens transform matrix, the result is*

$$G(i,j,\theta)\begin{bmatrix} x_i \\ x_j \end{bmatrix} = \begin{bmatrix} c & s \\ -s & c \end{bmatrix}\begin{bmatrix} x_i \\ x_j \end{bmatrix} = \begin{bmatrix} \sqrt{x_i^2 + x_j^2} \\ 0 \end{bmatrix}. \tag{4.7.5}$$

It can be seen that it is relatively easy to rotate one value in the vector to zero, since a $2 \times 2$ rotation matrix can be constructed with the above theorem. A MATLAB function givens() can be used to generate the rotation matrix $G$, with the syntax $G$=givens($x_i$, $x_j$). An illustrative example is given below to show the Givens transform method.

Note that, with givens() function, the rotation matrix is a $2\times2$ matrix, and it is not necessary to follow the form of (4.7.2). It is not necessary to transform the whole matrix. Transforming the relevant rows is sufficient. The following MATLAB commands can be used:

```
A([i,j],:)=G*A([i,j],:)
```

**Example 4.34.** Transform the following matrix with Givens rotation matrix such that the $(3, 2)$ element is transformed to zero:

$$A = \begin{bmatrix} 2 & 2 & 3 & 2 \\ 3 & 1 & 3 & 1 \\ 4 & 3 & 2 & 1 \\ 3 & 3 & 2 & 1 \end{bmatrix}.$$

**Solutions.** Since the target to change is in the third row, select $j = 3$. Also, $i$ can be selected arbitrarily from 1, 2, and 4, for instance, let $i = 1$. Since the target is the $(3, 2)$-element, the column is 2. Therefore, taking $x_i = 2$, $x_j = 3$, the following statements can be used to compute directly Givens transform matrices with numerical and analytical methods:

```
>> G1=givens(2,3), G2=givens(sym(2),3)
```

The obtained Givens transform matrices are respectively

$$G_1 = \begin{bmatrix} 0.5547 & 0.83205 \\ -0.83205 & 0.5547 \end{bmatrix}, \quad G_2 = \frac{\sqrt{13}}{13}\begin{bmatrix} 2 & 3 \\ -3 & 2 \end{bmatrix}.$$

With the transform matrices, the rotation results can be obtained.

```
>> A=[2,2,3,2; 3,1,3,1; 4,3,2,1; 3,3,2,1]; A1=A; A2=sym(A);
   A1([1,3],:)=G1*A1([1,3],:), A2([1,3],:)=G2*A2([1,3],:)
```

The results are given below, indicating that the Givens transformation can achieve the anticipated goals:

$$
A_1 = \begin{bmatrix} 4.4376 & 3.6056 & 3.3282 & 1.9415 \\ 3 & 1 & 3 & 1 \\ 0.5547 & 0 & -1.3868 & -1.1094 \\ 3 & 3 & 2 & 1 \end{bmatrix},
$$

$$
A_2 = \begin{bmatrix} 16\sqrt{13}/13 & \sqrt{13} & 12\sqrt{13}/13 & 7\sqrt{13}/13 \\ 3 & 1 & 3 & 1 \\ 2\sqrt{13}/13 & 0 & -5\sqrt{13}/13 & -4\sqrt{13}/13 \\ 3 & 3 & 2 & 1 \end{bmatrix}.
$$

If $i$ is selected as 2 or 4, different Givens transformation matrices can be designed, such that the element $A(3, 2)$ is mapped to zero. The readers can try these transforms and observe the results.

### 4.7.3 Householder transform

Householder transform is another kind of matrix transform in numerical linear algebra and it is widely used in the applications. In this section, the concept and implementation of Householder transform are presented.

**Definition 4.14.** If $u$ is an $n$-tuple column vector, the following Householder matrix is defined:

$$
H = I - 2\frac{vv^H}{v^Hv}. \tag{4.7.6}
$$

**Theorem 4.20.** *For the given column vector $x_1$, linear transform $H$ can be used to convert it into $x_2$, i. e., $x_2 = Hx_1$, and $\|x_1\| = \|x_2\|$, a column vector $v = x_2 - x_1$ can be created, and Householder transform matrix $H$ can be constructed.*

**Theorem 4.21.** *Householder matrix $H$ is a symmetric orthogonal matrix, and $H = H^{-1}$.*

**Example 4.35.** Consider matrix $A$ in Example 4.34. Design a Householder matrix $H$, such that $A_1 = HA$ can be used to transform the first column of $A_1$ into $x_2 = \alpha[1, 0, 0, 0]^T$.

**Solutions.** If Theorem 4.20 is used to construct $v$, then $\alpha = \|x_1\|$. The following statements can be used in selecting such a column vector $v$, then Householder matrix $H$ can be established, and the converted matrix $A_1$ is found

```
>> A=[2,2,3,2; 3,1,3,1; 4,3,2,1; 3,3,2,1]; A=sym(A);
   x1=A(:,1); x2=[1; 0; 0; 0]; v=norm(x1)*x2-x1;
   H=eye(4)-2*(v*v')/(v'*v), A1=H*A
```

The results obtained are given below, and through the selection of the Householder transform matrix, the first column in $A_1$ can be converted into the expected form:

$$
H = \begin{bmatrix}
\sqrt{38}/19 & 3\sqrt{38}/38 & 2\sqrt{38}/19 & 3\sqrt{38}/38 \\
3\sqrt{38}/38 & 25/34 - 9\sqrt{38}/646 & -6\sqrt{38}/323 - 6/17 & -9\sqrt{38}/646 - 9/34 \\
2\sqrt{38}/19 & -6\sqrt{38}/323 - 6/17 & 9/17 - 8\sqrt{38}/323 & -6\sqrt{38}/323 - 6/17 \\
3\sqrt{38}/38 & -9\sqrt{38}/646 - 9/34 & -6\sqrt{38}/323 - 6/17 & 25/34 - 9\sqrt{38}/646
\end{bmatrix},
$$

$$
A_1 = \begin{bmatrix}
\sqrt{38} & 14\sqrt{38}/19 & 29\sqrt{38}/38 & 7\sqrt{38}/19 \\
0 & 15\sqrt{38}/323 - 19/17 & 42\sqrt{38}/323 + 33/34 & 36\sqrt{38}/323 + 2/17 \\
0 & 20\sqrt{38}/323 + 3/17 & 56\sqrt{38}/323 - 12/17 & 48\sqrt{38}/323 - 3/17 \\
0 & 15\sqrt{38}/323 + 15/17 & 42\sqrt{38}/323 - 1/34 & 36\sqrt{38}/323 + 2/17
\end{bmatrix}.
$$

## 4.8 Problems

4.1 Test whether the following matrix is orthogonal or not:

$$
A = \begin{bmatrix}
0 & 2/\sqrt{6} & -1/\sqrt{3} \\
1/\sqrt{2} & 1/\sqrt{6} & 1/\sqrt{3} \\
-1/\sqrt{2} & 1/\sqrt{6} & 1/\sqrt{3}
\end{bmatrix}.
$$

4.2 Compute the inverse of the matrix, with elementary row transforms and the pivot version and compare the efficiency if

$$
A = \begin{bmatrix}
3 & 5 & 5 & 0 & 1 & 2 & 3 \\
3 & 2 & 5 & 4 & 6 & 2 & 5 \\
1 & 2 & 1 & 1 & 3 & 4 & 6 \\
3 & 5 & 1 & 5 & 2 & 1 & 2 \\
4 & 1 & 0 & 1 & 2 & 0 & 1 \\
-3 & -4 & -7 & 3 & 7 & 8 & 12 \\
1 & -10 & 7 & -6 & 8 & 1 & 5
\end{bmatrix}.
$$

4.3 Solve the following equation with Gaussian elimination method:

$$
\begin{bmatrix}
16 & 2 & 3 & 13 \\
5 & 11 & 10 & 8 \\
9 & 7 & 6 & 12 \\
4 & 14 & 15 & 1
\end{bmatrix} X = \begin{bmatrix} 1 \\ 3 \\ 4 \\ 7 \end{bmatrix}.
$$

4.4 Rewrite the Gaussian elimination method solver gauss_eq(), and add pivot selection in the function, then validate the results.

4.5 Compute triangular factorizations and singular value decompositions for the following matrices:

$$A = \begin{bmatrix} 8 & 0 & 1 & 1 & 6 \\ 9 & 2 & 9 & 4 & 0 \\ 1 & 5 & 9 & 9 & 8 \\ 9 & 9 & 4 & 7 & 9 \\ 6 & 9 & 8 & 9 & 6 \end{bmatrix}, \quad B = \begin{bmatrix} 1 & 2 & 2 & 2 \\ 1 & 1 & 2 & 0 \\ 1 & 1 & 1 & 0 \\ 0 & 0 & 2 & 0 \end{bmatrix}.$$

4.6 Test whether the following matrices are positive definite. If they are, compute the Cholesky factorization matrices for

$$A = \begin{bmatrix} 9 & 2 & 1 & 2 & 2 \\ 2 & 4 & 3 & 3 & 3 \\ 1 & 3 & 7 & 3 & 4 \\ 2 & 3 & 3 & 5 & 4 \\ 2 & 3 & 4 & 4 & 5 \end{bmatrix}, \quad B = \begin{bmatrix} 16 & 17 & 9 & 12 & 12 \\ 17 & 12 & 12 & 2 & 18 \\ 9 & 12 & 18 & 7 & 13 \\ 12 & 2 & 7 & 18 & 12 \\ 12 & 18 & 13 & 12 & 10 \end{bmatrix}.$$

4.7 Test whether the following matrices are positive definite. If they are, compute the Cholesky factorization matrices. If not, find the Cholesky factorization matrices and comment on the results when

$$A = \begin{bmatrix} 1 & 3 & 4 & 8 \\ 3 & 2 & 7 & 2 \\ 4 & 7 & 2 & 8 \\ 8 & 2 & 8 & 6 \end{bmatrix}, \quad B = \begin{bmatrix} 12 & 13 & 24 & 26 \\ 31 & 12 & 27 & 11 \\ 10 & 9 & 22 & 18 \\ 42 & 22 & 10 & 16 \end{bmatrix}.$$

4.8 Test whether the following quadratic forms are positive definite or not:

$$f_1(x) = x_1^2 + 2x_3^2 + 2x_1x_3 + 2x_2x_3, \quad f_2(x) = 2x_1^2 + x_2^2 + 5x_3^2 + 2x_1x_2 - 2x_2x_3.$$

4.9 Test whether the following quadratic forms are positive definite or not:

(1) $f_1(x) = 99x_1^2 - 12x_1x_2 + 48x_1x_3 + 130x_2^2 - 60x_2x_3 + 71x_3^2,$

(2) $f_2(x) = x_1^2 + x_2^2 + 4x_3^2 + 7x_4^2 + 6x_1x_3 + 4x_1x_4 - 4x_2x_3 + 2x_2x_4 + 4x_3x_4.$

4.10 Find Jordan canonical forms and transform matrices of

$$A = \begin{bmatrix} -2 & 0.5 & -0.5 & 0.5 \\ 0 & -1.5 & 0.5 & -0.5 \\ 2 & 0.5 & -4.5 & 0.5 \\ 2 & 1 & -2 & -2 \end{bmatrix}, \quad B = \begin{bmatrix} -2 & -1 & -2 & -2 \\ -1 & -2 & 2 & 2 \\ 0 & 2 & 0 & 3 \\ 1 & -1 & -3 & -6 \end{bmatrix}.$$

4.11 Find suitable transformation matrix such that matrix $A$ can be mapped into its companion matrix when

$$A = \begin{bmatrix} 2 & -2 & 2 & -1 & 0 \\ -2 & 1 & -1 & 2 & 1 \\ 1 & -1 & -1 & -2 & 2 \\ 1 & 1 & -1 & 2 & 2 \\ 1 & 0 & 2 & 2 & -2 \end{bmatrix}.$$

4.12 Test whether the matrix $A$ can be transformed into a diagonal matrix. If it can, what is the transformation matrix when

$$A = \begin{bmatrix} 2 & 2 & -2 & 7 & -2 \\ 38 & 15 & -28 & 56 & -10 \\ 17 & 8 & -15 & 24 & -5 \\ -4 & -2 & 2 & -9 & 2 \\ 20 & 10 & -16 & 31 & -8 \end{bmatrix}?$$

4.13 Find the eigenvalues and Jordan canonical form. If there exist complex eigenvalues, find the equivalent real Jordan form and the transformation matrix for

$$A = \begin{bmatrix} -5 & -2 & -4 & 0 & -1 & 0 \\ 1 & -2 & 2 & 0 & -1 & -2 \\ 2 & 2 & 0 & 3 & 2 & 0 \\ 1 & 3 & 1 & 0 & 3 & 1 \\ -1 & -2 & -3 & -4 & -4 & 1 \\ 3 & 4 & 3 & 1 & 2 & -1 \end{bmatrix}.$$

4.14 Convert the complex matrix with complex eigenvalues into Jordan canonical form. Is it possible to convert it into real Jordan form? How to perform this transform if

$$A = \begin{bmatrix} -2+5j & 0 & 0 & 6j & 0 \\ -3+41j & -2-1j & -3+5j & 56j & 3-1j \\ -1+17j & 0 & -3+2j & +24j & 1-1j \\ -4j & 0 & 0 & -2-5j & 0 \\ -2+22j & 0 & -2+2j & 30j & -1j \end{bmatrix}?$$

4.15 Compute the singular value decomposition of the following complex matrix and validate the results:

$$A = \begin{bmatrix} -2 & -2 & 1 & -1 \\ 0 & -1 & -2 & 2 \\ 0 & -2 & 2 & -1 \\ 1 & 0 & -2 & -1 \end{bmatrix} + j\begin{bmatrix} 2 & 0 & -2 & 0 \\ 2 & 1 & -2 & 1 \\ 0 & 0 & -1 & 2 \\ 0 & 0 & -1 & -1 \end{bmatrix}.$$

# 5 Solutions of matrix equations

The study of linear algebra and matrix theory has originated from the research on the solutions of linear equations. In this chapter, solutions of various matrix equations are explored.

In Section 5.1, numerical and analytical solutions to typical linear equations are presented, based on different types of linear equations. Equations with unique solutions, infinitely many solutions, and no solutions are addressed. In Section 5.2, various other simple linear equations are explored. In Section 5.3, various Lyapunov equations are studied, and immediate numerical solutions are presented, followed by Kronecker product-based analytical solutions and their MATLAB implementations. In Section 5.4, more general Sylvester equations are introduced, and their analytical solutions can be obtained with the MATLAB function provided in this book. In Section 5.5, nonlinear matrix equations are studied; Riccati equations are explored first, and we try to find all their solutions, real and complex. Different types of Riccati equations are also considered. A universal nonlinear matrix equation solver is formulated and implemented aiming at finding all the solutions. In Section 5.6, polynomial equation solutions are considered. Mainly, analytical Diophantine equation solutions are presented, and solutions of Bézout identity are discussed.

## 5.1 Linear equations

**Definition 5.1.** The general mathematical form of a system of linear equations is given by

$$\begin{cases} a_{11}x_1 + a_{12}x_2 + \cdots + a_{1n}x_n = b_1, \\ a_{21}x_1 + a_{22}x_2 + \cdots + a_{2n}x_n = b_1, \\ \vdots \\ a_{m1}x_1 + a_{m2}x_2 + \cdots + a_{mn}x_n = b_m. \end{cases} \tag{5.1.1}$$

If $b_i$ is a column vector, then the solution $x_i$ is also a column vector.

**Definition 5.2.** The matrix form of the linear equations given in Definition 5.1 is

$$Ax = B, \tag{5.1.2}$$

where matrices $A$ and $B$ are given as

$$A = \begin{bmatrix} a_{11} & a_{12} & \cdots & a_{1n} \\ a_{21} & a_{22} & \cdots & a_{2n} \\ \vdots & \vdots & \ddots & \vdots \\ a_{m1} & a_{m2} & \cdots & a_{mn} \end{bmatrix}, \quad B = \begin{bmatrix} b_{11} & b_{12} & \cdots & b_{1p} \\ b_{21} & b_{22} & \cdots & b_{2p} \\ \vdots & \vdots & \ddots & \vdots \\ b_{m1} & b_{m2} & \cdots & b_{mp} \end{bmatrix}. \tag{5.1.3}$$

https://doi.org/10.1515/9783110666991-005

**Definition 5.3.** From the given matrices $A$ and $B$, a solution judgement matrix $C$ is defined as

$$C = \begin{bmatrix} a_{11} & a_{12} & \cdots & a_{1n} & b_{11} & b_{12} & \cdots & b_{1p} \\ a_{21} & a_{22} & \cdots & a_{2n} & b_{21} & b_{22} & \cdots & b_{2p} \\ \vdots & \vdots & \ddots & \vdots & \vdots & \vdots & \ddots & \vdots \\ a_{m1} & a_{m2} & \cdots & a_{mn} & b_{m1} & b_{m2} & \cdots & b_{mp} \end{bmatrix}. \tag{5.1.4}$$

The solution judgement theorem describing the properties of linear equations is given below, without proofs[18].

**Theorem 5.1.** *The solutions of a linear equation $Ax = B$ should be considered in three cases:*

(1) *If $m = n$, and $\text{rank}(A) = n$, equations (5.1.2) have unique solution*

$$x = A^{-1}B. \tag{5.1.5}$$

(2) *If $\text{rank}(A) = \text{rank}(C) = r < n$, equations (5.1.2) have infinitely many solutions.*
(3) *If $\text{rank}(A) < \text{rank}(C)$, equations (5.1.2) are inconsistent, and there is no solution at all. Moore–Penrose generalized inverse can be used in finding the least squares solutions to the original equations.*

### 5.1.1 Unique solutions

In Chapter 4, several linear equation solvers in MATLAB are designed, and the numerical and analytical solutions to linear equations can be obtained. Here, the unique solution problem is discussed.

It is pointed out in Theorem 5.1(1) that if $A$ is a nonsingular matrix, the equation has a unique solution $x = A^{-1}B$. Function inv() in MATLAB can be used to solve the equation, with the syntax x=inv(A)*B. If $A$ is a double precision matrix, and it is singular or close to singular, erroneous solution may be obtained.

If Symbolic Math Toolbox is used, function inv() can be used to find analytical solutions of the equations. If solutions can be found, they are unique solutions, otherwise, other cases in Theorem 5.1 should be considered.

**Example 5.1.** Solve the following linear equations with low-level row transformation method:

$$\begin{bmatrix} 1 & 2 & 3 & 4 \\ 4 & 3 & 2 & 1 \\ 1 & 3 & 2 & 4 \\ 4 & 1 & 3 & 2 \end{bmatrix} X = \begin{bmatrix} 5 & 1 \\ 4 & 2 \\ 3 & 3 \\ 2 & 4 \end{bmatrix}.$$

**Solutions.** With the given matrices $A$ and $B$, the augmented matrix $C_1 = [A, B]$ can be constructed with the following statements:

```
>> A=[1 2 3 4; 4 3 2 1; 1 3 2 4; 4 1 3 2];
   B=[5 1; 4 2; 3 3; 2 4]; A=sym(A); C1=[A,B]
```

The augmented matrix $C_1$ is

$$C_1 = \begin{bmatrix} 1 & 2 & 3 & 4 & 5 & 1 \\ 4 & 3 & 2 & 1 & 4 & 2 \\ 1 & 3 & 2 & 4 & 3 & 3 \\ 4 & 1 & 3 & 2 & 2 & 4 \end{bmatrix}.$$

We assign $C_1$ to $C_2$, multiply its first row by −4, and then add to the second row. Next we subtract the third row from the first, multiply the first row by −4 and add it to the fourth row, to get the new $C_2$ matrix

```
>> C2=C1; C2(2,:)=C2(2,:)-4*C2(1,:);
   C2(3,:)=C2(3,:)-C2(1,:); C2(4,:)=C2(4,:)-4*C2(1,:)
```

The converted matrix $C_2$ is as follows. It can be seen that in the first column, the first element is 1, while others are all 0's.

$$C_2 = \begin{bmatrix} 1 & 2 & 3 & 4 & 5 & 1 \\ 0 & -5 & -10 & -15 & -16 & -2 \\ 0 & 1 & -1 & 0 & -2 & 2 \\ 0 & -7 & -9 & -14 & -18 & 0 \end{bmatrix}.$$

We continue by assigning $C_2$ to $C_3$. Since $C_2(2,2) = -5$, all the elements in the second row of $C_2$ should be divided by −5. Then the new second row can be multiplied by −2 and added to the first row; the third row is subtracted from the second row; finally, the second row is multiplied by 7 and added to the fourth row, and the new matrix $C_3$ is constructed

```
>> C3=C2; C3(2,:)=-C3(2,:)/5; C3(1,:)=C3(1,:)-2*C3(2,:);
   C3(3,:)=C3(3,:)-C3(2,:); C3(4,:)=C3(4,:)+7*C3(2,:)
```

This new matrix $C_3$ is

$$C_3 = \begin{bmatrix} 1 & 0 & -1 & -2 & -7/5 & 1/5 \\ 0 & 1 & 2 & 3 & 16/5 & 2/5 \\ 0 & 0 & -3 & -3 & -26/5 & 8/5 \\ 0 & 0 & 5 & 7 & 22/5 & 14/5 \end{bmatrix}.$$

Now the third row is multiplied by −1/3 and added to the first. Then multiplying the third row by −2, adding to the second row, and finally, multiplying the third row by −5 and adding to the fourth row, we get the new matrix $C_4$

```
>> C4=C3; C4(3,:)=-C4(3,:)/3; C4(1,:)=C4(1,:)+C4(3,:);
   C4(2,:)=C4(2,:)-2*C4(3,:); C4(4,:)=C4(4,:)-5*C4(3,:)
```

The result is

$$
C_4 = \begin{bmatrix} 1 & 0 & 0 & -1 & 1/3 & -1/3 \\ 0 & 1 & 0 & 1 & -4/15 & 22/15 \\ 0 & 0 & 1 & 1 & 26/15 & -8/15 \\ 0 & 0 & 0 & 2 & -64/15 & 82/15 \end{bmatrix}.
$$

The fourth row now is multiplied by 1/2 and added to the first. Next the second and third rows get subtracted by the fourth row, so that the needed matrix can finally be obtained

```
>> C5=C4; C5(4,:)=C5(4,:)/2; C5(1,:)=C5(1,:)+C5(4,:);
   C5(2,:)=C5(2,:)-C5(4,:); C5(3,:)=C5(3,:)-C5(4,:)
   X=C5(:,5:6), A*X-B
```

The results are given below.

It can be seen that the left-hand side of matrix $C_5$ is changed to an identity matrix. Therefore, the right-hand side of the matrix, i. e., the fifth and sixth columns form the solution $X$ of the equation. Substituting it back to the original equation, it can be seen that error matrix is zero, meaning that the solution is correct:

$$
C_5 = \begin{bmatrix} 1 & 0 & 0 & 0 & -9/5 & 12/5 \\ 0 & 1 & 0 & 0 & 28/15 & -19/15 \\ 0 & 0 & 1 & 0 & 58/15 & -49/15 \\ 0 & 0 & 0 & 1 & -32/15 & 41/15 \end{bmatrix}, \quad X = \begin{bmatrix} -9/5 & 12/5 \\ 28/15 & -19/15 \\ 58/15 & -49/15 \\ -32/15 & 41/15 \end{bmatrix}.
$$

**Example 5.2.** For the following $2 \times 2$ equation, derive its Kronecker product model.

$$
\begin{bmatrix} a_{11} & a_{12} \\ a_{21} & a_{22} \end{bmatrix} \begin{bmatrix} x_1 & x_3 \\ x_2 & x_4 \end{bmatrix} = \begin{bmatrix} b_1 & b_3 \\ b_2 & b_4 \end{bmatrix}.
$$

**Solutions.** To define symbolic variables, the following commands can be used:

```
>> A=sym('a%d%d',2);
   syms x1 x2 x3 x4 real; X=[x1 x3; x2 x4];
   syms b1 b2 b3 b4 real; B=[b1 b3; b2 b4]; M=A*X
```

and the equivalent form of the equation can be written as

$$
\begin{bmatrix} a_{11} & a_{12} & 0 & 0 \\ a_{21} & a_{22} & 0 & 0 \\ 0 & 0 & a_{11} & a_{12} \\ 0 & 0 & a_{21} & a_{22} \end{bmatrix} \begin{bmatrix} x_1 \\ x_2 \\ x_3 \\ x_4 \end{bmatrix} = \begin{bmatrix} b_1 \\ b_2 \\ b_3 \\ b_4 \end{bmatrix}.
$$

It can be seen that the coefficient matrix on the left-hand side of the equation can be written as $I \otimes A$, where $x$ and $b$ are the column vectors expanded in columns of the original matrices.

**Definition 5.4.** A given matrix $X \in \mathscr{C}^{n \times m}$ can be relabeled in a column-wise manner as

$$X = \begin{bmatrix} x_1 & x_{n+1} & \cdots & x_{(m-1)n+1} \\ x_2 & x_{n+2} & \cdots & x_{(m-1)n+2} \\ \vdots & \vdots & \ddots & \vdots \\ x_n & x_{2n} & \cdots & x_{mn} \end{bmatrix}, \tag{5.1.6}$$

and the thus expanded column vector $[x_1, x_2, \ldots, x_{nm}]^{\mathrm{T}}$ is denoted as vec($X$). If one wants to expand the matrix $X$ in a row-wise manner, then it is denoted as vec($X^{\mathrm{T}}$).

For a given matrix $C$, the expanded column vector $c = \text{vec}(C)$ can be obtained in MATLAB with the command $c=C(:)$. If one wants to restore an $n \times m$ matrix $C$ from the column vector $c$, the command $C=\texttt{reshape}(c,n,m)$ can be used.

**Theorem 5.2.** *The matrix equation* $AX = B$ *can be equivalently rewritten as*

$$(I_m \otimes A)x = b, \tag{5.1.7}$$

*where* $A \in \mathscr{C}^{n \times n}$, $B \in \mathscr{C}^{n \times m}$, *and* $x$ *and* $b$ *are the column-wise expanded forms of the matrices* $X$ *and* $B$, *i. e.,* $x = \text{vec}(X)$, $b = \text{vec}(B)$.

**Example 5.3.** Solve directly the linear equations in Example 5.1.

**Solutions.** The above equation can be expressed and solved with the following statements, and the accuracy of the solutions can be validated

```
>> A=[1 2 3 4; 4 3 2 1; 1 3 2 4; 4 1 3 2]; B=[5 1; 4 2; 3 3; 2 4];
   x=inv(A)*B, e1=norm(A*x-B), x1=inv(sym(A))*B
   e2=norm(A*x1-B) % validate the numerical solutions
   x2=inv(kron(eye(2),sym(A)))*B(:); x2=reshape(x2,4,2)
```

It can be seen that the numerical and analytical solutions of the equations are obtained as follows. When the numerical solutions are substituted back to the original equation, the norms of error matrices are respectively $e_1 = 8.4447 \times 10^{-15}$ and $e_2 = 0$. It can be seen that the analytical solution satisfies the original equation, with no errors. The results are exactly the same as those obtained by Kronecker products:

$$x = \begin{bmatrix} -1.8 & 2.4 \\ 1.8667 & -1.2667 \\ 3.8667 & -3.2667 \\ -2.1333 & 2.7333 \end{bmatrix}, \quad x_1 = x_2 = \begin{bmatrix} -9/5 & 12/5 \\ 28/15 & -19/15 \\ 58/15 & -49/15 \\ -32/15 & 41/15 \end{bmatrix}.$$

**Example 5.4.** Solve the equations in Example 5.1 with the elementary row transformation method.

**Solutions.** Several elementary row transform methods, such as Gaussian elimination and reduced row echelon form method, were discussed previously. The following statements can be used to compute the analytical solutions, and the results are the same as those in Example 5.1.

```
>> A=[1 2 3 4; 4 3 2 1; 1 3 2 4; 4 1 3 2]; A=sym(A);
   B=[5 1; 4 2; 3 3; 2 4]; x1=gauss_eq(A,B)
   C=[A,B]; C1=rref(C); x2=C1(:,5:6)
```

**Example 5.5.** Assess the efficiency of different algorithms when solving a 15×15 system of linear equations.

**Solutions.** A random symbolic linear system matrix can be established, and three different algorithms can be tried. The elapsed times are respectively 0.84, 1.3, and 0.34 seconds. It can be seen that the efficiency of function rref() is the highest of the three, then comes the inverse matrix method, and the slowest one is the low-level implementation of the Gaussian elimination method.

```
>> n=15; A=sym(rand(n)); B=sym(rand(n,5));
   tic, x1=inv(A)*B; toc, tic, x2=gauss_eq(A,B); toc
   tic, C=[A,B]; C1=rref(C); x3=C1(:,n+1:n+5); toc
   norm(A*x1-B), norm(A*x2-B), norm(A*x3-B)
```

### 5.1.2 Linear equations with infinitely many solutions

The method discussed earlier can only be used to solve equations when $A$ is a nonsingular square matrix. If $A$ is a rectangular or a singular matrix, other methods should be used. Infinitely many solutions mentioned in Theorem 5.1(2) can be constructed.

**Definition 5.5.** A nonzero vector $z$ satisfying $Az = 0$ is referred to as a null vector. A set of linearly independent null vectors span the null matrix $Z$.

In MATLAB, function null() can be used to compute a null matrix, with

$Z$=null($A$), or $Z$=null(sym($A$)).

The function can also be used in the numerical solution of the equations, where the number of columns of matrix $Z$ is $n - r$, while the vectors are also known as the basic set of solutions corresponding to matrix $A$.

**Theorem 5.3.** When rank($A$) = rank($C$) = $r < n$, the equations (5.1.2) have infinitely many solutions. The $n - r$ null vectors $z_i$, $i = 1, 2, \ldots, n - r$ can be constructed such that all solutions of the homogeneous equation $Az = 0$ are linear combinations of the null vectors $z_i$, that is, any solution is of the form

$$\hat{z} = \alpha_1 z_1 + \alpha_2 z_2 + \cdots + \alpha_{n-r} z_{n-r}, \tag{5.1.8}$$

where $\alpha_i, i = 1, 2, \ldots, n - r$ are arbitrary constants.

**Theorem 5.4.** When $\mathrm{rank}(A) = \mathrm{rank}(C) = r < n$, a particular solution of (5.1.2) is $x_0 = A^+B$.

**Theorem 5.5.** When $\mathrm{rank}(A) = \mathrm{rank}(C) = r < n$, the general solution of (5.1.2) can be written as

$$x = \alpha_1 z_1 + \alpha_2 z_2 + \cdots + \alpha_{n-r} z_{n-r} + x_0. \tag{5.1.9}$$

The solution of nonhomogeneous equation (5.1.2) is also very simple. If any one particular solution $x_0$ is found, the general solution is $x = \hat{x} + x_0$. In fact, the particular solution $x_0$ can easily be found in MATLAB, with $x_0$=pinv$(A)*B$.

**Example 5.6.** Solve the linear equation[2]

$$\begin{bmatrix} 1 & 4 & 0 & -1 & 0 & 7 & -9 \\ 2 & 8 & -1 & 3 & 9 & -13 & 7 \\ 0 & 0 & 2 & -3 & -4 & 12 & -8 \\ -1 & -4 & 2 & 4 & 8 & -31 & 37 \end{bmatrix} X = \begin{bmatrix} 3 \\ 9 \\ 1 \\ 4 \end{bmatrix}.$$

**Solutions.** The matrices $A$ and $B$ should be entered first, and then matrix $C$ can be constructed, such that the solvability of the equation can be judged.

```
>> A=[1,4,0,-1,0,7,-9;2,8,-1,3,9,-13,7;
      0,0,2,-3,-4,12,-8;-1,-4,2,4,8,-31,37];
   B=[3; 9; 1; 4]; C=[A B]; rank(A), rank(C) % compute the ranks
```

Through rank evaluation, the ranks of matrices $A$ and $C$ are checked to be identical, all of them being equal to 3, which is smaller than the number of columns of $A$, which is 7. It can be concluded that the original equation has infinitely many solutions. To solve the original equation, the null space $Z$ should be found first, and also a particular solution $x_0$ needs to be computed.

```
>> Z=null(sym(A)), x0=sym(pinv(A)*B) % null space and special solution
   a=sym('a%d',[4,1]); x=Z*a+x0, E=A*x-B % solve and validate
```

The null space $Z$ and a particular solution $x_0$ can be obtained as follows. For the selected constants $a_1, a_2, a_3$, and $a_4$, all the solutions can be constructed, and it can be seen that the error matrix is zero:

$$
Z = \begin{bmatrix}
-4 & -2 & -1 & 3 \\
1 & 0 & 0 & 0 \\
0 & -1 & 3 & -5 \\
0 & -2 & 6 & -6 \\
0 & 1 & 0 & 0 \\
0 & 0 & 1 & 0 \\
0 & 0 & 0 & 1
\end{bmatrix}, \quad
x_0 = \begin{bmatrix}
92/395 \\
368/395 \\
459/790 \\
-24/79 \\
347/790 \\
247/790 \\
303/790
\end{bmatrix},
$$

$$
x = \begin{bmatrix}
-4a_1 - 2a_2 - a_3 + 3a_4 + 92/395 \\
a_1 + 368/395 \\
-a_2 + 3a_3 - 5a_4 + 459/790 \\
-2a_2 + 6a_3 - 6a_4 - 24/79 \\
a_2 + 347/790 \\
a_3 + 247/790 \\
a_4 + 303/790
\end{bmatrix}.
$$

**Example 5.7.** Solve the matrix equation again in Example 5.6 with the reduced row echelon form method.

**Solutions.** With reduced row echelon form method, the equation is solved using

```
>> C=[A B]; D=rref(C) % augmentation the solve
```

with

$$
D = \begin{bmatrix}
1 & 4 & 0 & 0 & 2 & 1 & -3 & 4 \\
0 & 0 & 1 & 0 & 1 & -3 & 5 & 2 \\
0 & 0 & 0 & 1 & 2 & -6 & 6 & 1 \\
0 & 0 & 0 & 0 & 0 & 0 & 0 & 0
\end{bmatrix}.
$$

It can be seen that, if the free variables are selected as $x_2$, $x_5$, $x_6$, and $x_7$, they can be assigned to any constants. Letting $x_2 = b_1$, $x_5 = b_2$, $x_6 = b_3$, $x_7 = b_4$, from matrix $D$ the solutions of the matrix equation can be constructed as

$$
\begin{cases}
x_1 = -4b_1 - 2b_2 - b_3 + 3b_4 + 4, \\
x_3 = -b_2 + 3b_3 - 5b_4 + 2, \\
x_4 = -2b_2 + 6b_3 - 6b_4 + 1.
\end{cases}
$$

**Example 5.8.** Solve the linear equation

$$
\begin{bmatrix}
4 & 7 & 1 & 4 \\
3 & 7 & 4 & 6
\end{bmatrix} x = \begin{bmatrix} 3 \\ 4 \end{bmatrix}.
$$

**Solutions.** The equation can be solved directly with the following statements:

```
>> A=[4,7,1,4; 3,7,4,6]; B=[3; 4]; C=[A B];
   rank(A), rank(C)        % compute the ranks
```

Since the ranks of matrices **A** and **C** are the same, both of them being equal to 2, there are infinitely many solutions. The equation can be solved with the following two methods:

```
>> syms a1 a2 b1 b2; x1=null(sym(A))*[a1; a2]+sym(A\B)      % method 1
   a=rref(sym([A B])); x2=[a(:,3:5)*[-b1;-b2;1]; b1; b2]    % method 2
   e1=A*x1-B, e2=A*x2-B
```

and the analytical solutions are as follows. It can be seen that they both satisfy the original equation:

$$x_1 = \begin{bmatrix} a_1 \\ a_2 + 8/21 \\ 6a_1/5 + 7a_2/5 + 1/3 \\ -13a_1/10 - 21a_2/10 \end{bmatrix}, \quad x_2 = \begin{bmatrix} 3b_1 + 2b_2 - 1 \\ -13b_1/7 - 12b_2/7 + 1 \\ b_1 \\ b_2 \end{bmatrix}.$$

### 5.1.3 Inconsistent systems of equations

**Theorem 5.6.** *If* rank(**A**) < rank(**C**), *then equation (5.1.2) is inconsistent, with no solutions. The concept of Moore–Penrose generalized inverse can be used to compute the least squares solution,* $x = A^+B$.

Since pseudoinverse is involved, the command $x$=pinv($A$)*$B$ can be used to compute directly the least squares solution. The solution, of course, does not satisfy the original equations, however, it minimizes the norm of the error $\|Ax - B\|$.

**Example 5.9.** Solve the following linear equation:

$$\begin{bmatrix} 1 & 2 & 3 & 4 \\ 2 & 2 & 1 & 1 \\ 2 & 4 & 6 & 8 \\ 4 & 4 & 2 & 2 \end{bmatrix} X = \begin{bmatrix} 1 \\ 2 \\ 3 \\ 4 \end{bmatrix}.$$

**Solutions.** The two matrices can be entered first, a judgement matrix **C** can be constructed next, and then the ranks of both matrices can be computed

```
>> A=[1 2 3 4; 2 2 1 1; 2 4 6 8; 4 4 2 2]; B=[1:4]';
   C=[A B]; rank(A), rank(C)
```

It can be seen that rank(**A**) = 2 ≠ rank(**C**) = 3, therefore, the original equation is inconsistent, with no solution. Function pinv() can be used to compute Moore–Penrose generalized inverse, and the least squares solution can then be found.

```
>> X1=pinv(A)*B, e1=A*X1-B        % numerical results
   X2=pinv(sym(A))*B, e2=A*X2-B  % symbolic results
```

The numerical and symbolic solutions, as well as errors, are:

$$X_1 = \begin{bmatrix} 0.5466 \\ 0.4550 \\ 0.04427 \\ -0.0473 \end{bmatrix}, \quad e_1 = \begin{bmatrix} 0.4 \\ 0 \\ -0.2 \\ 0 \end{bmatrix}, \quad X_2 = \begin{bmatrix} 358/655 \\ 298/655 \\ 29/655 \\ -31/655 \end{bmatrix}, \quad e_2 = \begin{bmatrix} 2/5 \\ 0 \\ -1/5 \\ 0 \end{bmatrix}.$$

Two of the equations are satisfied, since $n - \text{rank}(A) = 2$. The solutions obtained minimize the norm of the error vectors.

### 5.1.4 Graphical interpretation of linear solutions

In linear equations with two unknown variables, each equation can be regarded as a straight line. If there are two equations, the intersection is regarded as the solution of the equations. If there are three or more equations, and they happen to intersect at the same point, it is the solution of the equations. If there is more than one intersection, and the lines are all different, the equations are inconsistent, and there is no solution in the equations.

Sometimes, although the equations are with two unknowns, there is no solution. It can be seen from Theorem 5.1 that when $A$ is not a full-rank matrix, and the ranks of $A$ and $C$ are different, the equation has no solution. Graphically, if two lines are parallel, while being not exactly the same, the two equations are inconsistent, and their system has no solutions. When the two lines are identical, it can be seen from Theorem 5.1 that, the two equations have infinitely many solutions, i. e., every point on the line is a solution.

**Example 5.10.** Find the solution of the chick–rabbit problem in Example 1.4.

**Solutions.** Let us consider the first equation $x+y = 35$. If one moves $x$ to the right-hand side of the equation, an explicit expression can be obtained, and fplot() function can be used to draw the straight line. If one does not want to move the term, the equation is an implicit function of $x$ and $y$. In this case, fimplicit() function can be used to draw the straight line. The lines of the equations can be drawn with the following statements, as shown in Figure 5.1, with the intersection at $x = 23$, $y = 12$. This is the solution of the simultaneous equations. For this particular example, the scales in $x$ and $y$ axes are assigned to $x, y \in [-80, 80]$.

```
>> syms x y; fimplicit(x+y==35,[-80,80])
   hold on; fimplicit(2*x+4*y==94,[-80,80])
```

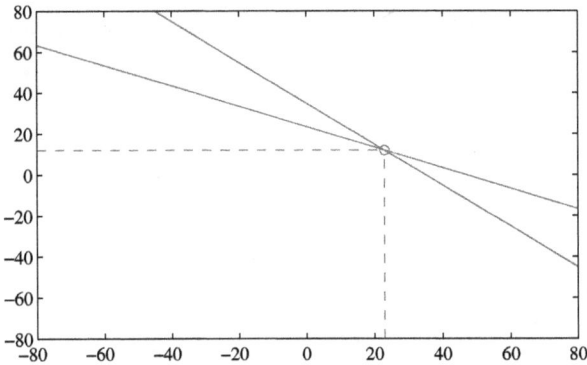

**Figure 5.1:** Graphical interpretation of the chick–rabbit problem.

For equations with three unknown variables, each equation can be regarded as a plane. Each pair of equations forms a straight line as the intersection of two planes. In this way, three straight lines can be formed. When the three lines meet at the same point, that point is the solution of the equations. If they fail to meet at the same point, the equations are inconsistent, and have no solution.

**Example 5.11.** Understand graphically the solutions of the following equations:

$$\begin{cases} 2x_1 + 3x_2 + x_3 = 2, \\ 3x_1 + 3x_2 + 2x_3 = 3, \\ 3x_1 + 2x_2 + 2x_3 = 4. \end{cases}$$

**Solutions.** Three-dimensional implicit functions can be drawn employing the `fimplicit3()` function. However, if the explicit form is known, then `surf()` function can be used to draw the planes, as shown in Figure 5.2. The intersections of the planes cannot be displayed clearly, and the solutions cannot be obtained graphically. For this ex-

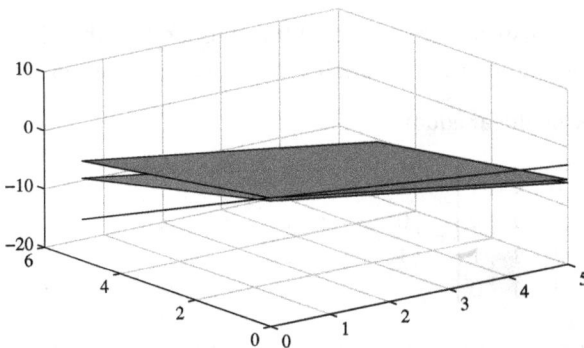

**Figure 5.2:** The intersection of three phases.

ample, the intersection of the three planes is the point $(4, -1, 3)$, which is the solution of the simultaneous equations.

```
>> [x1 x2]=meshgrid([0, 5]);
   z1=2-x1-3*x2; z2=1.5*(1-x1-x2); z3=2-1.5*x1-x2;
   surf(x1,x2,z1), hold on; surf(x1,x2,z2), surf(x1,x2,z3)
```

Equations with four and more variables cannot be solved graphically, and there is no graphical interpretations of their solutions.

## 5.2 Other forms of simple linear equations

The standard form of linear equations discussed so far was $AX = B$. In real applications, other forms of linear equations such as $XA = B$ and $AXB = C$, or their multiterm forms, may be encountered. In this section, the solutions to such linear equations are explored.

### 5.2.1 $XA = B$ equations

**Theorem 5.7.** *If a linear equation is described as*

$$XA = B, \tag{5.2.1}$$

*taking transpose to both sides, one has*

$$A^TZ = B^T, \tag{5.2.2}$$

*where $Z = X^T$, i. e., the equation can be converted to the case in (5.1.2). With the above algorithms, the original equations can be solved.*

It should be noted that the transpose here is a direct transpose, rather than Hermitian transpose, otherwise the solutions do not satisfy the original equations (5.2.1).

**Example 5.12.** Solve the following linear equation:

$$X \begin{bmatrix} 1 & 0 & 0 & 1 \\ 0 & 1 & 1 & 0 \\ 0 & 1 & 1 & 1 \\ 0 & 1 & 1 & 0 \end{bmatrix} = \begin{bmatrix} 0 & 2 & 2 & 1 \\ 1 & 2 & 2 & 2 \end{bmatrix}.$$

**Solutions.** The two matrices should be entered into MATLAB, and it can be found that the rank of $A$ is 3.

```
>> A=[1,0,0,1; 0,1,1,0; 0,1,1,1; 0,1,1,0]; A=sym(A);
   B=[0,2,2,1; 1,2,2,2]; rank(A)
```

A judgement matrix $C = [A^T, B^T]$ can be constructed, and it can be seen that the rank of $C$ is also 3, indicating that the equation has infinitely many solutions. With the following statements, the reduced row echelon form can be obtained, and the homogeneous equation solution can be obtained:

```
>> C=[A.', B.'], C1=rref(C), Z=C1(:,5:6)
```

It can be seen that the transformation matrix and homogeneous equation solutions are

$$C_1 = \begin{bmatrix} 1 & 0 & 0 & 0 & 0 & 1 \\ 0 & 1 & 0 & 1 & 1 & 1 \\ 0 & 0 & 1 & 0 & 1 & 1 \\ 0 & 0 & 0 & 0 & 0 & 0 \end{bmatrix}, \quad Z = \begin{bmatrix} 0 & 1 \\ 1 & 1 \\ 1 & 1 \\ 0 & 0 \end{bmatrix}.$$

Defining a free variable $x_4$ as $a$, the following commands can be used to solve the original equations:

```
>> syms a; Z(4,:)=[a a]; Z(2,:)=-[a a]+Z(2,:);
   x1=Z.', x1*A-B
```

The general solution of the equation can be obtained as follows. When it is substituted back to the original equation, it can be seen that the error is zero:

$$Z = \begin{bmatrix} 0 & 1 \\ 1-a & 1-a \\ 1 & 1 \\ a & a \end{bmatrix}, \quad x_1 = \begin{bmatrix} 0 & 1-a & 1 & a \\ 1 & 1-a & 1 & a \end{bmatrix}.$$

Besides, if the free variable is assigned as $b$, the following statements can be used to compute the general solution:

```
>> syms b; x=null(A.')*b+pinv(A.')*B.'; x2=x.', x2*A-B
```

It can be seen that the general solution obtained is as follows, which satisfies the original equation:

$$x_2 = \begin{bmatrix} 0 & 1/2-b & 1 & b+1/2 \\ 1 & 1/2-b & 1 & b+1/2 \end{bmatrix}.$$

**Theorem 5.8.** *Equation* $XA = B$ *can be converted to Kronecker product form*

$$(A^T \otimes I_n)x = b, \tag{5.2.3}$$

where $X \in \mathscr{C}^{m \times n}$, $A \in \mathscr{C}^{n \times n}$, and $x$ and $b$ are column-wise expanded column vectors of matrices $X$ and $B$, i. e., $x = \text{vec}(X)$, $b = \text{vec}(B)$.

Note that in (5.2.3) the direct rather than Hermitian transpose is used.

**Example 5.13.** Solve the following linear equation $XA = B$ with complex coefficients, and validate the results:

$$A = \begin{bmatrix} 5+j & 3j & 2+2j \\ 2 & 0 & 2j+4 \\ 3+2j & 2+6j & 6 \end{bmatrix}, \quad B = \begin{bmatrix} 0 & 2 & 1 \\ 0 & 2 & 1 \\ 1 & 0 & 0 \end{bmatrix}.$$

**Solutions.** The matrices can be input into MATLAB first, and then the solutions of the equation can be found

```
>> A=[5+1i,3i,2+2i; 2,0,4+2i; 3+2i,2+6i,6]; A=sym(A);
   B=[0,2,1; 0,2,1; 1,0,0]; X=inv(A.')*B.';
   X1=X.', X1*A-B
   X2=inv(kron(A.',eye(3))); X2=reshape(X2,3,3), X2*A-B
```

The two methods yield the same results. They are validated and it can be seen that the error is zero:

$$X_1 = X_2 = \frac{1}{3389} \begin{bmatrix} -1150 - 216j & 784 + 2823j/2 & 824 - 747j \\ -1150 - 216j & 784 + 2823j/2 & 824 - 747j \\ 802 + 186j & 78 - 227j & -333 - 204j \end{bmatrix}.$$

### 5.2.2 $AXB = C$ equations

We note that $AXB = C$ equations are more complicated than those presented earlier. Two cases should be considered for such equations:

(1) If in equation $AXB = C$, $A$ or $B$ is nonsingular, then the two sides can be left multiplied by $A^{-1}$ or right multiplied by $B^{-1}$, so that solvable forms of the equation can be obtained.

**Example 5.14.** Solve the following matrix equation:

$$\begin{bmatrix} 8 & 1 & 6 \\ 3 & 5 & 7 \\ 4 & 9 & 2 \end{bmatrix} X \begin{bmatrix} 0 & 1 & 0 & 0 & 1 \\ 1 & 0 & 1 & 2 & 2 \\ 1 & 2 & 0 & 0 & 2 \\ 0 & 0 & 1 & 1 & 1 \\ 1 & 0 & 0 & 2 & 1 \end{bmatrix} = \begin{bmatrix} 0 & 2 & 0 & 0 & 2 \\ 1 & 2 & 1 & 0 & 0 \\ 2 & 1 & 1 & 1 & 0 \end{bmatrix}.$$

**Solutions.** Since this equation can be expressed as $AXB = C$, and $A$ and $B$ are both nonsingular, the solution of the equation can be mathematically written as

$X = A^{-1}CB^{-1}$. The following commands can be used to find the numerical and analytic solutions:

```
>> B=[0,1,0,0,1; 1,0,1,2,2; 1,2,0,0,2; 0,0,1,1,1; 1,0,0,2,1];
   C=[0,2,0,0,2; 1,2,1,0,0; 2,1,1,1,0];
   A=[8,1,6; 3,5,7; 4,9,2]; X=inv(A)*C*inv(sym(B))
   A*X*B-C, X1=inv(A)*C*inv(B), norm(A*X1*B-C)
```

The numerical and analytical solutions can be found as follows. If the analytical solution is substituted back to the original equation, a zero error matrix is obtained. If the numerical solution is substituted to the equation, the norm of the error matrix is $2.3955 \times 10^{-15}$:

$$X = \begin{bmatrix} 257/360 & 7/15 & -29/90 & -197/360 & -29/180 \\ -179/180 & -8/15 & 23/45 & 119/180 & 23/90 \\ -163/360 & -8/15 & 31/90 & 223/360 & 31/180 \end{bmatrix},$$

$$X_1 = \begin{bmatrix} 0.71389 & 0.46667 & -0.32222 & -0.54722 & -0.16111 \\ -0.99444 & -0.53333 & 0.51111 & 0.66111 & 0.25556 \\ -0.45278 & -0.53333 & 0.34444 & 0.61944 & 0.17222 \end{bmatrix}.$$

It can be seen that if at least one of matrices $A$ and $B$ is nonsingular, the previously presented algorithms can be used to solve the equations. Next we discuss the case when both matrices are singular.

(2) If both matrices are singular, judgement should be made on whether there are solutions of the matrix equation. Only in the affirmative case, solutions can be constructed.

**Theorem 5.9.** *The equation* $AXB = C$ *has solutions if and only if*[23]

$$AA^- CB^- B = C. \tag{5.2.4}$$

**Theorem 5.10.** *If* $A \in \mathscr{C}^{m \times n}$, $B \in \mathscr{C}^{p \times q}$, $C \in \mathscr{C}^{m \times q}$, *the general solution of the equation* $AXB = C$ *can be constructed as*[23]

$$X = A^- CB^- + (Z - A^- AZBB^-), \tag{5.2.5}$$

*where* $Z \in \mathscr{C}^{n \times p}$ *is an arbitrary matrix.*

**Example 5.15.** Solve the matrix equation

$$\begin{bmatrix} 1 & 3 \\ 4 & 2 \end{bmatrix} X \begin{bmatrix} 16 & 2 & 3 & 13 \\ 5 & 11 & 10 & 8 \\ 9 & 7 & 6 & 12 \\ 4 & 14 & 15 & 1 \end{bmatrix} = \begin{bmatrix} 219 & 181 & 190 & 192 \\ 296 & 264 & 280 & 248 \end{bmatrix}.$$

**Solutions.** It can be recognized immediately that

$$A = \begin{bmatrix} 1 & 3 \\ 4 & 2 \end{bmatrix}, \quad B = \begin{bmatrix} 16 & 2 & 3 & 13 \\ 5 & 11 & 10 & 8 \\ 9 & 7 & 6 & 12 \\ 4 & 14 & 15 & 1 \end{bmatrix}, \quad C = \begin{bmatrix} 219 & 181 & 190 & 192 \\ 296 & 264 & 280 & 248 \end{bmatrix}.$$

In fact, if $A$ and $B$ matrices are not both singular, it is not necessary to test the conditions in Theorem 5.9, one may solve directly the equation with $X = A^+CB^+$, and then see whether the solution satisfies the original equation:

```
>> A=[1,3; 4,2]; C=[219,181,190,192; 296,264,280,248];
   B=[16,2,3,13; 5,11,10,8; 9,7,6,12; 4,14,15,1];
   X1=pinv(A)*C*pinv(B), e=norm(A*X1*B-C)
```

The solution obtained is given below. The norm of the error matrix is $e = 7.0451 \times 10^{-13}$, indicating the equation has infinitely many solutions.

$$X_1 = \begin{bmatrix} 1.85 & 0.55 & 0.45 & 2.15 \\ 2.3 & 0.9 & 1.1 & 1.7 \end{bmatrix}.$$

Since Moore–Penrose generalized inverse satisfies the generalized inverse conditions, it can be used to replace $A^-$ in Theorem 5.10 to configure the general solutions of the equation:

```
>> Z=sym('z%d%d',[2,4]); A=sym(A);
   Y=pinv(A)*C*pinv(B)+(Z-pinv(A)*A*Z*B*pinv(B)), A*Y*B-C
```

The general solution of the equation is as follows, and it has been validated that the original equation is satisfied:

$$Y = \frac{1}{20} \begin{bmatrix} z_{11} + 3z_{12} - 3z_{13} - z_{14} + 37 & 3z_{11} + 9z_{12} - 9z_{13} - 3z_{14} + 11 \\ z_{21} + 3z_{22} - 3z_{23} - z_{24} + 46 & 3z_{21} + 9z_{22} - 9z_{23} - 3z_{24} + 18 \end{bmatrix}$$

$$\begin{matrix} 9z_{13} - 9z_{12} - 3z_{11} + 3z_{14} + 9 & 3z_{13} - 3z_{12} - z_{11} + z_{14} + 43 \\ 9z_{23} - 9z_{22} - 3z_{21} + 3z_{24} + 22 & 3z_{23} - 3z_{22} - z_{21} + z_{24} + 34 \end{matrix}.$$

**Theorem 5.11.** *If the equation $AXB = C$ has no solution, the approximate solution with the least Frobenius norm is $X = A^+CB^+$.*

### 5.2.3 Kronecker product-based solutions

An alternative solution of equation $AXB = C$ is considered in this section. The equation can be converted to the typical form of $Ax = b$, with Kronecker product methods.

Then the methods previously presented can be adopted in finding the numerical and analytical solutions.

**Theorem 5.12.** *Equation $AXB = C$ can be converted to the linear equation*

$$(B^T \otimes A)x = c \tag{5.2.6}$$

*where $c = \text{vec}(C)$ and $x = \text{vec}(X)$ are the column-wise expanded column vectors of the matrices $C$ and $X$, respectively.*

**Example 5.16.** Solve the equation in Example 5.14 with Kronecker product method.

**Solutions.** The following statements can be used to find the analytical solution of the equation, and it is exactly the same as that in Example 5.14.

```
>> B=[0,1,0,0,1; 1,0,1,2,2; 1,2,0,0,2; 0,0,1,1,1; 1,0,0,2,1];
   C=[0,2,0,0,2; 1,2,1,0,0; 2,1,1,1,0];
   A=[8,1,6; 3,5,7; 4,9,2]; A=sym(A); c=C(:);
   x=inv(kron(B.',A))*c; X=reshape(x,3,5), A*X*B-C
```

### 5.2.4 Multiterm $AXB = C$ equations

**Definition 5.6.** Multiterm $AXB = C$ equation can be expressed as

$$A_1 X B_1 + A_2 X B_2 + \cdots + A_k X B_k = -C, \tag{5.2.7}$$

where $A_i \in \mathscr{C}^{n \times n}$, $B_i \in \mathscr{C}^{m \times m}$, $i = 1, 2, \ldots, k$, and $C, X \in \mathscr{C}^{n \times m}$.

**Theorem 5.13.** *Multiterm $AXB = C$ equation can be converted into the following linear equation:*

$$(B_1^T \otimes A_1 + B_2^T \otimes A_2 + \cdots + B_k^T \otimes A_k)x = -c, \tag{5.2.8}$$

*where $c = \text{vec}(C)$ and $x = \text{vec}(X)$ are column-wise expanded vectors of matrices $C$ and $X$, respectively.*

**Example 5.17.** Solve the multiterm equation $AXB = C$ for the matrices given below:

$$A_1 = \begin{bmatrix} 2 & 0 \\ 2 & 2 \end{bmatrix}, \quad A_2 = \begin{bmatrix} 1 & 0 \\ 0 & 1 \end{bmatrix}, \quad C = -\begin{bmatrix} 17 & 9 & 9 & 7 \\ 21 & 17 & 13 & 13 \end{bmatrix},$$

$$B_1 = \begin{bmatrix} 2 & 2 & 1 & 1 \\ 2 & 1 & 2 & 0 \\ 0 & 2 & 2 & 2 \\ 2 & 0 & 2 & 2 \end{bmatrix}, \quad B_2 = \begin{bmatrix} 2 & 1 & 0 & 2 \\ 2 & 0 & 0 & 0 \\ 2 & 2 & 0 & 2 \\ 1 & 0 & 2 & 0 \end{bmatrix}.$$

**Solutions.** The following statements can be used to solve the equation

```
>> A1=[2,0; 2,2]; A2=[1,0; 0,1]; A1=sym(A1);
   C=-[17,9,9,7; 21,17,13,13]; c=C(:);
   B1=[2,2,1,1; 2,1,2,0; 0,2,2,2; 2,0,2,2];
   B2=[2,1,0,2; 2,0,0,0; 2,2,0,2; 1,0,2,0];
   x=inv(kron(B1.',A1)+kron(B2.',A2))*c;
   X=reshape(x,2,4), A1*X*B1+A2*X*B2-C
```

The solution can be obtained as follows, and it can be seen that the equation is satisfied:

$$X = \begin{bmatrix} -279/182 & -369/364 & 3/26 & -71/182 \\ -23\,923/16\,562 & 1\,405/8\,281 & -477/1\,183 & -1\,321/8\,281 \end{bmatrix}.$$

In fact, multiterm equation can also be regarded as the simultaneous one in the form of $A_iXB_i = -C_i$, $i = 1, 2, \ldots, k$, where $C = C_1 + C_2 + \cdots + C_k$.

## 5.3 Lyapunov equations

The equations discussed so far were linear equations as $AX = B$, or other simple forms. In practical applications, other types of equations may be encountered, for instance, Lyapunov equations. Special solution algorithms should be used. In this section, solutions of Lyapunov equations are presented.

### 5.3.1 Continuous Lyapunov equations

**Definition 5.7.** Continuous Lyapunov equation can be expressed as

$$AX + XA^H = -C, \tag{5.3.1}$$

with matrices $A, C, X \in \mathscr{C}^{n\times n}$.

Lyapunov equation is named after Russian mathematician Aleksandr Mikhailovich Lyapunov (1857–1918). Lyapunov equations originate from the stability theory of ordinary differential equations. It was normally assumed that $-C$ is an $n \times n$ symmetric matrix. It can be shown that the solution $X$ is also an $n \times n$ symmetric matrix. In real applications, $C$ can be any matrix.

It is not easy to solve such equation directly. With the use of MATLAB and its toolboxes, the solutions are easier to find. For instance, the function $\texttt{lyap}()$ provided in the Control System Toolbox can be used to solve the equations numerically, with the syntax $X=\texttt{lyap}(A,C)$. If the matrices $A$ and $C$ in Lyapunov equation are given, the numerical solution $X$ can be obtained immediately. Examples are given below to demonstrate the solutions of Lyapunov equations.

**Example 5.18.** Assume that matrices $A$ and $C$ in (5.3.1) are given below, solve the Lyapunov equation and check the accuracy of the solution:

$$A = \begin{bmatrix} 1 & 2 & 3 \\ 4 & 5 & 6 \\ 7 & 8 & 0 \end{bmatrix}, \quad C = - \begin{bmatrix} 10 & 5 & 4 \\ 5 & 6 & 7 \\ 4 & 7 & 9 \end{bmatrix}.$$

**Solutions.** For the given matrices, the following MATLAB commands can be used:

```
>> A=[1 2 3;4 5 6; 7 8 0]; C=-[10,5,4; 5,6,7; 4,7,9]; % the matrices
   X=lyap(A,C), norm(A*X+X*A'+C)   % solve and validate equations
```

The numerical solutions are given below, and it is found that $\|AX + XA^T + C\| = 2.3211 \times 10^{-14}$, meaning the solution $X$ satisfies the original equation, and the accuracy is high:

$$X = \begin{bmatrix} -3.9444444444442 & 3.8888888888887 & 0.38888888888891 \\ 3.8888888888887 & -2.7777777777775 & 0.22222222222221 \\ 0.38888888888891 & 0.22222222222221 & -0.11111111111111 \end{bmatrix}.$$

**Example 5.19.** In classical Lyapunov equations, it is always assumed that $C$ is a real symmetric positive definite matrix. Explore whether the conditions are necessary or not.

**Solutions.** Influenced by the stability study of differential equations, it seems that the condition for Lyapunov equation to have solutions is that $-C$ matrix is real positive definite. In fact, if this condition is not satisfied, the linear algebraic equation in (5.3.1) may also have unique solutions. For instance, matrix $A$ in Example 5.18 can be kept unchanged, and the $C$ matrix can be modified to a complex asymmetric matrix

$$C = - \begin{bmatrix} 1+1j & 3+3j & 12+10j \\ 2+5j & 6 & 11+6j \\ 5+2j & 11+j & 2+12j \end{bmatrix}.$$

Then the following statements can be used to solve the equation:

```
>> A=[1 2 3;4 5 6; 7 8 0]; % input the known matrices
   C=-[1+1i, 3+3i, 12+10i; 2+5i, 6, 11+6i; 5+2i, 11+1i, 2+12i];
   X=lyap(A,C), norm(A*X+X*A'+C)
```

The result obtained is as follows. When it is substituted back to the original equation, the norm of the error matrix is $2.3866 \times 10^{-14}$:

$$X = \begin{bmatrix} -0.0490 + 1.5871j & 0.8824 - 0.8083j & -0.1993 + 0.3617j \\ 0.2353 - 1.3638j & -0.1961 + 0.3486j & 0.7516 + 1.3224j \\ -0.1797 + 0.3617j & -0.1699 - 0.4553j & 1.3268 + 0.7832j \end{bmatrix}.$$

It can be concluded that, if the physical interpretations of stability of linear systems are not considered, conventional Lyapunov equations can be extended to further including the case where any complex $C$ matrices can be handled. Function `lyap()` can be applied to directly solve Lyapunov equations of any sort.

### 5.3.2 2 × 2 Lyapunov equations

A $2 \times 2$ Lyapunov equation is discussed here, and its Kronecker product-based conversions are implemented. The results can easily be applied to handle Lyapunov equations of any sizes.

**Example 5.20.** For a given $2 \times 2$ Lyapunov equation, formulate an analytical method.

**Solutions.** The matrices $X$ and $C$ can be column-wise expanded, and the following statements can be used to configure the matrices, then the left-hand side of the equation can be simplified

```
>> A=sym('a%d%d',2); assume(A,'real');
   syms x1 x2 x3 x4 c1 c2 c3 c4;
   X=[x1 x3; x2 x4]; C=[c1 c3; c2 c4]; simplify(A*X+X*A')
```

The left-hand side can be derived as

$$\begin{bmatrix} 2a_{11}x_1 + a_{12}x_2 + a_{12}x_3 & a_{21}x_1 + (a_{11} + a_{22})x_3 + a_{12}x_4 \\ a_{21}x_1 + (a_{11} + a_{22})x_2 + a_{12}x_4 & a_{21}x_2 + a_{21}x_3 + 2a_{22}x_4 \end{bmatrix} = - \begin{bmatrix} c_1 & c_3 \\ c_2 & c_4 \end{bmatrix}.$$

The equation can be rewritten as

$$\begin{bmatrix} 2a_{11} & a_{12} & a_{12} & 0 \\ a_{21} & a_{11} + a_{22} & 0 & a_{12} \\ a_{21} & 0 & a_{11} + a_{22} & a_{12} \\ 0 & a_{21} & a_{21} & 2a_{22} \end{bmatrix} \begin{bmatrix} x_1 \\ x_2 \\ x_3 \\ x_4 \end{bmatrix} = - \begin{bmatrix} c_1 \\ c_2 \\ c_3 \\ c_4 \end{bmatrix}.$$

It can be seen that the coefficient matrix on the left-hand side can be written as $I \otimes A + A \otimes I$, and the original equation can be converted to the form of $AX = B$, so that the solution can then be found:

```
>> I=eye(2); A0=kron(I,A)+kron(A,I)
   x=-inv(A0)*[c1;c3;c2;c4]; simplify(x)
```

For this particular equation, the analytical vector solution is

$$x = \frac{1}{\Delta} \begin{bmatrix} (a_{12}a_{21} - a_{11}a_{22} - a_{22}^2)c_1 + a_{12}a_{22}c_2 + a_{12}a_{22}c_3 - a_{12}^2c_4 \\ a_{21}a_{22}c_1 + (a_{12}a_{21} - 2a_{11}a_{22})c_2 - a_{12}a_{21}c_3 + a_{11}a_{12}c_4 \\ a_{21}a_{22}c_1 - a_{12}a_{21}c_2 + (a_{12}a_{21} - 2a_{11}a_{22})c_3 + a_{11}a_{12}c_4 \\ -a_{21}^2c_1 + a_{11}a_{21}c_2 + a_{11}a_{21}c_3 + (a_{12}a_{21} - a_{11}a_{22} - a_{11}^2)c_4 \end{bmatrix},$$

where $\Delta = 2(a_{11} + a_{22})(a_{11}a_{22} - a_{12}a_{21})$. If $\Delta = 0$, there is no solution.

### 5.3.3 Analytical solution of Lyapunov equations

From the results for the $2 \times 2$ Lyapunov equation discussed earlier, analytical solution of Lyapunov equation of any size can be formulated.

**Theorem 5.14.** *With Kronecker product representation, Lyapunov equation can be written as*

$$(I \otimes A + A \otimes I)x = -c, \tag{5.3.2}$$

*where $A \otimes B$ represents the Kronecker product of matrices $A$ and $B$, while $x$ and $c$ are the column-wise expanded column vectors of matrices $X$ and $C$, i. e., $c = \mathrm{vec}(C)$, $x = \mathrm{vec}(X)$.*

It can be seen that the condition to have a unique solution is no longer that $-C$ is a symmetric positive definite matrix. The condition in (5.3.1) is that $(I \otimes A + A \otimes I)$ is a nonsingular square matrix. It can be seen that, if $A$ is nonsingular, the equation has a unique solution.

**Example 5.21.** Solve analytically the Lyapunov equation in Example 5.18.

**Solutions.** The analytical solution can be obtained directly with the following statements and the solution can also be validated:

```
>> A=[1 2 3;4 5 6; 7 8 0]; C=-[10,5,4; 5,6,7; 4,7,9];
   A0=sym(kron(eye(3),A)+kron(A,eye(3)));
   c=C(:); x0=-inv(A0)*c; x=reshape(x0,3,3)
   norm(A*x+x*A'+C) % validation of the solution
```

The analytical solution $x$ is given below, and it can be seen that there is no error in the solution:

$$x = \begin{bmatrix} -71/18 & 35/9 & 7/18 \\ 35/9 & -25/9 & 2/9 \\ 7/18 & 2/9 & -1/9 \end{bmatrix}.$$

**Example 5.22.** Solve analytically the Lyapunov equation in Example 5.19.

**Solutions.** Matrices $A$ and $C$ should be entered into MATLAB, and then the complex solution of the original equation can be found:

```
>> A=[1 2 3;4 5 6; 7 8 0]; % input the matrices
   C=-[1+1i, 3+3i, 12+10i; 2+5i, 6, 11+6i; 5+2i, 11+1i, 2+12i];
   A0=sym(kron(eye(3),A)+kron(A,eye(3)));
   c=C(:); x0=-inv(A0)*c; x=reshape(x0,3,3)
   norm(A*x+x*A'+C)          % validate the solutions
```

The analytical solution $x$ is given below, and it can be seen that there is no error in the solution

$$
x = \begin{bmatrix}
-5/102 + j1\,457/918 & 15/17 - j371/459 & -61/306 + j166/459 \\
4/17 - j626/459 & -10/51 + j160/459 & 115/153 + j607/459 \\
-55/306 + j166/459 & -26/153 - j209/459 & 203/153 + j719/918
\end{bmatrix}.
$$

### 5.3.4 Stein equations

**Definition 5.8.** The general form of Stein equation is

$$AXB - X + Q = 0, \tag{5.3.3}$$

where $A \in \mathscr{C}^{n \times n}$, $B \in \mathscr{C}^{m \times m}$, $X, Q \in \mathscr{C}^{n \times m}$.

**Theorem 5.15.** *Stein equation can be converted into the linear equation*

$$(I_{nm} - B^H \otimes A)x = q, \tag{5.3.4}$$

*where $q$ and $x$ are respectively the column-wise expanded column vectors of matrices $Q$ and $X$, i. e., $q = \mathrm{vec}(Q)$, $x = \mathrm{vec}(X)$.*

**Example 5.23.** Solve analytically Stein equation

$$
\begin{bmatrix}
-2 & 2 & 1 \\
-1 & 0 & -1 \\
1 & -1 & 2
\end{bmatrix} X
\begin{bmatrix}
-2 & -1 & 2 \\
1 & 3 & 0 \\
3 & -2 & 2
\end{bmatrix} - X +
\begin{bmatrix}
0 & -1 & 0 \\
-1 & 1 & 0 \\
1 & -1 & -1
\end{bmatrix} = 0.
$$

**Solutions.** The following statements can be used to directly solve the equation:

```
>> A=[-2,2,1;-1,0,-1;1,-1,2]; B=[-2,-1,2;1,3,0;3,-2,2];
   Q=[0,-1,0;-1,1,0;1,-1,-1]; x=inv(sym(eye(9))-kron(B.',A))*Q(:);
   X=reshape(x,3,3), norm(A*X*B-X+Q)
```

The following analytical solution can be obtained:

$$
X = \begin{bmatrix}
4\,147/47\,149 & 3\,861/471\,490 & -40\,071/235\,745 \\
-2\,613/94\,298 & 2\,237/235\,745 & -43\,319/235\,745 \\
20\,691/94\,298 & 66\,191/235\,745 & -10\,732/235\,745
\end{bmatrix}.
$$

### 5.3.5 Discrete Lyapunov equation

**Definition 5.9.** Discrete Lyapunov equation is expressed as

$$AXA^H - X + Q = 0, \tag{5.3.5}$$

where $A, Q, X \in \mathscr{C}^{n \times n}$.

**Theorem 5.16.** *The Kronecker-product form of discrete Lyapunov equation is expressed as*

$$(I_{n^2} - A \otimes A)x = q, \tag{5.3.6}$$

*where* $x = \text{vec}(X)$, $q = \text{vec}(Q)$ *are expanded vectors of matrices* $X$ *and* $Q$.

This equation is a special case of Stein equation. It can be numerically solved with function dlyap() provided in Control System Toolbox. The syntax of the function is $X$=dlyap$(A,Q)$.

**Example 5.24.** Solve the discrete Lyapunov equation

$$\begin{bmatrix} 8 & 1 & 6 \\ 3 & 5 & 7 \\ 4 & 9 & 2 \end{bmatrix} X \begin{bmatrix} 8 & 1 & 6 \\ 3 & 5 & 7 \\ 4 & 9 & 2 \end{bmatrix}^T - X + \begin{bmatrix} 16 & 4 & 1 \\ 9 & 3 & 1 \\ 4 & 2 & 1 \end{bmatrix} = 0.$$

**Solutions.** The equation can be solved directly with function dlyap() using

```
>> A=[8,1,6; 3,5,7; 4,9,2]; Q=[16,4,1; 9,3,1; 4,2,1];
   X=dlyap(A,Q), norm(A*X*A'-X+Q)   % precision validation
```

and the numerical solution is obtained as follows. The error of the solution is $1.7909 \times 10^{-14}$ and

$$X = \begin{bmatrix} -0.1647 & 0.0691 & -0.0168 \\ 0.0528 & -0.0298 & -0.0062 \\ -0.102 & 0.045 & -0.0305 \end{bmatrix}.$$

With Theorem 5.15, the analytical solution of discrete Lyapunov equation

```
>> x=inv(sym(eye(9))-kron(A,A))*Q(:);
   X=reshape(x,3,3), norm(A*X*A'-X+Q)
```

and the analytical solution can be obtained below, with the norm of the error matrix being equal to zero and

$$X = \begin{bmatrix} -22\,912\,341/139\,078\,240 & 48\,086\,039/695\,391\,200 & -11\,672\,009/695\,391\,200 \\ 36\,746\,487/695\,391\,200 & -20\,712\,201/695\,391\,200 & -4\,279\,561/695\,391\,200 \\ -70\,914\,857/695\,391\,200 & 31\,264\,087/695\,391\,200 & -4\,247\,541/139\,078\,240 \end{bmatrix}.$$

## 5.4 Sylvester equations

Observing the Lyapunov equation, it can be seen that the left-hand side is composed of two terms, $AX$ and $XA^H$. If the latter term is replaced with the free term $XB$, the equation can be extended into a Sylvester equation. Sylvester equation is named after British mathematician James Joseph Sylvester (1814–1897). In this section, numerical and analytical solutions of Sylvester equations are presented.

### 5.4.1 Numerical solutions of Sylvester equations

**Definition 5.10.** The mathematic form of Sylvester equation is

$$AX + XB = -C, \tag{5.4.1}$$

where $A \in \mathscr{C}^{n \times n}$, $B \in \mathscr{C}^{m \times m}$, $C, X \in \mathscr{C}^{n \times m}$. The equation is also known as a generalized Lyapunov equation.

The function $\mathrm{lyap}()$ in Control System Toolbox can be used to solve Sylvester equation, with $X=\mathrm{lyap}(A,B,C)$. Schur decomposition is used in finding the numerical solutions. Besides, MATLAB function $\mathrm{sylvester}()$ can also be used to solve Sylvester equations numerically, employing the command $X=\mathrm{sylvester}(A,B,-C)$. Note that $-C$ is used here, since the standard form of the equation is $AX + XB = C$.

**Example 5.25.** Solve the Sylvester equation

$$\begin{bmatrix} 8 & 1 & 6 \\ 3 & 5 & 7 \\ 4 & 9 & 2 \end{bmatrix} X + X \begin{bmatrix} 16 & 4 & 1 \\ 9 & 3 & 1 \\ 4 & 2 & 1 \end{bmatrix} = \begin{bmatrix} 1 & 2 & 3 \\ 4 & 5 & 6 \\ 7 & 8 & 0 \end{bmatrix}.$$

**Solutions.** Function $\mathrm{lyap}()$ can be used in finding the numerical solutions

```
>> A=[8,1,6; 3,5,7; 4,9,2]; B=[16,4,1; 9,3,1; 4,2,1]; % input matrices
   C=-[1,2,3; 4,5,6; 7,8,0]; X=lyap(A,B,C)
   norm(A*X+X*B+C) % validating the solution
```

The numerical solutions obtained are as follows. It can be seen that the error norm is $7.5409 \times 10^{-15}$ and

$$X = \begin{bmatrix} 0.074871873700251 & 0.089913433762636 & -0.432920003296282 \\ 0.008071644736313 & 0.481441768049986 & -0.216033912855526 \\ 0.019577082629844 & 0.182643828725430 & 1.157921439176529 \end{bmatrix}.$$

Function $\mathrm{sylvester}()$ can also be used, and the error is $9.6644 \times 10^{-15}$. Note the calling syntax:

```
>> x=lyapsym(A,B,C), norm(A*x+x*B+C)
   X=sylvester(A,B,-C), norm(A*X+X*B+C)
```

### 5.4.2 Analytical solution of Sylvester equation

If the analytical solution of the Sylvester equation is expected, similar to the case in Lyapunov equation, Kronecker product can be used.

**Theorem 5.17.** *Sylvester equation can be converted to the following:*

$$(I_m \otimes A + B^T \otimes I_n)x = -c \tag{5.4.2}$$

*where* $c = \text{vec}(C)$ *and* $x = \text{vec}(X)$ *are column-wise expanded column vectors of the respective matrices.*

*If* $(I_m \otimes A + B^T \otimes I_n)$ *is nonsingular, Sylvester equation has a unique solution.*

Note that in (5.4.2) the direct transpose is used, rather than Hermitian one.

With the above algorithm, a universal analytical solver of Sylvester equation can be implemented in `lyapsym()` function:

```
function X=lyapsym(A,B,C)
if nargin==2, C=B; B=A'; end % If two input arguments
[nr,nc]=size(C); A0=kron(eye(nc),A)+kron(B.',eye(nr));
if rank(A0)==nr*nc, x0=-inv(A0)*C(:); X=reshape(x0,nr,nc);
else, error('singular matrix found.'), end
```

**Theorem 5.18.** *Consider the discrete Lyapunov equation in (5.3.5). If both sides are multiplied by* $(A^H)^{-1}$, *the equation can be rewritten as*

$$AX + X[-(A^H)^{-1}] = -Q(A^H)^{-1}. \tag{5.4.3}$$

Letting $B = -(A^H)^{-1}$, $C = Q(A^H)^{-1}$, the discrete Lyapunov equation can be converted into Sylvester equation in (5.4.1). Therefore, function `lyapsym()` can be used to find its analytical solutions.

The syntaxes of the function are

$X$=lyapsym(sym($A$),$C$), %continuous Lyapunov equation

$X$=lyapsym(sym($A$),-inv($A'$),$Q$*inv($A'$)), %discrete equation

$X$=lyapsym(sym($A$),-inv($B$),$Q$*inv($B$)), % Stein equation

$X$=lyapsym(sym($A$),$B$,$C$), %Sylvester equation

**Example 5.26.** Solve the Sylvester equation in Example 5.25.

**Solutions.** With function `lyapsym()`, the analytical solution can be obtained

```
>> A=[8,1,6; 3,5,7; 4,9,2]; B=[16,4,1; 9,3,1; 4,2,1]; % known matrices
   C=-[1,2,3; 4,5,6; 7,8,0]; x=lyapsym(sym(A),B,C)
   norm(A*x+x*B+C) % validate the results
```

The solution of the equation is as follows. It can be shown that the solution is analytical:

$$x = \begin{bmatrix} 1\,349\,214/18\,020\,305 & 648\,107/7\,208\,122 & -15\,602\,701/36\,040\,610 \\ 290\,907/36\,040\,610 & 3\,470\,291/7\,208\,122 & -3\,892\,997/18\,020\,305 \\ 70\,557/3\,604\,061 & 1\,316\,519/7\,208\,122 & 8\,346\,439/7\,208\,122 \end{bmatrix}.$$

Of course, function lyapsym() can still be used to solve Sylvester equations numerically.

**Example 5.27.** Solve analytically the discrete Lyapunov equation in Example 5.24.

**Solutions.** The following statements can be used to solve the equation:

```
>> A=[8,1,6; 3,5,7; 4,9,2]; Q=[16,4,1; 9,3,1; 4,2,1]; % given matrices
   x=lyapsym(sym(A),-inv(A'),Q*inv(A'))
   norm(A*x*A'-x+Q) % validate the results
```

The analytical solution of the equation found is exactly the same as that obtained in Example 5.24.

**Example 5.28.** Solve the Sylvester equation with the following matrices:

$$A = \begin{bmatrix} 8 & 1 & 6 \\ 3 & 5 & 7 \\ 4 & 9 & 2 \end{bmatrix}, \quad B = \begin{bmatrix} 2 & 3 \\ 4 & 5 \end{bmatrix}, \quad C = -\begin{bmatrix} 1 & 2 \\ 3 & 4 \\ 5 & 6 \end{bmatrix}.$$

**Solutions.** It is not necessary to assume that matrix $C$ in Sylvester equation is square. The following statements can be used to compute the numerical solution of the equation. Also, the new function lyapsym() can be used in finding the analytical solution:

```
>> A=[8,1,6; 3,5,7; 4,9,2]; B=[2,3; 4,5]; C=-[1,2; 3,4; 5,6]
   X=lyapsym(sym(A),B,C), norm(A*X+X*B+C) % analytical solution
```

The result obtained is as follows. It can be validated that the solution obtained is analytical:

$$X = \begin{bmatrix} -2\,853/14\,186 & -11\,441/56\,744 \\ -557/14\,186 & -8\,817/56\,744 \\ 9\,119/14\,186 & 50\,879/56\,744 \end{bmatrix}.$$

Function solve() can also be used in MATLAB to symbolically solve the matrix equations. The solution procedure is rather complicated, and an example is given below to demonstrate it.

**Example 5.29.** Solve again the equation in Example 5.28, and validate the results.

**Solutions.** The following statements can be used to solve the matrix equation once more, and the analytical solution can be obtained, which is exactly the same as that in Example 5.28.

```
>> A=[8,1,6; 3,5,7; 4,9,2]; B=[2,3; 4,5]; C=-[1,2; 3,4; 5,6];
   X=sym('x%d%d',[3,2]); X=solve(A*X+X*B==-C,X)
   X=[X.x11 X.x12; X.x21 X.x22; X.x31 X.x32], A*X+X*B+C
```

**Example 5.30.** Solve the Sylvester equation and validate the result for

$$A = \begin{bmatrix} 5 & 0 & 2 \\ 2 & 0 & 4 \\ 3 & 2 & 6 \end{bmatrix}, \quad B = \begin{bmatrix} 1 & 3 & 2 \\ 0 & 0 & 2 \\ 2 & 6 & 0 \end{bmatrix}, \quad C = \begin{bmatrix} 0 & 2 & 1 \\ 0 & 2 & 1 \\ 1 & 0 & 0 \end{bmatrix}.$$

**Solutions.** The matrices should be input in MATLAB, and the rank can be obtained. It can be seen that $B$ is a singular matrix. The following statements can be entered to solve the given Sylvester equation:

```
>> A=[5,0,2; 2,0,4; 3,2,6]; A=sym(A);
   B=[1,3,2; 0,0,2; 2,6,0]; rank(A), rank(B)
   C=[0,2,1; 0,2,1; 1,0,0];
   X=lyapsym(A,B,C), norm(A*X+X*B+C)
```

The solutions are obtained below, and it can be validated that the original equation is satisfied. It can be seen from the example that if $A$ and $B$ are not both singular, the Sylvester solution can be solved, with the same analytical solution:

$$X = \begin{bmatrix} 1/13 & -1/52 & -7/26 \\ -45/52 & 21/52 & 29/104 \\ 1/26 & -27/104 & 3/26 \end{bmatrix}.$$

### 5.4.3 Sylvester equations with constant parameters

If the original matrices contain symbolic variables, the functions `lyap()` and `sylvester()` provided in MATLAB cannot be used in solving the equations. The new function `lyapsym()` should be used instead to find the solutions. An example is given to demonstrate the analytical solution of the Sylvester equation.

**Example 5.31.** If $b_{21}$ is set to $b_{21} = a$ in Example 5.28, solve Sylvester equation again.

**Solutions.** If the (2,1)th element in matrix $B$ is changed to a symbolic variable $a$, then the following statements can be used to solve the Sylvester equation:

```
>> syms a real; A=[8,1,6; 3,5,7; 4,9,2];
   B=[2,3; a,5]; C=-[1,2; 3,4; 5,6];
   X=simplify(lyapsym(A,B,C)), norm(A*X+X*B+C)
```

The result obtained is as follows:

$$X = \frac{1}{\Delta} \begin{bmatrix} 6(3a^3 + 155a^2 - 2\,620a + 200) & -(513a^2 - 10\,716a + 80\,420) \\ 4(9a^3 - 315a^2 + 314a + 980) & -3(201a^2 - 7\,060a + 36\,780) \\ 2(27a^3 - 1\,869a^2 + 25\,472a - 760) & -477a^2 + 4\,212a + 194\,300 \end{bmatrix},$$

where $\Delta = 27a^3 - 3\,672a^2 + 69\,300a + 6\,800$. Therefore, when $\Delta = 0$, the equation has no solutions.

**Example 5.32.** Solve again the Sylvester equation in Example 5.31.

**Solutions.** The following MATLAB statements can also be used to solve the Sylvester equation with symbolic variables, and the result is exactly the same as that obtained in the previous example:

```
>> syms a real; A=[8,1,6; 3,5,7; 4,9,2];
   B=[2,3; a,5]; C=-[1,2; 3,4; 5,6];
   X=sym('x%d%d',[3,2]); X=solve(A*X+X*B==-C,X)
   X=[X.x11 X.x12; X.x21 X.x22; X.x31 X.x32]
   simplify(A*X+X*B+C)
```

### 5.4.4 Multiterm Sylvester equations

In the Sylvester equations discussed earlier, there is only one term $AX + XB$. More generally, multiterm Sylvester equation can also be solved with Kronecker product method.

**Definition 5.11.** Multiterm Sylvester equation is expressed as

$$A_1 X + X B_1 + A_2 X + X B_2 + \cdots + A_k X + X B_k = -C, \tag{5.4.4}$$

where $A_i \in \mathscr{C}^{n \times n}$, $B_i \in \mathscr{C}^{m \times m}$, $i = 1, 2, \ldots, k$, and $C, X \in \mathscr{C}^{n \times m}$.

In fact, multiterm Sylvester equation can be understood as the simultaneous equations $A_i X + X B_i = -C_i$, $i = 1, 2, \ldots, k$, where $C = C_1 + C_2 + \cdots + C_k$.

**Theorem 5.19.** *Multiterm Sylvester equation can be converted to the following linear equation:*

$$[I_m \otimes (A_1 + A_2 + \cdots + A_k) + (B_1 + B_2 + \cdots + B_k)^{\mathrm{T}} \otimes I_n]x = -c, \tag{5.4.5}$$

*where $x = \mathrm{vec}(X)$ and $c = \mathrm{vec}(C)$ are column-wise expanded column vectors of matrices $X$ and $C$, respectively.*

**Example 5.33.** If the matrices are given in Example 5.17, find the analytical solution of multiterm Sylvester equation.

**Solutions.** The following commands can be used to directly solve the multiterm Sylvester equations:

```
>> A1=[2,0; 2,2]; A2=[1,0; 0,1]; A1=sym(A1);
   C=-[17,9,9,7; 21,17,13,13]; c=C(:);
   B1=[2,2,1,1; 2,1,2,0; 0,2,2,2; 2,0,2,2];
   B2=[2,1,0,2; 2,0,0,0; 2,2,0,2; 1,0,2,0];
   c=C(:); x=inv(kron(eye(4),A1+A2)+kron((B1+B2).',eye(2)))*c;
   X=reshape(x,2,4), A1*X+X*B1+A2*X+X*B2-C
```

The validated analytical solution can be obtained as

$$X = \begin{bmatrix} -14/33 & -31/12 & 43/66 & -5/3 \\ 391/2178 & -1487/396 & 721/1089 & -509/198 \end{bmatrix}.$$

## 5.5 Nonlinear matrix equations

The algebraic equations discussed so far were essentially linear equations. Therefore numerical and analytical solutions may all be computed. In this section, nonlinear matrix equations are considered. Riccati algebraic equation is presented first, and then variations of Riccati equations are also discussed. Nonlinear matrix equations will be explored, aiming at finding all or as many as possible solutions of such equations.

### 5.5.1 Riccati equations

In optimal control theory, Riccati algebraic equation is introduced, with some algorithms. In this section, numerical solutions of Riccati equations are mainly discussed.

**Definition 5.12.** Riccati algebraic equation is a well known quadratic equation, whose mathematical form is given by

$$A^H X + XA - XBX + C = 0. \tag{5.5.1}$$

Riccati equation is named after Italian mathematician Jacopo Francesco Riccati (1676–1754).

Since there is a quadratic form of matrix $X$ in the equation, the solutions of Riccati equations are much more complicated than for the Lyapunov equations discussed earlier.

**Example 5.34.** For a $2 \times 2$ matrix Riccati equation, derive the corresponding simultaneous algebraic equations.

**Solutions.** Similar to the case in Example 5.20, the simultaneous algebraic equations can be derived with the following statements:

```
>> A=sym('a%d%d',2); assume(A,'real');
   syms x1 x2 x3 x4 c1 c2 c3 c4; B=sym('b%d%d',2);
   X=[x1 x3; x2 x4]; C=[c1 c3; c2 c4];
   simplify(A'*X+X*A-X*B*X+C)
```

The following algebraic equations can be found:

$$\begin{cases} c_1 + 2a_{11}x_1 + a_{21}x_2 + a_{21}x_3 - x_1(b_{11}x_1 + b_{21}x_3) - x_2(b_{12}x_1 + b_{22}x_3) = 0, \\ c_2 + a_{12}x_1 + (a_{11} + a_{22})x_2 + a_{12}x_4 - x_1(b_{11}x_2 + b_{21}x_4) - x_2(b_{12}x_2 + b_{22}x_4) = 0, \\ c_3 + a_{12}x_1 + (a_{11} + a_{22})x_3 + a_{21}x_4 - x_3(b_{11}x_1 + b_{21}x_3) - x_4(b_{12}x_1 + b_{22}x_3) = 0, \\ c_4 + a_{12}x_2 + a_{12}x_3 + 2a_{22}x_4 - x_3(b_{11}x_2 + b_{21}x_4) - x_4(b_{12}x_2 + b_{22}x_4) = 0. \end{cases}$$

It can be seen that there are many $x_i x_j$ terms, and from the well-known Abel–Ruffini theorem one knows that there is no analytical solution for the Riccati equation of size higher than 2. Numerical solutions must be involved.

A MATLAB function `are()` is provided in the Control System Toolbox, which can be used to solve directly the algebraic equation in (5.5.1). The syntax of the function is $X$=`are`$(A,B,C)$.

**Example 5.35.** Consider the Riccati equation in (5.5.1), where

$$A = \begin{bmatrix} -2 & 1 & -3 \\ -1 & 0 & -2 \\ 0 & -1 & -2 \end{bmatrix}, \quad B = \begin{bmatrix} 2 & 2 & -2 \\ -1 & 5 & -2 \\ -1 & 1 & 2 \end{bmatrix}, \quad C = \begin{bmatrix} 5 & -4 & 4 \\ 1 & 0 & 4 \\ 1 & -1 & 5 \end{bmatrix}.$$

Compute the numerical solution of the equation and validate it.

**Solutions.** The following statements can be used to solve the equation directly:

```
>> A=[-2,1,-3; -1,0,-2; 0,-1,-2]; B=[2,2,-2; -1 5 -2; -1 1 2];
   C=[5 -4 4; 1 0 4; 1 -1 5]; X=are(A,B,C)
   norm(A'*X+X*A-X*B*X+C) % validate the result
```

The solution obtained is shown below, and when substituted back to the original equation, the norm of the error matrix is $1.4370 \times 10^{-14}$, indicating that the solution satisfies the original equation very well:

$$X = \begin{bmatrix} 0.987394908497906 & -0.798327696888299 & 0.418868996625638 \\ 0.577405649554727 & -0.130792336490926 & 0.577547768361485 \\ -0.284045000180512 & -0.073036978332803 & 0.692411488305715 \end{bmatrix}.$$

**Example 5.36.** If matrix $B$ in Example 5.35 is changed to the following one, solve again the Riccati equation, and validate the result:

$$B = \begin{bmatrix} 2 & 1 & -1 \\ 1 & 2 & 0 \\ -1 & 0 & -4 \end{bmatrix}.$$

**Solutions.** The following statements can be tried to solve the Riccati equation

```
>> A=[-2,1,-3; -1,0,-2; 0,-1,-2]; B=[2,1,-1; 1,2,0; -1,0,-4];
   C=[5 -4 4; 1 0 4; 1 -1 5]; X=are(A,B,C)
```

Unfortunately, an error message "No solution: $(A,B)$ may be uncontrollable or no solution exists" is displayed. It can be seen that matrix $B$ is not positive definite, so function are() failed to find a solution.

### 5.5.2 Solutions of general nonlinear equations

It is well known that a quadratic equation with one unknown usually has two roots. Riccati equations are also quadratic equations, but usually with more than one unknown. There are no theoretical results about how many solutions ordinary Riccati equations have, let alone what all the roots are.

Besides, if one of the solutions can be obtained with a certain algorithm, is it possible to find other solutions? Or is it possible to find all the solutions of the matrix equations? A general purpose and widely applicable nonlinear matrix equation solver is needed. In this section, general purpose numerical solvers of matrix equations are discussed.

A MATLAB function more_sols() is provided in [32] for finding as many as possible solutions of some nonlinear matrix equations.

```
function more_sols(f,X0,varargin)
[A,tol,tlim]=default_vals({1000,eps,30},varargin{:});
if length(A)==1, a=-0.5*A; b=0.5*A; else, a=A(1); b=A(2); end
ar=real(a); br=real(b); ai=imag(a); bi=imag(b);
ff=optimset; ff.Display='off'; [n,m,i]=size(X0);
ff1=ff; ff.TolX=tol; ff.TolFun=tol; X=X0;
try, err=evalin('base','err');
catch, err=0; end, if i<=1; err=0; end, tic
while (1), % infinite loop,    Ctrl+C keys to terminate the loop
    x0=ar+(br-ar)*rand(n,m); % generate initial search matrix
    if abs(imag(A))>1e-5, x0=x0+(ai+(bi-ai)*rand(n,m))*1i; end
    [x,aa,key]=fsolve(f,x0,ff1); t=toc; if t>tlim, break; end
```

```
    if key>0, N=size(X,3); % if the solution is recorded, discard it
        for j=1:N, if norm(X(:,:,j)-x)<1e-5; key=0; break; end, end
        if key>0, [x1,aa,key]=fsolve(f,x,ff); % if new, refine it
            if norm(x-x1)<1e-5 & key>0; X(:,:,i+1)=x1; % record it
                assignin('base','X',X); err=max([norm(aa),err]);
                assignin('base','err',err); i=i+1, tic % update information
end, end, end, end
```

The idea of the function is to generate an initial random matrix from the region of interest, and solve the equations with a searching algorithm. If a solution is found, then it should be checked whether the solution has been recorded. If it is, discard it, otherwise record the solution and continue the solution process. Repeat the solution process until all the solutions are found. If no new solutions are found within a certain period of time, terminate the solution process.

The syntax of the function more_sols() is

more_sols$(f, X_0, A, \epsilon, t_{\lim})$

where $\epsilon$ is the error tolerance, with a default value of eps; the default value of $t_{\lim}$ is 30, meaning that if no new solution is found within 30 seconds, the solver is terminated. Alternatively, the user may terminate the solution process at any time by pressing the Ctrl+C keys.

The low-level function default_vals() is given below

```
function varargout=default_vals(vals,varargin)
if nargout~=length(vals), error('number of arguments mismatch');
else, nn=length(varargin)+1;
    varargout=varargin; for i=nn:nargout, varargout{i}=vals{i};
end, end, end
```

and it can be used to assign default values of input arguments.

In the syntax, $A$ indicates the region of interested, which can be flexibly selected. If $A$ is a scalar, then the region of interest for all the unknowns is $(-A/2, A/2)$, with a default value of $A = 1\,000$, indicating a search in a large interval; if $A$ is a vector $[a, b]$, the region of interest is $[a, b]$; while if $A$ is complex, complex solutions are also expected.

The solution of the equations is returned as a three-dimensional array $X$, with $X(:,:,i)$ storing the $i$th solution.

Compared with other functions, this one is special. An infinite loop structure is used, and one may terminate the execution of the function at any time by pressing Ctrl+C keys. If so, normal returned arguments cannot be ensured. Therefore, it is not appropriate to write the returned arguments in the leading sentence. Instead, if a new solution is found, the three-dimensional array is written back to MATLAB workspace with the `assignin()` function. If there is no new solution found within certain time, the function is completed normally. Function `evalin()` is also used in the function to load the variables in MATLAB workspace into the function. An alternative MATLAB variable `err` is used to store the maximum norm of the errors.

If one wants to continue finding other solutions after interruption, and in MATLAB workspace, the three-dimensional array $X$ is not cleared, the command `more_sols(f,X)` can be used.

**Example 5.37.** Consider the Riccati equation in Example 5.35. Find all of its solutions.

**Solutions.** With function `more_sols()`, the solution of Riccati equation is rather simple. The original matrix equation can be expressed directly with anonymous function in MATLAB, then the following statements can be employed directly:

```
>> A=[-2,1,-3; -1,0,-2; 0,-1,-2]; B=[2,2,-2; -1 5 -2; -1 1 2];
   C=[5 -4 4; 1 0 4; 1 -1 5]; f=@(X)A'*X+X*A-X*B*X+C
   more_sols(f,zeros(3,3,0),1000), X, err
```

All eight real solutions of the matrix equation can be found directly, with a maximum error of $1.1237 \times 10^{-12}$. Please note that since random initial points were assigned, the orders of the solutions may be different each time the function is called:

$$X_1 = \begin{bmatrix} 0.9874 & -0.7983 & 0.4189 \\ 0.5774 & -0.1308 & 0.5775 \\ -0.2840 & -0.0730 & 0.6924 \end{bmatrix}, \quad X_2 = \begin{bmatrix} 1.2213 & -0.4165 & 1.9775 \\ 0.3578 & -0.4894 & -0.8863 \\ -0.7414 & -0.8197 & -2.3560 \end{bmatrix},$$

$$X_3 = \begin{bmatrix} 0.6665 & -1.3223 & -1.720 \\ 0.3120 & -0.5640 & -1.191 \\ -1.2273 & -1.6129 & -5.594 \end{bmatrix}, \quad X_4 = \begin{bmatrix} -2.1032 & 1.2978 & -1.9697 \\ -0.2467 & -0.3563 & -1.4899 \\ -2.1494 & 0.7190 & -4.5465 \end{bmatrix},$$

$$X_5 = \begin{bmatrix} -0.1538 & 0.1087 & 0.4623 \\ 2.0277 & -1.7437 & 1.3475 \\ 1.9003 & -1.7513 & 0.5057 \end{bmatrix}, \quad X_6 = \begin{bmatrix} 0.8878 & -0.9609 & -0.2446 \\ 0.1072 & -0.8984 & -2.5563 \\ -0.0185 & 0.3604 & 2.4620 \end{bmatrix},$$

$$X_7 = \begin{bmatrix} 23.9467 & -20.6673 & 2.4529 \\ 30.1460 & -25.9830 & 3.6699 \\ 51.9666 & -44.9108 & 4.6410 \end{bmatrix}, \quad X_8 = \begin{bmatrix} -0.7619 & 1.3312 & -0.8400 \\ 1.3183 & -0.3173 & -0.1719 \\ 0.6371 & 0.7885 & -2.1996 \end{bmatrix}.$$

**Example 5.38.** Solve the Riccati equation in Example 5.35 again, and find all the solutions, including complex ones.

**Solutions.** If the argument $A$ is a real number, then real matrix solutions are found, however, if $A$ is a complex number, for instance, letting $A = 1000 + 1000\text{j}$, the real and imaginary axes are both set to the interval of $[-500, 500]$. The following statements can be written, and a total of 20 solutions can be found:

```
>> A=[-2,1,-3; -1,0,-2; 0,-1,-2]; B=[2,2,-2; -1 5 -2; -1 1 2];
   C=[5 -4 4; 1 0 4; 1 -1 5]; f=@(X)A'*X+X*A-X*B*X+C
   more_sols(f,zeros(3,3,0),1000+1000i), X, err
```

The maximum error obtained is $3.6088 \times 10^{-13}$. Since the result is quite complicated, only one of the complex solution is shown below:

$$X_9 = \begin{bmatrix} -0.1767 + 0.3472j & 0.1546 - 0.6981j & 0.4133 + 0.7436j \\ 1.6707 - 0.3261j & -1.0258 + 0.6557j & 0.5828 - 0.6984j \\ 1.9926 - 0.6791j & -1.9368 + 1.3654j & 0.7033 - 1.4544j \end{bmatrix}.$$

**Example 5.39.** An attempt was made in Example 5.36 to solve a Riccati equation, when **B** is not positive definite, but it failed. Does this mean that the Riccati equation has no solutions?

**Solutions.** Although are() function call failed, function more_sols() can be used in the same way as in the previous example:

```
>> A=[-2,1,-3; -1,0,-2; 0,-1,-2]; B=[2,1,-1; 1,2,0; -1,0,-4];
   C=[5 -4 4; 1 0 4; 1 -1 5]; f=@(X)A'*X+X*A-X*B*X+C;
   more_sols(f,zeros(3,3,0),1000), X, err
```

It is found that the equation has eight real solutions again, with the maximum error of $1.6118 \times 10^{-14}$:

$$X_1 = \begin{bmatrix} -1.0819 & 0.9858 & 1.0125 \\ 2.3715 & -2.2627 & 1.4691 \\ -0.5022 & 0.5495 & -0.2243 \end{bmatrix}, \quad X_2 = \begin{bmatrix} 0.9674 & -2.6868 & 4.4395 \\ 0.1991 & 1.6306 & -2.1639 \\ -0.5083 & 0.5604 & -0.2345 \end{bmatrix},$$

$$X_3 = \begin{bmatrix} 0.9397 & -0.6723 & 0.1514 \\ 0.2284 & -0.5050 & 2.3819 \\ -0.5082 & 0.5544 & -0.2217 \end{bmatrix}, \quad X_4 = \begin{bmatrix} 1.3202 & 5.0545 & 2.8973 \\ 2.2258 & 0.0431 & 0.7047 \\ 3.0469 & 3.1350 & 3.4330 \end{bmatrix},$$

$$X_5 = \begin{bmatrix} 5.5119 & -4.2335 & 0.1801 \\ 3.8740 & -3.6091 & -0.3637 \\ 6.6868 & -4.9302 & 1.0736 \end{bmatrix}, \quad X_6 = \begin{bmatrix} 4.4723 & -1.9300 & 0.8540 \\ 2.3535 & -0.2399 & 0.6220 \\ 5.3631 & -1.9973 & 1.9316 \end{bmatrix},$$

$$X_7 = \begin{bmatrix} 6.2488 & -5.8664 & -0.2976 \\ 3.4292 & -2.6235 & -0.0754 \\ 8.3125 & -8.5326 & 0.0197 \end{bmatrix}, \quad X_8 = \begin{bmatrix} -6.9247 & 4.7691 & -7.0558 \\ 8.5655 & -6.2734 & 10.0220 \\ -0.4848 & 0.5382 & -0.2003 \end{bmatrix}.$$

If complex solutions are expected, the following statements can be used, and the number of solutions is again 20, with the maximum error of $1.1709 \times 10^{-14}$:

```
>> more_sols(f,X,1000+1000i); X, err
```

The `vpasolve()` function in MATLAB can be used to solve some quadratic equations under the symbolic framework. The difference between this function and `solve()` is that it can be used to find high precision numerical solutions. An example is given to demonstrate the use of the function in solving Riccati equations.

**Example 5.40.** Compute high precision solutions for the equation in Example 5.39.

**Solutions.** An unknown matrix should be defined first, then the symbolic expression of the equation can be established, from which the equation can be solved directly with function `solve()`. After 77.03 seconds of waiting, all the 20 complex solutions can be obtained. It can be seen that the method cannot be used to find real solutions alone:

```
>> A=[-2,1,-3; -1,0,-2; 0,-1,-2]; B=[2,1,-1; 1,2,0; -1,0,-4];
   C=[5 -4 4; 1 0 4; 1 -1 5]; X=sym('x%d%d',[3 3]);
   F=A'*X+X*A-X*B*X+C; tic, X1=vpasolve(F), toc
```

Letting $k = 3$, the third solution matrix can be extracted. When substituted back into the original equation, it can be seen that the error is $3.0099 \times 10^{-35}$, much smaller than that obtained under the double precision framework:

```
>> k=3;
   X2=[X1.x11(k),X1.x12(k),X1.x13(k);
       X1.x21(k),X1.x22(k),X1.x23(k);
       X1.x31(k),X1.x32(k),X1.x33(k)];
   F0=A'*X2+X2*A-X2*B*X2+C, norm(F0)
```

### 5.5.3 Variations of Riccati equations

If the classical Riccati equation is extended, the following variations of it can be achieved. In this section, we concentrate on a sample of these equations.

**Definition 5.13.** Generalized Riccati equation is expressed as

$$AX + XD - XBX + C = 0. \tag{5.5.2}$$

**Definition 5.14.** Quasi-Riccati equation is mathematically given by

$$AX + XD - XBX^{\mathrm{H}} + C = 0. \tag{5.5.3}$$

These variations of Riccati equations can be solved directly with function `more_sols()`. Normally, all the numerical solutions can be obtained.

**Example 5.41.** Solve the quasi-Riccati equation $AX + XD - XBX^H + C = 0$. Find all the solutions and validate the results, where

$$A = \begin{bmatrix} 2 & 1 & 9 \\ 9 & 7 & 9 \\ 6 & 5 & 3 \end{bmatrix}, \quad B = \begin{bmatrix} 0 & 3 & 6 \\ 8 & 2 & 0 \\ 8 & 2 & 8 \end{bmatrix}, \quad C = \begin{bmatrix} 7 & 0 & 3 \\ 5 & 6 & 4 \\ 1 & 4 & 4 \end{bmatrix}, \quad D = \begin{bmatrix} 3 & 9 & 5 \\ 1 & 2 & 9 \\ 3 & 3 & 0 \end{bmatrix}.$$

**Solutions.** The following commands can be used to find the solutions of the equation. All 16 real solutions can be found, with the maximum error of $2.713 \times 10^{-13}$:

```
>> A=[2 1 9; 9 7 9; 6 5 3]; B=[0 3 6; 8 2 0; 8 2 8];
   C=[7 0 3; 5 6 4; 1 4 4]; D=[3 9 5; 1 2 9; 3 3 0];
   f=@(X)A*X+X*D-X*B*X'+C; more_sols(f,zeros(3,3,0)); X, err
```

If complex solutions are also expected, the following commands can be issued. However, no more new solutions are found, meaning that for this equation, there are only real solutions.

```
>> more_sols(f,X,1000+1000i); X, err
```

If the equation is changed to $AX + XD - XBX^T + C = 0$, the following statements can be used. It can be seen that all 38 complex solutions can be obtained. It can be seen that if the form of the equation is changed, the description can be modified with no new effort. The new equation can be directly solved.

```
>> A=[2 1 9; 9 7 9; 6 5 3]; B=[0 3 6; 8 2 0; 8 2 8];
   C=[7 0 3; 5 6 4; 1 4 4]; D=[3 9 5; 1 2 9; 3 3 0];
   f=@(X)A*X+X*D-X*B*X.'+C;
   more_sols(f,zeros(3,3,0),1000+1000i); X, err
```

### 5.5.4 Numerical solutions of nonlinear matrix equations

**Definition 5.15.** A nonlinear matrix equation is expressed as $F(X) = 0$.

A nonlinear matrix equation can be described by an anonymous function, or by other function format, then the equation can be solved directly with function `more_sols()`. Note that when the equation is described, only one input argument is allowed. Sometimes, not all solutions are obtained in one run of the solver, so the function `more_sols(f, X)` can be called again and again to find more solutions.

**Example 5.42.** For the given nonlinear matrix equation

$$AX^3 + X^4D - X^2BX + CX - I = 0,$$

assume that the matrices **A**, **B**, **C**, and **D** are given in Example 5.41. Find as many solutions as possible.

**Solutions.** The following statements can be used to solve the equation directly, and with repeated calls and validation, it can be found that there are 70 real solutions. If complex solutions are also expected, thousands of solutions may be found. So far, 3 322 complex solutions are obtained and saved in file data542c.mat.

```
>> A=[2 1 9; 9 7 9; 6 5 3]; B=[0 3 6; 8 2 0; 8 2 8];
   C=[7 0 3; 5 6 4; 1 4 4]; D=[3 9 5; 1 2 9; 3 3 0];
   f=@(X)A*X^3+X^4*D-X^2*B*X+C*X-eye(3);
   more_sols(f,zeros(3,3,0)); X, err
```

The following statements can be used to continuously solve the equation, and new solutions can be found. The function may be called thousands of times to find more solutions, if they exist:

```
>> load data542c; more_sols(f,X,50+50i);
```

**Example 5.43.** Solve the following nonlinear matrix equation

$$e^{0.05AX} \sin BX - CX + D = 0,$$

where the matrices **A**, **B**, **C**, and **D** are given in Example 5.41, find all the real solutions of the nonlinear matrix equation.

**Solutions.** The following statements can be used to describe and solve numerically the complicated nonlinear matrix equation. After many trials, altogether 39 real solutions can be found, with the maximum error of $1.2136 \times 10^{-12}$:

```
>> A=[2 1 9; 9 7 9; 6 5 3]; B=[0 3 6; 8 2 0; 8 2 8];
   C=[7 0 3; 5 6 4; 1 4 4]; D=[3 9 5; 1 2 9; 3 3 0];
   f=@(X)expm(0.05*A*X)*funm(B*X,@sin)-C*X+D;
   more_sols(f,zeros(3,3,0),10); X, err
```

It can be seen from the example that function more_sols() can be used to solve matrix solutions of any complexity.

## 5.6 Diophantine equations

Matrix equations were the main topic in the chapter so far. In this section, polynomial equations are considered. The concept of polynomial coprimeness is presented first, then analytical solutions of Diophantine polynomial equations are presented.

### 5.6.1 Coprimeness of polynomials

**Definition 5.16.** For two polynomials $A(s)$ and $B(s)$, if the greatest common divisor is not a polynomial, the two polynomials are coprime.

Powerful facilities are provided in MATLAB for handling polynomials. Function gcd() can be used directly in finding the greatest common divisor of two symbolic polynomials, with $d$=gcd$(A,B)$. If the divisor $d$ is not a polynomial, the two polynomials are coprime, otherwise they are not coprime.

**Example 5.44.** Two polynomials are given as

$$A(s) = 2s^4 + 16s^3 + 36s^2 + 32s + 10, \quad B(s) = s^5 + 12s^4 + 55s^3 + 120s^2 + 124s + 48.$$

Test whether they are coprime or not. If they are not, what is the common factor?

**Solutions.** The two polynomials can be input into MATLAB as symbolic expressions. Then the greatest common divisor $d$ can be obtained. It can be found that $d = s + 1$, indicating the two polynomials are not coprime.

```
>> syms s; A=2*s^4+16*s^3+36*s^2+32*s+10;
   B=s^5+12*s^4+55*s^3+120*s^2+124*s+48;
   d=gcd(A,B)
```

### 5.6.2 Diophantine polynomial equations

**Definition 5.17.** Consider the following polynomial equation:

$$A(s)X(s) + B(s)Y(s) = C(s), \tag{5.6.1}$$

where $A(s)$, $B(s)$, and $C(s)$ are given polynomials

$$
\begin{aligned}
A(s) &= a_1 s^n + a_2 s^{n-1} + a_3 s^{n-2} + \cdots + a_n s + a_{n+1}, \\
B(s) &= b_1 s^m + b_2 s^{m-1} + b_3 s^{m-2} + \cdots + b_m s + b_{m+1}, \\
C(s) &= c_1 s^k + c_2 s^{k-1} + c_3 s^{k-2} + \cdots + c_k s + c_{k+1}.
\end{aligned}
\tag{5.6.2}
$$

Such a polynomial equation is referred to as a Diophantine equation.

Diophantine equation is named after ancient Greek mathematician Diophantus of Alexandria (second century CE). The original form was an indeterminate equation. Later the equation was extended to polynomial equations.

It is known from the orders of polynomials $A(s)$ and $B(s)$ that the orders of polynomials $X(s)$ and $Y(s)$ are respectively $m-1$ and $n-1$, and the polynomials are denoted as

$$
\begin{aligned}
X(s) &= x_1 s^{m-1} + x_2 s^{m-2} + x_3 s^{m-3} + \cdots + x_{m-1} s + x_m, \\
Y(s) &= y_1 s^{n-1} + y_2 s^{n-2} + y_3 s^{n-3} + \cdots + y_{n-1} s + y_n.
\end{aligned}
\tag{5.6.3}
$$

**Theorem 5.20.** *If $A(s)$ and $B(s)$ are coprime polynomials, the matrix form of Diophantine equation can be written as*

$$
\begin{bmatrix}
a_1 & 0 & \cdots & 0 & b_1 & 0 & \cdots & 0 \\
a_2 & a_1 & \ddots & 0 & b_2 & b_1 & \ddots & 0 \\
a_3 & a_2 & \ddots & 0 & b_3 & b_2 & \ddots & 0 \\
\vdots & \vdots & \ddots & a_1 & \vdots & \vdots & \ddots & b_1 \\
a_{n+1} & a_n & \ddots & a_2 & \cdot & \cdot & \ddots & b_2 \\
0 & a_{n+1} & \ddots & a_3 & \cdot & \cdot & \ddots & b_3 \\
\vdots & \vdots & \ddots & \vdots & \vdots & \vdots & \ddots & \vdots \\
0 & 0 & \cdots & a_{n+1} & 0 & 0 & \cdots & b_{m+1}
\end{bmatrix}
\begin{bmatrix}
x_1 \\ x_2 \\ \vdots \\ x_m \\ y_1 \\ y_2 \\ \vdots \\ y_n
\end{bmatrix}
=
\begin{bmatrix}
0 \\ 0 \\ \vdots \\ 0 \\ c_1 \\ c_2 \\ \vdots \\ c_{k+1}
\end{bmatrix}. \tag{5.6.4}
$$

$\underbrace{\phantom{aaaaaaaaaaa}}_{m \text{ columns}}\underbrace{\phantom{aaaaaaaaaaa}}_{n \text{ columns}}$

*The coefficient matrix in this equation is the transpose of a Sylvester matrix.*

It can be shown that when polynomials $A(s)$ and $B(s)$ are coprime, such Sylvester matrix is nonsingular, so that the equation has a unique solution. The coprimeness of the two polynomials is automatically checked.

The following MATLAB function can be used to create a Sylvester matrix:

```
function S=sylv_mat(A,B)
n=length(B)-1; m=length(A)-1; S=[];
A1=[A(:); zeros(n-1,1)]; B1=[B(:); zeros(m-1,1)];
for i=1:n, S=[S A1]; A1=[0; A1(1:end-1)]; end
for i=1:m, S=[S B1]; B1=[0; B1(1:end-1)]; end; S=S.';
```

Based on this function, the function `diophantine()` is written to solve Diophantine equation. The coprimeness check is embedded in the function. If the processed $C(s)$ is no longer a polynomial, an error message is given, indicating that there is no solution for the equation.

```
function [X,Y]=diophantine(A,B,C,x)
d=gcd(A,B); A=simplify(A/d); B=simplify(B/d); C=simplify(C/d);
A1=polycoef(A,x); B1=polycoef(B,x); C1=polycoef(C,x);
n=length(B1)-1; m=length(A1)-1; S=sylv_mat(A1,B1);
C2=zeros(n+m,1); C2(end-length(C1)+1:end)=C1(:); x0=inv(S.')*C2;
X=poly2sym(x0(1:n),x); Y=poly2sym(x0(n+1:end),x);
```

**Example 5.45.** For the following polynomials, solve the Diophantine equation:

$$
A(s) = s^4 - \frac{27s^3}{10} + \frac{11s^2}{4} - \frac{1\,249s}{1\,000} + \frac{53}{250},
$$

$$
B(s) = 3s^2 - \frac{6s}{5} + \frac{51}{25}, \quad C(s) = 2s^2 + \frac{3s}{5} - \frac{9}{25}.
$$

**Solutions.** The following statements can be used to solve directly Diophantine equation

```
>> syms s; A=s^4-27*s^3/10+11*s^2/4-1249*s/1000+53/250;
   B=3*s^2-6*s/5+51/25; C=2*s^2+3*s/5-9/25;        % three polynomials
   [X,Y]=diophantine(A,B,C,s), simplify(A*X+B*Y-C) % solve and check
```

The solutions obtained are as follows, and when substituted back to the original Diophantine equation, it can be seen that the error is zero. It can be shown that the solutions obtained are correct:

$$X(s) = \frac{4\,280\,s}{4\,453} + \frac{9\,480}{4\,453},$$

$$Y(s) = -\frac{4\,280s^3}{13\,359} + \frac{364s^2}{13\,359} + \frac{16\,882s}{13\,359} - \frac{1771}{4\,453}.$$

If the Diophantine equation is changed to

$$A(s)X(s) + s^d B(s)Y(s) = C(s), \tag{5.6.5}$$

the $s^d B(s)$ term can be regarded as a single polynomial. With similar statements, the corresponding Diophantine equation can be solved, and the analytical solution is validated.

**Example 5.46.** Change the equation of Example 5.45 into the form in (5.6.5), with $d = 2$, and solve the equation.

**Solutions.** The two polynomials can be entered into MATLAB, and then the polynomial $B(s)$ can be modified to the new one, so that the new Diophantine equation can be solved:

```
>> syms s; A=s^4-27*s^3/10+11*s^2/4-1249*s/1000+53/250;
   B=s^2*(3*s^2-6*s/5+51/25); C=2*s^2+3*s/5-9/25;
   [X,Y]=diophantine(A,B,C,s), simplify(A*X+B*Y-C) % solve and check
```

The solution can be obtained as follows. It can be validated that the solution satisfies the original equation:

$$X(s) = -\frac{382\,119\,125}{25\,016\,954}s^3 + \frac{6\,026\,575}{12\,508\,477}s^2 - \frac{40\,305}{5\,618}s - \frac{90}{53},$$

$$Y(s) = \frac{382\,119\,125}{75\,050\,862}s^3 - \frac{593\,951\,425}{50\,033\,908}s^2 + \frac{862\,183\,205}{100\,067\,816}s - \frac{234\,765\,227}{200\,135\,632}.$$

**Example 5.47.** Bézout identity is a commonly used polynomial equation, with

$$A(x)X(x) + B(x)Y(x) = 1.$$

For $A(x) = x^3 + 4x^2 + 5x + 2$ and $B(x) = x^4 + 13x^3 + 63x^2 + 135x + 108$, solve and validate the Bézout identity.

**Solutions.** Bézout identity is a special form of Diophantine equation, named after French mathematician Étienne Bézout (1730–1783), where $C(s) = 1$. The following statements can be used to solve the equation above:

```
>> syms x; A=x^3+4*x^2+5*x+2; B=x^4+13*x^3+63*x^2+135*x+108;
   [X,Y]=diophantine(A,B,1,x), simplify(A*X+Y*B-1)
```

The solution obtained is as follows, and it can be validated that the error is zero:

$$X(x) = -\frac{55x^3}{144} - \frac{33x^2}{8} - \frac{239x}{16} - \frac{73}{4},$$
$$Y(x) = \frac{55x^2}{144} + \frac{11x}{16} + \frac{25}{72}.$$

### 5.6.3 Solutions of pseudo-polynomial equations

Pseudo-polynomial equations are special cases of nonlinear matrix equations, since the unknown matrix is, in fact, a scalar. The definition and solution methods are given below for pseudo-polynomial equations.

**Definition 5.18.** The general form of a pseudo-polynomial is

$$p(s) = c_1 s^{\alpha_1} + c_2 s^{\alpha_2} + \cdots + c_{n-1} s^{\alpha_{n-1}} + c_n s^{\alpha_n} \tag{5.6.6}$$

It can be seen that pseudo-polynomial equations are direct extensions of polynomial equations. The solutions of these equations may be far more complicated that those of the ordinary polynomial equations. In this section, different methods are explored.

**Example 5.48.** Solve the following pseudo-polynomial equation[28]:

$$x^{2.3} + 5x^{1.6} + 6x^{1.3} - 5x^{0.4} + 7 = 0.$$

**Solutions.** A natural way is to introduce a new variable $z = x^{0.1}$, such that the original equation can be mapped into a polynomial equation of $z$, as follows:

$$f_1(z) = z^{23} + 5z^{16} + 6z^{13} - 5z^4 + 7.$$

There are 23 solutions for the equation. When the expression $x = z^{10}$ is used, we can find all the solutions in the equation. The idea can be implemented in the following MATLAB statements:

```
>> syms x z; f1=z^23+5*z^16+6*z^13-5*z^4+7;
   p=sym2poly(f1); r=roots(p);
   f=x^2.3+5*x^1.6+6*x^1.3-5*x^0.4+7;
   r1=r.^10, double(subs(f,x,r1))
```

Unfortunately, when the solutions are substituted back to the original equation, it can be found that most of the solutions do not satisfy the original pseudo-polynomial equation. How many solutions are there in the original equation? There are only two solutions $x$ satisfying the original equation, namely $x = -0.1076 \pm j0.5562$. The remaining 21 solutions are extraneous roots. The two genuine solutions can also be obtained with the following statements:

```
>> f=@(x)x.^2.3+5*x.^1.6+6*x.^1.3-5*x.^0.4+7;
   more_sols(f,zeros(1,1,0),100+100i), x0=X(:)
```

It can be seen from the mathematics viewpoint that the genuine solutions are located on the first Riemann sheet, with the remaining solutions (extraneous roots) being located on other Riemann sheets.

**Example 5.49.** Solve the following pseudo-polynomial equation:

$$s^{\sqrt{5}} + 25s^{\sqrt{3}} + 16s^{\sqrt{2}} - 3s^{0.4} + 7 = 0.$$

**Solutions.** Since some of the orders are changed to irrational numbers, there is no way to convert this equation into an ordinary polynomial equation. Therefore, $\texttt{more\_sols()}$ function becomes the only solver for such equation. The following statements can be used to solve it directly. The pseudo-polynomial equation with irrational orders has only two solutions, $s = -0.0812 \pm 0.2880j$:

```
>> f=@(s)s^sqrt(5)+25*s^sqrt(3)+16*s^sqrt(2)-3*s^0.4+7;
   more_sols(f,zeros(1,1,0),100+100i); x0=X(:)
```

It can be seen from this example that, even though the orders are changed to irrational numbers, there is no extra computation load added to the solution process. The computational complexity is the same as that in the previous example.

## 5.7 Problems

5.1 Check whether the following linear equations have solutions or not:

$$\begin{bmatrix} 16 & 2 & 3 & 13 \\ 5 & 11 & 10 & 8 \\ 9 & 7 & 6 & 12 \\ 4 & 14 & 15 & 1 \end{bmatrix} X = \begin{bmatrix} 1 \\ 3 \\ 4 \\ 7 \end{bmatrix}.$$

5.2 Solve analytically the following linear equations and validate the result:

$$\begin{bmatrix} 2 & 9 & 4 & 12 & 5 & 8 & 6 \\ 12 & 2 & 8 & 7 & 3 & 3 & 7 \\ 3 & 0 & 3 & 5 & 7 & 5 & 10 \\ 3 & 11 & 6 & 6 & 9 & 9 & 1 \\ 11 & 2 & 1 & 4 & 6 & 8 & 7 \\ 5 & -18 & 1 & -9 & 11 & -1 & 18 \\ 26 & -27 & -1 & 0 & -15 & -13 & 18 \end{bmatrix} X = \begin{bmatrix} 1 & 9 \\ 5 & 12 \\ 4 & 12 \\ 10 & 9 \\ 0 & 5 \\ 10 & 18 \\ -20 & 2 \end{bmatrix}.$$

5.3 Solve the following linear equations:

$$\begin{cases} x_1 + 2x_2 + x_3 + 2x_5 + x_6 + x_8 = 1, \\ 2x_1 + x_2 + 4x_3 + 4x_4 + 4x_5 + x_6 + 3x_7 + 3x_8 = 1, \\ 2x_1 + x_3 + x_4 + 3x_5 + 2x_6 + 2x_7 + 2x_8 = 0, \\ 2x_1 + 4x_2 + 2x_3 + 4x_5 + 2x_6 + 2x_8 = 2. \end{cases}$$

5.4 Find the basic set of solutions of the following matrix, and find the general solution to the homogeneous equation $AZ = 0$.

$$A = \begin{bmatrix} 3 & 1 & 2 & 1 & 1 \\ 1 & 2 & 3 & 3 & 2 \\ 4 & 1 & 4 & 2 & 1 \\ 4 & 4 & 4 & 4 & 4 \\ 3 & 2 & 1 & 1 & 2 \end{bmatrix}.$$

5.5 Solve the following matrix equation:

$$\begin{bmatrix} 1 & 2 & 3 \\ 3 & 5 & 7 \\ 4 & 9 & 2 \end{bmatrix} X \begin{bmatrix} 0 & 1 & 0 & 0 & 1 \\ 1 & 0 & 1 & 2 & 2 \\ 1 & 2 & 0 & 0 & 2 \\ 0 & 0 & 1 & 1 & 1 \\ 1 & 0 & 0 & 2 & 1 \end{bmatrix} = \begin{bmatrix} 1 & 1 & 2 & 2 & 2 \\ 1 & 1 & 2 & 2 & 2 \\ 1 & 1 & 2 & 2 & 2 \end{bmatrix}.$$

5.6 Check whether the following equation has solutions or not. If it does, find all the solutions.

$$\begin{bmatrix} -1 & -1 & 0 & 0 & -1 & 0 \\ 1 & 1 & -1 & 0 & -1 & -1 \\ 1 & 1 & 0 & 0 & 1 & 0 \end{bmatrix} X \begin{bmatrix} 0 & 1 & -1 & -1 & 0 \\ 0 & 1 & -1 & -1 & 0 \\ 0 & -1 & 0 & 0 & 1 \end{bmatrix} = \begin{bmatrix} 0 & -6 & 21 & 21 & -15 \\ 0 & -30 & 27 & 27 & 3 \\ 0 & 6 & -21 & -21 & 15 \end{bmatrix}.$$

5.7 Solve the equation $A_1 X B_1 + A_2 X B_2 = C$ and validate the results, where

$$A_1 = \begin{bmatrix} 4+4j & 1+j \\ 1+4j & 4+2j \end{bmatrix}, \quad A_2 = \begin{bmatrix} 3+j & 2+2j \\ 1+2j & 4+2j \end{bmatrix},$$

$$B_1 = \begin{bmatrix} 3 & 4 & 1 & 1 \\ 2 & 4 & 1 & 1 \\ 1 & 2 & 1 & 2 \\ 4 & 3 & 1 & 2 \end{bmatrix}, \quad B_2 = \begin{bmatrix} 2 & 1 & 4 & 4 \\ 3 & 4 & 2 & 3 \\ 4 & 1 & 4 & 3 \\ 2 & 3 & 2 & 3 \end{bmatrix},$$

$$C = \begin{bmatrix} 141 + 47j & 77 + 3j & 98 + 27j & 122 + 37j \\ 115 + 58j & 72 + 4j & 93 + 34j & 106 + 46j \end{bmatrix}.$$

5.8 If the $A_1$ and $B_1$ matrices in Problem 5.7 are changed to the following singular matrices, does the original equation have solutions? If it does, solve the equation and validate the results when

$$A_1 = \begin{bmatrix} 1 & 3 \\ 4 & 2 \end{bmatrix}, \quad B_1 = \begin{bmatrix} 16 & 2 & 3 & 13 \\ 5 & 11 & 10 & 8 \\ 9 & 7 & 6 & 12 \\ 4 & 14 & 15 & 1 \end{bmatrix}.$$

5.9 Solve numerically and analytically the following Sylvester equation, and validate the solutions:

$$\begin{bmatrix} 3 & -6 & -4 & 0 & 5 \\ 1 & 4 & 2 & -2 & 4 \\ -6 & 3 & -6 & 7 & 3 \\ -13 & 10 & 0 & -11 & 0 \\ 0 & 4 & 0 & 3 & 4 \end{bmatrix} X + X \begin{bmatrix} 3 & -2 & 1 \\ -2 & -9 & 2 \\ -2 & -1 & 9 \end{bmatrix} = \begin{bmatrix} -2 & 1 & -1 \\ 4 & 1 & 2 \\ 5 & -6 & 1 \\ 6 & -4 & -4 \\ -6 & 6 & -3 \end{bmatrix}.$$

5.10 Solve analytically the following matrix equation and validate the results. Find the value $a$ such that there is no solution.

$$\begin{bmatrix} -2 & 2 & c \\ -1 & 0 & -1 \\ 1 & -1 & 2 \end{bmatrix} X + X \begin{bmatrix} -2 & -1 & 2 \\ a & 3 & 0 \\ b & -2 & 2 \end{bmatrix} + \begin{bmatrix} 0 & -1 & 0 \\ -1 & 1 & 0 \\ 1 & -1 & -1 \end{bmatrix} = 0.$$

5.11 Solve numerically and analytically the discrete Lyapunov equation $AXA^T - X + Q = 0$, where

$$A = \begin{bmatrix} -2 & -1 & 0 & -3 \\ -2 & -2 & -1 & -3 \\ 2 & 2 & -3 & 0 \\ -3 & 1 & 1 & -3 \end{bmatrix}, \quad Q = \begin{bmatrix} -12 & -16 & 14 & -8 \\ -20 & -25 & 11 & -20 \\ 3 & 1 & -16 & 1 \\ -4 & -10 & 21 & 10 \end{bmatrix}.$$

5.12 Solve the multiterm Sylvester equation, whose matrices are given in Example 5.7, and validate the result. If matrices $A_1$ and $B_1$ are changed to the singular ones in Problem 5.8, solve again the multiterm Sylvester equation, and validate the results.

5.13  A Riccati equation is given by $PA + A^{\mathrm{T}}P - PBR^{-1}B^{\mathrm{T}}P + Q = 0$, with

$$A = \begin{bmatrix} -27 & 6 & -3 & 9 \\ 2 & -6 & -2 & -6 \\ -5 & 0 & -5 & -2 \\ 10 & 3 & 4 & -11 \end{bmatrix}, \quad B = \begin{bmatrix} 0 & 3 \\ 16 & 4 \\ -7 & 4 \\ 9 & 6 \end{bmatrix},$$

$$Q = \begin{bmatrix} 6 & 5 & 3 & 4 \\ 5 & 6 & 3 & 4 \\ 3 & 3 & 6 & 2 \\ 4 & 4 & 2 & 6 \end{bmatrix}, \quad R = \begin{bmatrix} 4 & 1 \\ 1 & 5 \end{bmatrix}.$$

Solve the equation for matrix $P$, and check the precision.

5.14  Solve and validate the extended Riccati equation $AX + XD - XBX + C = 0$, where

$$A = \begin{bmatrix} 2 & 1 & 9 \\ 9 & 7 & 9 \\ 6 & 5 & 3 \end{bmatrix}, \quad B = \begin{bmatrix} 0 & 3 & 6 \\ 8 & 2 & 0 \\ 8 & 2 & 8 \end{bmatrix}, \quad C = \begin{bmatrix} 7 & 0 & 3 \\ 5 & 6 & 4 \\ 1 & 4 & 4 \end{bmatrix}, \quad D = \begin{bmatrix} 3 & 9 & 5 \\ 1 & 2 & 9 \\ 3 & 3 & 0 \end{bmatrix}.$$

5.15  Find high precision numerical solutions of the equation in Problem 5.14, and also the accuracy of the solution.

5.16  If the equation in Problem 5.14 is changed to $AX + XD - XBX^{\mathrm{T}} + C = 0$, find all the solutions, and find the high precision solutions as well.

5.17  Find all the complex solutions of the equation in Problem 5.43.

5.18  Find all the real solutions in the interval $-\pi \leqslant x, y \leqslant \pi$ for the following nonlinear simultaneous equations:

$$\begin{cases} x^2 e^{-xy^2/2} + e^{-x/2}\sin(xy) = 0, \\ y^2\cos(x + y^2) + x^2 e^{x+y} = 0. \end{cases}$$

5.19  Solve the pseudo-polynomial equation $x^{2.3} + 5x^{1.6} + 6x^{1.3} - 5x^{0.4} + 7 = 0$.

5.20  Solve the following Diophantine equations and validate the results:
   (1)  $A(x) = 1 - 0.7x$, $B(x) = 0.9 - 0.6x$, $C(x) = 2x^2 + 1.5x^3$,
   (2)  $A(x) = 1 + 0.6x - 0.08x^2 + 0.152x^3 + 0.0591x^4 - 0.0365x^5$,
        $B(x) = 5 - 4x - 0.25x^2 + 0.42x^3$, $C(x) = 1$.

# 6 Matrix functions

Matrix functions are very important in matrix analysis. They play important roles in many scientific and engineering applications.

The term "matrix function" may have different meanings in mathematics[12]. In this chapter, the following definition is mainly used.

**Definition 6.1.** Given a scalar function $f(x)$ and a complex square matrix $A$, if the independent variable $x$ is substituted by matrix $A$, $f(A)$ is a matrix of the same size, then $f(A)$ is referred to as a matrix function.

It can be seen that matrix function is an extension of scalar function. It is itself a function, and can be used to map one matrix into another through matrix function. An example below is given to demonstrate the extension from scalar functions.

**Example 6.1.** Extend the scalar function $f(x) = (1 + x^2)/(1 - x)$ into a matrix function.

**Solutions.** Addition, subtraction and multiplication in the scalar function can be extended to matrix computations directly, and number 1 can be replaced by an identity matrix. The handling of divisions in the function is complicated, it can be extended to left or right multiplication of inverse matrices. For instance, the scalar function can be extended to a matrix function, $f(A) = (I - A^2)(I - A)^{-1}$.

Apart from the extension of algebraic computation demonstrated here, transcendental functions can also be extended to matrix functions by corresponding Taylor series computation, and this topic will be presented in detail later in the chapter.

Before introducing matrix functions, nonlinear computations of matrix elements are presented in Section 6.1. Such nonlinear functions are similar to those in dot operations studied earlier. It is a kind of element-by-element computation of matrices. Integer rounding and rationalization of matrix elements are summarized first, followed by transcendental function evaluations and discrete mathematics computations such as sorting and finding maximum or minimum values.

In the remaining sections of the chapter, the concept and MATLAB solutions of matrix functions are studied. In Section 6.2, the commonly used matrix exponential functions are introduced. Several numerical algorithms are implemented in MATLAB, and the limitations of the algorithms are demonstrated. The universal MATLAB function is recommended for computing numerically and analytically the matrix exponentials. In Section 6.3, logarithmic and square root functions of matrices are presented, and in Section 6.4, trigonometric functions of matrices are introduced. In Section 6.5, a universal matrix function solver in MATLAB is provided, which can be used to evaluate matrix functions of any complexity. In Section 6.6, the matrices of $A^k$ and $k^A$ are also presented.

https://doi.org/10.1515/9783110666991-006

## 6.1 Nonlinear computation of matrix elements

Taking nonlinear computation to a matrix in an element-by-element manner, also known as element-wise manner, the result is also a matrix, while such function is also a matrix function, but this is not the mainstream matrix function to be discussed in this chapter.

In this section, integer rounding and rationalization of data in matrices are presented. Also the evaluation of transcendental functions is introduced. Finally, vector-based data processing such as sorting, maximum and minimum value extraction, and other facilities are also addressed.

### 6.1.1 Integer rounding and rationalization

A set of integer rounding functions is provided in MATLAB to take integers in different directions. The related functions are shown in Table 6.1. Their syntaxes are also listed in the table. Examples are given to demonstrate the use and results of the functions.

**Table 6.1:** Integer rounding and conversion functions.

| name | syntaxes | function explanations |
|------|----------|----------------------|
| floor() | $n$=floor($x$) | rounding $x$ towards $-\infty$ and getting the vector $n$, denoted as $n = \lfloor x \rfloor$ |
| ceil() | $n$=ceil($x$) | rounding $x$ towards $+\infty$ and getting the vector $n = \lceil x \rceil$ |
| round() | $n$=round($x$) | rounding $x$ to the nearest integer and finding $n$ |
| fix() | $n$=fix($x$) | rounding $x$ towards 0 and getting vector $n$ |
| rat() | [$n,d$]=rat($x$) | find rational approximations to $x$, to get two integer vectors $n$ and $d$, the numerator and denominator matrices |
| rem() | $B$=rem($A,C$) | remainder after division $A-n.*C$, where, $n$=fix($A./C$) |
| mod() | $B$=mod($A,C$) | modulus after division, $A-n.*C$, where $n$=floor($A./C$) |

**Example 6.2.** For a set of data −0.2765, 0.5772, 1.4597, 2.1091, 1.191, −1.6187, round them to integers with different functions and observe the results. Also understand better the facilities provided in the functions.

**Solutions.** The given data can be expressed first with a vector. The following statements can be used to round the data:

```
>> A=[-0.2765,0.5772,1.4597,2.1091,1.191,-1.6187];
   v1=floor(A), v2=ceil(A), v3=round(A),
   v4=fix(A) % with different rounding functions for different results
```

With the rounding functions, the following results can be obtained:

$$v_1 = [-1, 0, 1, 2, 1, -2], \quad v_2 = [0, 1, 2, 3, 2, -1], \quad v_3 = [0, 1, 1, 2, 1, -2], \quad v_4 = [0, 0, 1, 2, 1, -1].$$

**Example 6.3.** For a $3 \times 3$ Hilbert matrix, find the rational representations of the matrix elements.

**Solutions.** The following statements can be used and, with function $A$=hilb(3), the matrix will be created, and the rationalization results can be obtained

```
>> A=hilb(3); [n,d]=rat(A) % extract numerator and denominator matrices
```

The two integer matrices obtained are respectively

$$n = \begin{bmatrix} 1 & 1 & 1 \\ 1 & 1 & 1 \\ 1 & 1 & 1 \end{bmatrix}, \quad d = \begin{bmatrix} 1 & 2 & 3 \\ 2 & 3 & 4 \\ 3 & 4 & 5 \end{bmatrix}.$$

**Example 6.4.** From the given vector $v = [5.2, 0.6, 7, 0.5, 0.4, 5, 2, 6.2, -0.4, -2]$, find and display the integers.

**Solutions.** The function isinteger() provided in MATLAB cannot be used to test whether a number is an integer or not, it can only be used to test if a number is presented using integer data type or not. For the elements in the vector, they are all stored in the double precision rather than integer format. If integer entities are expected, the remainders must be used. If the remainders are zeros, the numbers are integers, otherwise they are not integers. The following statements can be issued, and it is found that $i = [3, 6, 7, 10]$, meaning the values at those positions are integers. The integers at these positions are $[7, 5, 2, -2]$.

```
>> v=[5.2 0.6 7 0.5 0.4 5 2 6.2 -0.4 -2];
   i=find(rem(v,1)==0), v(i)   % they can be simplified to v(rem(v,1)==0)
```

In fact, it is not a reliable way to check whether a number is an integer or not in this way since the use of double precision framework may produce imprecise terms as 5.000000000000001. Therefore, a better condition below is usually adopted.

```
>> i=find(rem(v,1)<=1e-12), v(i)   % or use other small numbers
```

### 6.1.2 Transcendental function computations

The definition of transcendental functions is presented first. Exponential, logarithmic, and trigonometric functions, as well as their inverses, are discussed. Element-wise functions of matrices are also presented.

**Definition 6.2.** Transcendental functions are usually functions which cannot be evaluated with finitely many addition, subtraction, multiplication, division, and exponentiation operations.

For instance, exponential, logarithmic and trigonometric functions are all transcendental functions.

Generally, there are MATLAB functions which can be used to directly evaluate these functions, and the syntax is unified as $y$=fun($x$), where fun is the name of the function, $x$ is the independent variable, $y$ is the value of the function. The argument $x$ can be a scalar, vector, matrix or even other data type such as multidimensional array. It can also be a symbolic variable. The data type of the returned argument $y$ is exactly the same as that of $x$. The facilities of the function are similar to the dot operation in MATLAB.

(1) Exponential and logarithmic functions. Normally, the power function $a^x$ can be evaluated directly with $y$=a.^x. The commonly used exponential function $e^x$ can be evaluated with $y$=exp($x$) directly.

**Example 6.5.** For the given matrix $A$, compute its exponential if

$$A = \begin{bmatrix} 2 & 2 & -1 & 2 & 2 \\ 2 & 1 & 0 & -2 & 0 \\ -2 & -2 & 2 & 2 & 2 \end{bmatrix}.$$

**Solutions.** The following rectangular matrix can be entered into MATLAB directly, and exp() function can be used to compute the exponential element-by-element. There is no restrictions to the size of matrix $A$.

```
>> A=[2,2,-1,2,2; 2,1,0,-2,0; -2,-2,2,2,2]; A1=exp(A)
```

The exponentials obtained are

$$A_1 = \begin{bmatrix} 7.3891 & 7.3891 & 0.3679 & 7.3891 & 7.3891 \\ 7.3891 & 2.7183 & 1 & 0.1353 & 1 \\ 0.1353 & 0.1353 & 7.3891 & 7.3891 & 7.3891 \end{bmatrix}.$$

Logarithmic function $\ln x$ can be evaluated with log(), while common logarithmic function $\lg x$ can be evaluated with log10(); the logarithmic function with base 2, $\log_2 x$, can be evaluated using log2(); the logarithmic function with base $a$, $\log_a x$, can be evaluated with the formula of change of base of logarithms, log($x$)/log($a$).

**Example 6.6.** For the results obtained in Example 6.5, evaluate ln() function to restore the original matrix, and then compute $\log_3(A)$.

**Solutions.** The following statements can be used directly:

```
>> A=[2,2,-1,2,2; 2,1,0,-2,0; -2,-2,2,2,2]; A1=exp(A);
   A2=log(A1), A3=log(A)/log(3)
```

and the matrix $A_2$ is exactly the same as the original matrix $A$, while the logarithmic matrix $A_3$ is obtained as

$$A_3 = \begin{bmatrix} 0.6309 & 0.6309 & 0+2.8596j & 0.6309 & 0.6309 \\ 0.6309 & 0 & -\infty & 0.6309+2.8596j & -\infty \\ 0.6309+2.8596j & 0.6309+2.8596j & 0.6309 & 0.6309 & 0.6309 \end{bmatrix}.$$

(2) Trigonometric functions. Sine, cosine, tangent, and cotangent functions are commonly used trigonometric functions. Secant, cosecant, hyperbolic sine and cosine are also commonly used. The definitions of these functions are given below.

**Definition 6.3.** The secant function is the reciprocal of cosine, $\sec x = 1/\cos x$, while the cosecant function is the reciprocal of sine, $\csc x = 1/\sin x$.

**Definition 6.4.** The hyperbolic sine function is defined as $\sinh x = (e^x - e^{-x})/2$, while the hyperbolic cosine function is $\cosh x = (e^x + e^{-x})/2$.

Sine, cosine, tangent, and cotangent functions can be evaluated with sin(), cos(), tan(), and cot() functions directly; secant and cosecant functions can be evaluated with sec() and csc() functions; hyperbolic sine, $\sinh x$, and hyperbolic cosine function, $\cosh x$, can be evaluated with functions sinh() and cosh(). The unit of trigonometric functions is radian. If degree is used, it should be converted to radian with y=pi*x/180. Alternatively, functions like sind() can be used.

In fact, the functions such as sin() are similar to dot operations discussed earlier. The sine function of each element in the input matrix is evaluated individually so that the size of the output argument is the same as that of the input argument.

**Example 6.7.** For the matrix in Example 6.5, compute the nonlinear functions sin($A$) and cos($A$) of the matrix elements.

**Solutions.** Input the original matrix to the computer, and the following statements can be used to compute directly the nonlinear functions:

```
>> A=[2,2,-1,2,2; 2,1,0,-2,0; -2,-2,2,2,2];
   A1=sin(A), A2=cos(A)
```

The results obtained are

$$A_1 = \begin{bmatrix} 0.9093 & 0.9093 & -0.8415 & 0.9093 & 0.9093 \\ 0.9093 & 0.8415 & 0 & -0.9093 & 0 \\ -0.9093 & -0.9093 & 0.9093 & 0.9093 & 0.9093 \end{bmatrix},$$

$$A_2 = \begin{bmatrix} -0.4162 & -0.4162 & 0.5403 & -0.4162 & -0.4162 \\ -0.4162 & 0.5403 & 1 & -0.4162 & 1 \\ -0.4162 & -0.4162 & -0.4162 & -0.4162 & -0.4162 \end{bmatrix}.$$

(3) Inverse trigonometric function. Arcsine, arccosine, arctangent, arccotangent, arcsecant, arccosecant are commonly used inverse trigonometric functions. Also, arc hyperbolic sine and cosine functions are commonly applied. Based on the definitions of the functions, MATLAB based solutions are presented and examples are shown.

**Definition 6.5.** Inverse trigonometric functions, also known as antitrigonometric functions or arcus functions, are the inverse functions of trigonometric functions. For instance, if $\sin x = y$, the inverse function is referred to as arcsine function, denoted as $x = \sin^{-1} y$, or $x = \arcsin y$. The latter notation is adopted in the book.

Since the trigonometric function $y = \sin x$ is periodic, its inverse function is not single-valued. If $x$ is the value of its arcsine function, the values $x + 2k\pi$ are also valid arcsine function values, where $k$ is any integer. The range of the function is $[-\pi/2, \pi/2]$. Similarly, the ranges of arctangent, arccotangent and arcsecant functions are $[-\pi/2, \pi/2]$. The ranges of arcosine and arcsecant functions are $[0, \pi]$.

A letter a is placed in front of the trigonometric function to form an inverse trigonometric function. For instance, asin() can be used to evaluate arcsine function. Similar MATLAB functions are acos(), atan(), acot(), asec(), acsc(), asinh(), and acosh(). The unit of the functions is radian. If degrees are expected, command y=180*x/pi can be used for the conversion. Alternatively, functions like asind() could be used.

**Example 6.8.** Compute the inverse trigonometric functions of the results in Example 6.7, and see whether the original functions can be restored.

**Solutions.** Compute the trigonometric functions first in Example 6.7, then for the results the inverse functions can be calculated to see whether the original functions can be restored.

```
>> A=[2,2,-1,2,2; 2,1,0,-2,0; -2,-2,2,2,2];
   A1=sin(A); A2=cos(A); A3=asin(A1), A, A4=acos(A2)
```

The results obtained are as follows:

$$A_3 = \begin{bmatrix} 1.1416 & 1.1416 & -1 & 1.1416 & 1.1416 \\ 1.1416 & 1 & 0 & -1.1416 & 0 \\ -1.1416 & -1.1416 & 1.1416 & 1.1416 & 1.1416 \end{bmatrix},$$

$$A = \begin{bmatrix} 2 & 2 & -1 & 2 & 2 \\ 2 & 1 & 0 & -2 & 0 \\ -2 & -2 & 2 & 2 & 2 \end{bmatrix}, \quad A_4 = \begin{bmatrix} 2 & 2 & 1 & 2 & 2 \\ 2 & 1 & 0 & 2 & 0 \\ 2 & 2 & 2 & 2 & 2 \end{bmatrix}.$$

It can be seen that some of the values cannot be restored with inverse trigono-metric functions, since the inverse functions have their own ranges. For the arcsine function, the range is $[-\pi/2, \pi/2]$. If the range is exceeded, for instance, the original value is $\pm 2$, it can be converted to $\pm 1.1416$, since $1.1416 + 2 = \pi$. The inverse cosine func-tion is defined in the range $(0, \pi)$, therefore, the ranges of the inverse functions should be carefully considered.

### 6.1.3 Vector sorting, minimization, and maximization

MATLAB function sort() can be used to sort a vector in the ascending order. The func-tion can be sorted with v=sort(a) and [v,k]=sort(a). Using the former syntax, a sorted vector v is returned, while using the latter, the index vector k is also returned. If one wants to sort a vector in the descending order, the vector −a can be used in-stead. Alternatively, the option 'descend' can be used, with the syntax sort(a,'de-scend').

**Example 6.9.** If a is a matrix, function sort() can also be used to sort it. Each column is individually sorted. Sort a given $4 \times 4$ magic matrix and observe the results.

**Solutions.** If one wants to sort a matrix, the following statements can be used:

```
>> A=magic(4), [a k]=sort(A)
```

The results and the original magic matrix are all given below. The readers can under-stand the results by comparisons:

$$A = \begin{bmatrix} 16 & 2 & 3 & 13 \\ 5 & 11 & 10 & 8 \\ 9 & 7 & 6 & 12 \\ 4 & 14 & 15 & 1 \end{bmatrix}, \quad a = \begin{bmatrix} 4 & 2 & 3 & 1 \\ 5 & 7 & 6 & 8 \\ 9 & 11 & 10 & 12 \\ 16 & 14 & 15 & 13 \end{bmatrix}, \quad k = \begin{bmatrix} 4 & 1 & 1 & 4 \\ 2 & 3 & 3 & 2 \\ 3 & 2 & 2 & 3 \\ 1 & 4 & 4 & 1 \end{bmatrix}.$$

If a is a matrix, then each column is sorted individually. If one wants to sort each row, two methods can be used, the first one is to sort the matrix $a^T$, while the other is to use the command sort(a,2), where 2 indicates sorting according to rows. If option 1 is used, the default sorting of columns is performed. The following statements can be used to sort the rows:

```
>> A, [a k]=sort(A,2)
```

The following statements can be used to sort the matrix:

$$A = \begin{bmatrix} 16 & 2 & 3 & 13 \\ 5 & 11 & 10 & 8 \\ 9 & 7 & 6 & 12 \\ 4 & 14 & 15 & 1 \end{bmatrix}, \quad a = \begin{bmatrix} 2 & 3 & 13 & 16 \\ 5 & 8 & 10 & 11 \\ 6 & 7 & 9 & 12 \\ 1 & 4 & 14 & 15 \end{bmatrix}, \quad k = \begin{bmatrix} 2 & 3 & 4 & 1 \\ 1 & 4 & 3 & 2 \\ 3 & 2 & 1 & 4 \\ 4 & 1 & 2 & 3 \end{bmatrix}.$$

If one wants to sort all the elements in a matrix, the command sort(**A**(:)) can be used. In fact, this method also applies to the case where **A** is a multidimensional array.

```
>> [v,k]=sort(A(:))     % sort all the elements in a matrix
```

The sorted index vector **k** is $k^{\mathrm{T}} = [16, 5, 9, 4, 2, 11, 7, 14, 3, 10, 6, 15, 13, 8, 12, 1]$.

The functions for finding the maximum and minimum values in MATLAB are max() and min(). The functions sum() and prod() can be used for finding the sum and product of all the elements in a vector. If the input argument is a matrix, then column-wise maximum, minimum, sum, and product are computed.

### 6.1.4 Means, variances, and standard deviations

If a set of data is given, statistical analysis of the data is usually expected. For instance, the statistical quantities such as mean and variance can be evaluated. In this section, the definitions of these statistical quantities are given first, then MATLAB solutions are summarized.

**Definition 6.6.** If a vector $A = [a_1, a_2, \ldots, a_n]$ is known, the mean $\mu$, variance $\sigma^2$, and standard deviation $s$ are respectively defined as

$$\mu = \frac{1}{n}\sum_{k=1}^{n} a_k, \quad \sigma^2 = \frac{1}{n}\sum_{k=1}^{n}(a_k - \mu)^2, \quad s = \sqrt{\frac{1}{n-1}\sum_{k=1}^{n}(a_k - \mu)^2}. \tag{6.1.1}$$

If **A** is a vector, the function $\mu$=mean(**A**), $c$=cov(**A**), and $s$=std(**A**) can directly be used to evaluate the mean, variance, and standard deviation of the elements in vector **A**.

If **A** is a matrix, every column of **A** is computed individually, for instance, mean() can be used to find the mean of each column, and the result is a vector. The syntaxes of these functions are similar to those of functions such as max(). Each column of matrix **A** is regarded as a sample of a signal. Function cov() can be used to evaluate the covariance matrix.

**Example 6.10.** With function randn(3000,4), four input signals from standard normal distribution can be generated, each signal having 3 000 samples. Find the mean of each signal, and also the covariance matrix of the four signals.

**Solutions.** With the powerful tool such as MATLAB, the following statements can be used to solve the problem:

```
>> R=randn(3000,4); m=mean(R), C=cov(R)
```

The means of the signals are $m = [-0.00388, 0.0163, -0.00714, -0.0108]$, and the covariance matrix is

$$C = \begin{bmatrix} 0.98929 & -0.01541 & -0.012409 & 0.011073 \\ -0.01541 & 0.99076 & -0.003764 & 0.003588 \\ -0.012409 & -0.003764 & 1.0384 & 0.013633 \\ 0.011073 & 0.003588 & 0.013633 & 0.99814 \end{bmatrix}.$$

## 6.2 Matrix exponentials

Matrix exponential function is the most widely used matrix function. The definition of the matrix exponential is introduced, and some numerical algorithms and MATLAB implementations are presented, including series truncation and Cayler–Hamilton theorem-based algorithms, and the functions provided in MATLAB are also presented. The Jordan transformation matrix is demonstrated in an example.

### 6.2.1 Definitions and properties of matrix functions

Before considering matrix functions, consider a scalar function $f(x)$. Matrix functions can be extended from scalar ones, then matrix exponential functions are introduced, and the computation of the functions is presented.

**Definition 6.7.** For a scalar function $f(x)$, the Taylor series expansion is

$$f(x) = f(0) + \frac{1}{1!}f'(0)x + \frac{1}{2!}f''(0)x^2 + \cdots + \frac{1}{k!}f^{(k)}(0)x^k + \cdots. \tag{6.2.1}$$

Matrix functions are extensions of scalar functions to the field of matrices. The independent scalar variables can be extended to matrices. The definition of a matrix function is given first, then matrix exponential functions are proposed, with various algorithms presented to compute matrix exponentials.

**Definition 6.8.** The Taylor series expansion of a matrix function is

$$f(A) = f(0)I + \frac{1}{1!}f'(0)A + \frac{1}{2!}f''(0)A^2 + \cdots + \frac{1}{k!}f^{(k)}(0)A^k + \cdots. \tag{6.2.2}$$

Some of the matrix function properties are summarized without proofs[12]:

**Theorem 6.1.** For a matrix $A$, function $f(A)$ and $A$ satisfy $f(A)A = Af(A)$.

**Theorem 6.2.** *Matrix function $f(A)$ satisfies*

$$f(A^\mathrm{T}) = f^\mathrm{T}(A), \quad f(XAX^{-1}) = Xf(A)X^{-1}. \tag{6.2.3}$$

**Theorem 6.3.** *If $A = \mathrm{diag}(A_1, A_2, \ldots, A_m)$ is a block-diagonal matrix, then*

$$f(A) = \mathrm{diag}(f(A_1), f(A_2), \ldots, f(A_m)). \tag{6.2.4}$$

Matrix exponential functions are the most commonly used matrix functions, since they are applied in many fields. For instance, they are used in the analytical solutions of linear systems. In this section, the definition of the matrix exponential function is presented, and two numerical algorithms are introduced and implemented. General solvers provided in MATLAB are also demonstrated.

### 6.2.2 Computing matrix exponentials

Apart from the element-based exponential computation, sometimes the nonlinear functions of the whole matrices are needed. In this section, matrix exponentials are introduced.

**Definition 6.9.** The exponential function of $A$ matrix can be expressed in an infinite series as

$$e^A = \sum_{i=0}^{\infty} \frac{1}{i!} A^i = I + A + \frac{1}{2!} A^2 + \frac{1}{3!} A^3 + \cdots + \frac{1}{m!} A^m + \cdots. \tag{6.2.5}$$

In [21], 19 different algorithms for computing matrix exponentials are presented. Two low-level algorithms are introduced, with MATLAB implementations provided. Also, universal functions for the evaluation of exponential functions are discussed.

### 6.2.3 Taylor series-based truncation algorithm

For the Taylor series in (6.2.5), it can be seen that loop structures are suitable for the implementation. High-order powers of matrix $A$ need to be computed, which is time consuming, and may add accumulative errors. Therefore, a better way for computing the general term $F_k = A^k/k!$ is the iterative algorithm. From the $(k + 1)$th term divided by the $k$th term (for the convenience of presentation, division is used, rather than the correct inverse matrices), it can be seen that

$$\frac{F_{k+1}}{F_k} = \frac{A^{k+1}/(k+1)!}{A^k/k!} = \frac{1}{k+1} A. \tag{6.2.6}$$

An iterative formula of the general term can be obtained as

$$F_{k+1} = \frac{1}{k+1} AF_k, \quad k = 0, 1, 2, \ldots, \tag{6.2.7}$$

where $F_0$ is an identity matrix. If $\|F_{k+1}\| \leqslant \epsilon$, and $\epsilon$ is a predefined error tolerance, the loop structure can be terminated, and the accumulation is considered successful. The matrix obtained can be regarded as the matrix exponential of the given matrix.

Based on the above ideas, it is easily seen that the following MATLAB implementation can be written to compute the matrix exponentials. If necessary, the number $k$ of accumulations can also be returned.

```
function [A1,k]=exp_taylor(A)
A1=zeros(size(A)); F=eye(size(A)); k=0;
while norm(F)>eps, A1=A1+F; F=A*F/(k+1); k=k+1; end
```

**Example 6.11.** For the matrix $A$ below, compute the matrix exponential $e^A$ if

$$A = \begin{bmatrix} -3 & -1 & -1 \\ 0 & -3 & -1 \\ 1 & 2 & 0 \end{bmatrix}.$$

**Solutions.** Matrix $A$ can be entered into MATLAB workspace first, then the following statements can be used:

```
>> A=[-3,-1,-1; 0,-3,-1; 1,2,0]; [A1,k]=exp_taylor(A)
```

The result is obtained as follows, and the number of iteration is $k = 26$:

$$A_1 = \begin{bmatrix} -0.000000000000000 & -0.135335283236613 & -0.135335283236613 \\ -0.067667641618306 & -0.067667641618307 & -0.203002924854919 \\ 0.203002924854919 & 0.338338208091532 & 0.473673491328144 \end{bmatrix}.$$

**Example 6.12.** For matrix $A$ below, compute matrix exponential $e^A$ if

$$A = \begin{bmatrix} -21 & 19 & -20 \\ 19 & -21 & 20 \\ 40 & -40 & -40 \end{bmatrix}.$$

**Solutions.** The matrix exponential can be evaluated directly with

```
>> A=[-21,19,-20; 19,-21,20; 40,-40,-40];
   [A1,k]=exp_taylor(A)
```

The result obtained is as follows, with the number of iteration $k = 184$:

$$A_1 = \begin{bmatrix} -0.598980906452256 & 0.598980906452256 & -1.005954603300903 \\ 0.598980906452256 & -0.598980906452256 & 1.005954603300903 \\ 2.860007351570487 & -4.536904952352106 & -0.114550879637390 \end{bmatrix} \times 10^7.$$

It will be demonstrated later that the above result is incorrect, although the convergence condition of the general term is satisfied. This is because the sign of the general term is alternating all the time, and when taking the sum of the series, the terms are canceling each other. Although the general term satisfies the convergence condition, the sum itself has not converged. Therefore, this algorithm is sometimes not reliable, and may lead to wrong results.

### 6.2.4 A Cayley–Hamilton theorem-based algorithm

It can be seen from the Taylor series expansion that the sum of high-order power of $A$ is involved, and this method may introduce errors. Other algorithms may be considered.

It is known from Cayley–Hamilton theorem that $A^n$ can be expressed as a linear combination of $I, A, A^2, \ldots, A^{n-1}$ as follows:

$$A^n = -a_1 A^{n-1} - a_2 A^{n-2} - \cdots - a_{n-1} A - a_n I. \tag{6.2.8}$$

Taylor series expansion can be divided into two parts

$$e^A = \sum_{k=0}^{n-1} \frac{1}{k!} A^k + \sum_{j=0}^{\infty} \frac{1}{(n+j)!} A^{n+j} \tag{6.2.9}$$

Applying (6.2.8) repeatedly, it can be seen that $A^{n+j}$ can also be expressed as a linear combination of $I, A, \ldots, A^{n-1}$:

$$B_{n+j} = \frac{1}{(n+j)!} A^{n+j}, \quad j = 0, 1, 2, \ldots \tag{6.2.10}$$

Matrix $B_{n+j}$ can also be expressed as the linear combination of $I, A, \ldots, A^{n-1}$. Denote

$$B_{n+j} = \beta_{1,j} I + \beta_{2,j} A + \cdots + \beta_{n,j} A^{n-1}. \tag{6.2.11}$$

From $B_{n+j+1} = A B_{n+j}/(n+j+1)$, it can be found that

$$B_{n+j+1} = \frac{1}{n+j+1} (\beta_{1,j} A + \beta_{2,j} A^2 + \cdots + \beta_{n,j} A^n). \tag{6.2.12}$$

The following recursive formula can be established:

$$\beta_{1,j+1} = -\frac{a_n \beta_{n,j}}{n+j}, \quad \beta_{i,j+1} = \frac{1}{n+j} (\beta_{i-1,j} - \beta_{n,j} a_{n-i}), \tag{6.2.13}$$

where $i = 2, 3, \ldots, n$, and $\beta_{i,1} = -a_{n+1-i}/n!, i = 1, 2, \ldots, n$. It follows that[33]

$$e^A = \sum_{k=0}^{n-1} \left( \frac{1}{k!} + \sum_{j=1}^{\infty} \beta_{k+1,j} \right) A^k. \tag{6.2.14}$$

If the coefficients $\beta_{.j}$ are small enough, the subsequent terms can be truncated, the matrix exponential can then be approximately computed. Based on the above algorithm, the following MATLAB function can be written:

```
function [A1,k]=exp_ch(A)
[n,m]=size(A); p=poly1(A); F=eye(n); A1=zeros(n);
a=p(2:end).'; bet=-a(end:-1:1)/factorial(n); k=1;
while (1), b0=bet(:,end);
    b1=-a(n)*b0(n); b2=(b0(1:n-1)-b0(n)*a(n-1:-1:1));
    b=[b1; b2]/(n+k); k=k+1; bet=[bet, b];
    if norm(b)<eps; break; end
end
for j=1:n, A1=A1+(1/factorial(j-1)+sum(bet(j,:)))*F; F=A*F; end
```

**Example 6.13.** For matrix $A$ in Example 6.12, compute $e^A$.

**Solutions.** The following statements can be used to compute matrix exponential. However, with this example, it is seen that the algorithm here still yields large errors, and the number of iterations is $k = 176$. Effective algorithms and MATLAB functions are needed to compute matrix exponentials.

```
>> A=[-21,19,-20; 19,-21,20; 40,-40,-40];
   [A1,k]=exp_ch(A)
```

### 6.2.5 Direct MATLAB solver

For matrix exponentials, an effective MATLAB function `expm()` can be used with the syntax $E$=`expm`$(A)$, where $A$ can be double precision or symbolic matrices. The function can also be used for computing functions $e^{At}$. This function cannot be evaluated with numerical methods.

**Example 6.14.** Compute the matrix exponential $e^A$ of the matrix $A$ in Example 6.12.

**Solutions.** The result obtained in Example 6.12 was not validated. Here, the new function can be used to compute the numerical and analytical solutions. The numerical solutions can be obtained with the commands

```
>> A=[-21,19,-20; 19,-21,20; 40,-40,-40]; A1=expm(A)
```

The exact solution obtained is as follows, and it can be seen that the result is completely different from that obtained in Example 6.12:

$$A_1 = \begin{bmatrix} 0.0676676416183064 & 0.0676676416183064 & 9.0544729577 \times 10^{-20} \\ 0.0676676416183064 & 0.0676676416183064 & 3.256049339 \times 10^{-18} \\ 4.0095395715 \times 10^{-19} & -5.9300552112 \times 10^{-18} & -2.8333891522 \times 10^{-18} \end{bmatrix}.$$

The analytical solution can also be computed with the following commands:

```
>> A2=expm(sym(A)), A3=vpa(A2,15)
```

The analytical solution of the matrix exponential is

$$A_2 = \begin{bmatrix} e^{-2}/2 + e^{-40}\cos 40/2 & e^{-2}/2 - e^{-40}\cos 40/2 & -e^{-40}\sin 40/2 \\ e^{-2}/2 - e^{-40}\cos 40/2 & e^{-2}/2 + e^{-40}\cos 40/2 & e^{-40}\sin 40/2 \\ e^{-40}\sin 40 & -e^{-40}\sin 40 & e^{-40}\cos 40 \end{bmatrix}.$$

From the analytical solution, it can be seen that a reliable numerical solution can be found, and it can be seen that the error level is $10^{-16}$, meaning the result obtained by expm() function is accurate:

$$A_3 = \begin{bmatrix} 0.0676676416183063 & 0.0676676416183063 & -1.582752333 \times 10^{-18} \\ 0.0676676416183063 & 0.0676676416183063 & 1.582752333 \times 10^{-18} \\ 3.16550466560 \times 10^{-18} & -3.1655046660 \times 10^{-18} & -2.8333891522 \times 10^{-18} \end{bmatrix}.$$

**Example 6.15.** For the given matrix $A$ in Example 6.11, compute function $e^{e^{At}t}$.

**Solutions.** Even with such a composite matrix function, nested calls of the function expm() can be used to compute the function directly:

```
>> syms t; A=[-3,-1,-1; 0,-3,-1; 1,2,0]; A=sym(A);
   A1=simplify(expm(expm(A*t)*t))
```

and the result obtained is

$$A_1 = \begin{bmatrix} e^{te^{-2t}-2t}(e^{2t}-t^2) \\ -t^3 e^{-2t}e^{te^{-2t}}/2 - t^4 e^{-4t}e^{te^{-2t}}/2 \\ t^2 e^{-2t}e^{te^{-2t}} + t^3 e^{-2t}e^{te^{-2t}}/2 + t^4 e^{-4t}e^{te^{-2t}}/2 \end{bmatrix}$$

$$\begin{array}{c} -t^2 e^{-2t}e^{te^{-2t}} \\ e^{te^{-2t}} - t^2 e^{-2t}e^{te^{-2t}} - t^3 e^{-2t}e^{te^{-2t}}/2 - t^4 e^{-4t}e^{te^{-2t}}/2 \\ 2t^2 e^{-2t}e^{te^{-2t}} + t^3 e^{-2t}e^{te^{-2t}}/2 + t^4 e^{-4t}e^{te^{-2t}}/2 \end{array}$$

$$\begin{array}{c} -t^2 e^{-2t}e^{te^{-2t}} \\ -t^2 e^{-2t}e^{te^{-2t}} - t^3 e^{-2t}e^{te^{-2t}}/2 - t^4 e^{-4t}e^{te^{-2t}}/2 \\ e^{te^{-2t}} + 2t^2 e^{-2t}e^{te^{-2t}} + t^3 e^{-2t}e^{te^{-2t}}/2 + t^4 e^{-4t}e^{te^{-2t}}/2 \end{array} \Bigg].$$

### 6.2.6 Jordan transform-based approach

Although the function expm() can be used to compute the matrix exponential of any matrix, an alternative algorithm – Jordan transformation matrix-based algorithm – is yet to be introduced. There is no need to use the method in computing matrix exponential, however, the algorithm is useful in computing other matrix functions. Here, an example is given to demonstrate the Jordan transformation matrix-based algorithm.

**Example 6.16.** For the matrix $A$ in Example 6.11, compute $e^A$ and $e^{At}$.

**Solutions.** Matrix exponential and exponential function can both be evaluated directly with the function expm().

```
>> syms t; A=[-3,-1,-1; 0,-3,-1; 1,2,0];    % input matrix
   A1=expm(A), A2=expm(sym(A))      % numerical and analytical solutions
   simplify(expm(A*t))              % matrix exponential function
```

The numerical result obtained is exactly the same as that obtained in Example 6.11. The analytical solution can also be obtained as

$$A_2 = \begin{bmatrix} 0 & -e^{-2} & -e^{-2} \\ -e^{-2}/2 & -e^{-2}/2 & -3e^{-2}/2 \\ 3e^{-2}/2 & 5e^{-2}/2 & 7e^{-2}/2 \end{bmatrix}.$$

Moreover, the matrix exponential function can also be obtained as

$$e^{At} = \begin{bmatrix} -e^{-2t}(-1+t) & -te^{-2t} & -te^{-2t} \\ -t^2 e^{-2t}/2 & -e^{-2t}(-1+t+t^2/2) & -te^{-2t}(2+t/2) \\ te^{-2t}/2 & te^{-2t}(2+t/2) & e^{-2t}(1+2t+t^2/2) \end{bmatrix}.$$

A Jordan transformation matrix-based algorithm for computing $e^{At}$ is demonstrated below:

```
>> [V,J]=jordan(A)    % Jordanian transform
```

The Jordan matrix $J$ and transformation matrix $V$ can be found as

$$V = \begin{bmatrix} 0 & -1 & 1 \\ -1 & 0 & 0 \\ 1 & 1 & 0 \end{bmatrix}, \quad J = \begin{bmatrix} -2 & 1 & 0 \\ 0 & -2 & 1 \\ 0 & 0 & -2 \end{bmatrix}.$$

It is known that for Jordan matrix $J$, one may directly write down $e^{Jt}$ as

$$e^{Jt} = e^{-t} \begin{bmatrix} 1 & t & t^2/2 \\ 0 & 1 & t \\ 0 & 0 & 1 \end{bmatrix}.$$

Therefore, with the Jordan transformation technique and Theorem 6.2, the matrix exponential function can also be found, and the result is exactly the same as that obtained above:

```
>> J1=exp(-2*t)*[1 t t^2/2; 0 1 t; 0 0 1];
   A1=simplify(V*J1*inv(V))
```

In fact, the computation of matrix exponential with such an algorithm is not our real objective, since the function expm() is powerful enough. The Jordan transformation-based method can be explored to compute matrix sine functions, or other functions to be discussed later.

## 6.3 Matrix logarithmic and square root functions

The matrix logarithmic and square root functions can be evaluated, and the algorithm in [12] can be considered. Examples are given to demonstrate the matrix computation.

### 6.3.1 Matrix logarithmic function

It is known from calculus that functions such as $\ln x$ cannot be expanded as a Taylor series about point $x = 0$. One may expand the function $\ln(1+x)$, or alternatively expand the function about $x = 1$. Therefore, Taylor series expansion of a logarithmic function can be written below as

$$\ln x = (x - 1) - \frac{1}{2}(x - 1)^2 + \frac{1}{3}(x - 1)^3 + \cdots + \frac{(-1)^{k+1}}{k}(x - 1)^k + \cdots, \tag{6.3.1}$$

and the convergence region is $|x - 1| < 1$.

**Theorem 6.4.** *Taylor series expansion of the natural logarithmic function* $\ln A$ *is*

$$\ln A = A - I - \frac{1}{2}(A - I)^2 + \frac{1}{3}(A - I)^3 + \cdots + \frac{(-1)^{k+1}}{k}(A - I)^k + \cdots, \tag{6.3.2}$$

*with the convergence region* $\|A - I\| < 1$.

Alternatively, MATLAB function $C$=logm($A$) can be used to evaluate the matrix logarithmic function directly. Also, $A$ can be expressed as a symbolic matrix.

**Example 6.17.** Compute numerically matrix exponential $e^A$ from Example 6.16. Compute the logarithmic function of the result and see whether the original matrix can be restored.

**Solutions.** The following commands can be used to input the matrix, then compute respectively the matrix exponential and logarithmic functions. The final result is compared with the original matrix:

```
>> syms t; A=[-3,-1,-1; 0,-3,-1; 1,2,0];
   A1=expm(A); A2=logm(A1), err=norm(A-A2)
```

The result obtained is given below, and the restoration error is $4.4979 \times 10^{-15}$:

$$A_2 = \begin{bmatrix} -3.000000000000000 & -1.000000000000001 & -1.000000000000000 \\ -0.000000000000003 & -3.000000000000002 & -1.000000000000002 \\ 1.000000000000001 & 2.000000000000000 & 0.000000000000002 \end{bmatrix}.$$

**Example 6.18.** Compute analytically the matrix exponential $e^A$ from Example 6.14. Compute the logarithmic function of/ the result and see whether the original matrix can be restored.

**Solutions.** The following statements can be used to compute the matrix exponential, and from the result, the matrix logarithm can be obtained:

```
>> A=[-21,19,-20; 19,-21,20; 40,-40,-40];
   A1=expm(A); A2=logm(A1)
```

The final result can be obtained as follows. Since the elements in the last row and last column of $A_1$ are very close to zero, the original matrix $A$ cannot be restored correctly with the function logm():

$$A_2 = \begin{bmatrix} -20.6379406437531 & 18.6379406437531 & -1.13770229508543 \\ 18.6379406437531 & -20.6379406437531 & 1.13770229508546 \\ 2.27540459565457 & -2.27540459565457 & -41.3125571195314 \end{bmatrix}.$$

Even with the symbolic framework, the original matrix $A$ cannot successfully be restored:

```
>> A1=expm(sym(A)); A2=simplify(logm(A1)), A3=vpa(A2)
```

The erroneous result is

$$A_3 = \begin{bmatrix} -21.0 & 19.0 & -1.15044407846124 \\ 19.0 & -21.0 & 1.15044407846124 \\ 2.30088815692248 & -2.30088815692248 & -40.0 \end{bmatrix}.$$

In fact, observing the first column of matrix $A_2$, it is easily found that the correct result is

$$A_2(:,1) = \begin{bmatrix} \ln(\cos 40 - j\sin 40)/4 + \ln(\cos 40 + j\sin 40)/4 - 21 \\ 19 - \ln(\cos 40 + j\sin 40)/4 - \ln(\cos 40 - j\sin 40)/4 \\ (\ln(\cos 40 - j\sin 40)j)/2 - (\ln(\cos 40 + j\sin 40)j)/2 \end{bmatrix},$$

since with Euler's formula, a series of manual simplifications are possible:

$$\ln(\cos -j\sin 40)/4 + \ln(\cos 40 + j\sin 40)/4 - 21 \rightarrow -40j/4 + 40j/4 - 21 \rightarrow -21$$

$$19 - \ln(\cos 40 + j\sin 40)/4 - \ln(\cos 40 - j\sin 40)/4 \rightarrow 19 - 40j/4 - (-40j)/4 \rightarrow 19$$

$$(\ln(\cos 40 - j\sin 40)j)/2 - (\ln(\cos 40 + j\sin 40)j)/2 \rightarrow (-40j)j/2 - (40j)j/2 \rightarrow 40.$$

The other columns can also be manipulated in the same way. For this particular example, the result obtained by vpa() function is not correct.

### 6.3.2 Square root of a matrix

Similar to the case of logarithmic function, there is no Taylor series expansion about $x = 0$ for a root function. The expansion about $x = 1$ should be used instead. In this section, a MATLAB solver of the square root function is demonstrated with an example.

The Taylor series expansion about $x = 1$ can be written as

$$\sqrt[q]{x^p} = 1 + \frac{p}{q}(x-1) + \frac{p(p-q)}{q \times 2q}(x-1)^2 + \frac{p(p-q)(p-2q)}{q \times 2q \times 3q}(x-1)^3 + \cdots. \tag{6.3.3}$$

The square root function is just a special case of the root function discussed above. If Taylor series expansion of a square root function is expected, it may be rather complicated to write down a general term. Symbolic computation in MATLAB can be used to compute the Taylor series formula.

**Example 6.19.** Find the Taylor series expansion about $x = 1$ for the function $\sqrt{x}$.

**Solutions.** If a Taylor series expansion is needed, the following MATLAB statements should be used:

```
>> syms x positive
   F=taylor(sqrt(x),x,1,'Order',7)
```

The first seven terms in the expansion can be obtained as

$$1 + \frac{x-1}{2} - \frac{(x-1)^2}{8} + \frac{(x-1)^3}{16} - \frac{5(x-1)^4}{128} + \frac{7(x-1)^5}{256} - \frac{21(x-1)^6}{1024}.$$

MATLAB function $A_1$=sqrtm($A$) can be used to compute the matrix square root function directly. An example is given below, and the accuracy can be tested with the function.

**Example 6.20.** Consider the matrix $A$ in Example 6.14. Compute $A^2$, then find its square root and validate the result.

**Solutions.** Computing $A^2$ first, the square root of it can be evaluated with sqrtm() function. The result can be compared with the original matrix $A$ with the following statements:

```
>> A=[-21,19,-20; 19,-21,20; 40,-40,-40];
   A0=A^2; A1=sqrtm(A0), A1=sqrtm(sym(A0))
```

The result is

$$A_1 = \begin{bmatrix} 20.999999999999883 & -19.000000000000071 & 19.999999999999932 \\ -19.000000000000046 & 20.999999999999940 & -20.000000000000004 \\ -40.000000000000000 & 40.000000000000043 & 40.000000000000050 \end{bmatrix}.$$

In fact, one of the square roots of matrix $A^2$ is found here, the other square root is $-A_1$.

In fact, apart from the method given above, there are different ways to compute the roots, such as the method of $A\hat{\,}(1/2)$, since with such a method, other root can be found, rather than just one square root. Multiple solutions of arbitrary roots of a matrix can also be found, with details given in Section 2.5.3.

## 6.4 Matrix trigonometric functions

Matrix trigonometric functions are also widely used. Numerical methods of matrix trigonometric function evaluation are introduced in this section first. Then, with the use of Euler's formula, numerical and analytical values of matrix trigonometric functions can be evaluated through exponential functions.

### 6.4.1 Computation of matrix trigonometric functions

In the current versions of MATLAB, there is no function for computing matrix trigonometric functions. Instead, a universal function funm() can be used to evaluate arbitrary matrix functions, with the syntax:

$A_1$=funm($A$, 'function_name'), for instance, $B$=funm($A$,@sin)

where "function_name" can be a function name in the form of quoted string or function handle. In the new versions of MATLAB, the funm($A*t$,'sin') or funm($A*t$,@sin) commands can also be used in computing sin $At$.

**Example 6.21.** Consider matrix $A$ in Example 6.11, compute sin $A$ and sin $At$.

**Solutions.** If one wants to compute the matrix sine of matrix $A$, the following statements can be used:

```
>> A=[-3,-1,-1; 0,-3,-1; 1,2,0];
   B1=funm(A,@sin), B2=funm(sym(A),@sin)
   syms t; C=simplify(funm(A*t,@sin))     % matrix sine function
   norm(B1-double(B2))
```

The numerical and analytical values of matrix sinusoidal function can be evaluated directly as

$$
B_1 = \begin{bmatrix}
-0.493150590278543 & 0.416146836547141 & 0.416146836547141 \\
-0.454648713412841 & -0.947799303691380 & -0.038501876865700 \\
0.038501876865700 & -0.377644959681444 & -1.286942386507125
\end{bmatrix},
$$

$$
B_2 = \begin{bmatrix}
-\cos 2 - \sin 2 & -\cos 2 & -\cos 2 \\
-\sin 2/2 & -\cos 2 - 3\sin 2/2 & -\cos 2 - \sin 2/2 \\
\cos 2 + \sin 2/2 & 2\cos 2 + \sin 2/2 & 2\cos 2 - \sin 2/2
\end{bmatrix},
$$

and the matrix function $\sin At$ is

$$
C = \begin{bmatrix}
-\sin 2t - t\cos 2t & -t\cos 2t & -t\cos 2t \\
-t^2 \sin 2t/2 & -\sin 2t - t\cos 2t - t^2 \sin 2t/2 & -t\cos 2t - t^2 \sin 2t/2 \\
t\cos 2t + t^2 \sin 2t/2 & 2t\cos 2t + t^2 \sin 2t/2 & 2t\cos 2t - \sin 2t + t^2 \sin 2t/2
\end{bmatrix}.
$$

It is also seen that the error of the numerical solution is $4.2028 \times 10^{-15}$.

### 6.4.2 Power series-based trigonometric function evaluation

Power series-based matrix exponential algorithm and its MATLAB implementation were discussed earlier. If fact, matrix trigonometric functions can also be expressed as Taylor series, and hence similar algorithms can be established. In this section, matrix sine function is presented, followed by its MATLAB implementation.

**Theorem 6.5.** *The matrix sinusoidal function* $\sin A$ *can be expressed in Taylor series as*

$$
\sin A = \sum_{k=0}^{\infty} (-1)^k \frac{A^{2k+1}}{(2k+1)!} = A - \frac{1}{3!}A^3 + \frac{1}{5!}A^5 + \cdots. \tag{6.4.1}
$$

**Theorem 6.6.** *The matrix cosine function* $\cos A$ *can be expressed in Taylor series as*

$$
\cos A = I - \frac{1}{2!}A^2 + \frac{1}{4!}A^4 - \frac{1}{6!}A^6 + \cdots + \frac{(-1)^n}{(2n)!}A^{2n} + \cdots. \tag{6.4.2}
$$

**Theorem 6.7.** *The matrix arcsine function* $\arcsin A$ *can be expressed in Taylor series as*

$$
\arcsin A = A + \frac{1}{2 \cdot 3}A^3 + \frac{1 \cdot 3}{2 \cdot 4 \cdot 5}A^5 + \frac{1 \cdot 3 \cdot 5}{2 \cdot 4 \cdot 6 \cdot 7}A^7
$$

$$
+ \frac{1 \cdot 3 \cdot 5 \cdot 7}{2 \cdot 4 \cdot 6 \cdot 8 \cdot 9}A^9 + \cdots + \frac{(2n)!}{2^{2n}(n!)^2(2n+1)}A^{2n+1} + \cdots. \tag{6.4.3}
$$

If power series is used to find the numerical values of a matrix function, the general term can be evaluated recursively by finding the radio of the next and current terms. Then a loop structure can be used to accumulate the sum. For this example, the $k$th term is expressed as

$$F_k = (-1)^k \frac{A^{2k+1}}{(2k+1)!}, \quad k = 0, 1, 2, \ldots \tag{6.4.4}$$

Therefore, the next term divided by the current term can be evaluated as follows (for simplicity, the division notation is used rather than the correct form of inverse matrices):

$$\frac{F_{k+1}}{F_k} = \frac{(-1)^{k+1} A^{2(k+1)+1}/(2(k+1)+1)!}{(-1)^k A^{2k+1}/(2k+1)!} = -\frac{A^2}{(2k+3)(2k+2)}. \tag{6.4.5}$$

It can be seen that the following recursive formula can be obtained:

$$F_{k+1} = -\frac{A^2 F_k}{(2k+3)(2k+2)}, \tag{6.4.6}$$

where $k = 0, 1, 2, \ldots$, and the initial term is $F_0 = A$.

Similarly, the recursive formulas for other matrix trigonometric functions can be formulated.

**Example 6.22.** In fact, nonlinear matrix functions can easily be evaluated with power series. Write down a general purpose MATLAB function to compute the matrix sine function.

**Solutions.** The accumulation process can be implemented in a loop structure. When the general term is small enough, the accumulation process can be terminated. Here, the convergence condition is expressed mathematically as $\|E + F - E\|_1 > 0$, meaning that, it the amount in $F$, when added to $E$, does not change the value of $E$ in the double precision framework, the general term is small enough. Note that the condition cannot be modified to $\|F\|_1 > 0$.

```
function E=sinm1(A)
F=A; E=A; k=0; % if the general term is negligible, terminate the loop
while norm(E+F-E,1)>0,
    F=-A^2*F/(2*k+3)/(2*k+2); E=E+F; k=k+1;
end
```

It can be seen from the MATLAB function that the seemingly rather complicated algorithm for the matrix sine function evaluation can be implemented with a few lines of code. With the new function $A_1$=sinm1 ($A$), the sine matrix $A_1$ can immediately be evaluated.

**Example 6.23.** Consider again the matrix given in Example 6.11. Compute $\sin A$ with the new function and validate the results.

**Solutions.** The matrix function can be evaluated with the $\mathtt{sinm1()}$ function

```
>> A=[-3,-1,-1; 0,-3,-1; 1,2,0];
   A1=sinm1(A), norm(A1-double(funm(sym(A),@sin)))
```

and the result obtained is as follows. It can be seen that, compared with the theoretical result, the error is $2.4022 \times 10^{-16}$. The result in Example 6.21 is more reliable:

$$A_1 = \begin{bmatrix} -0.493150590278539 & 0.416146836547142 & 0.416146836547142 \\ -0.454648713412841 & -0.947799303691380 & -0.038501876865698 \\ 0.038501876865698 & -0.377644959681444 & -1.286942386507125 \end{bmatrix}.$$

### 6.4.3 Analytical evaluations of matrix trigonometric functions

The analytical computations of matrix trigonometric functions are presented first, with Euler's formula. It is named after Swiss mathematician Leonhard Euler (1707–1783), who showed that $e^{j\theta} = \cos\theta + j\sin\theta$.

**Theorem 6.8.** *If $A$ is a square matrix, from the well-known Euler's formulas $e^{jA} = \cos A + j\sin A$ and $e^{-jA} = \cos A - j\sin A$, it can be found that*

$$\sin A = \frac{1}{j2}(e^{jA} - e^{-jA}), \quad \cos A = \frac{1}{2}(e^{jA} + e^{-jA}). \tag{6.4.7}$$

From the reliable $\mathtt{expm()}$ function, the analytical values of matrix trigonometric functions can easily be computed. Matrix trigonometric functions with complicated structures can also be computed.

**Example 6.24.** For the matrix $A$ in Example 6.11 compute $\sin A$.

**Solutions.** With the existing function $\mathtt{expm()}$, matrix sine function can easily be evaluated from

```
>> A=[-3,-1,-1; 0,-3,-1; 1,2,0]; j=sqrt(-1);
   A1=(expm(A*j)-expm(-A*j))/(2*j)
```

It can be seen that, the result obtained here is exactly the same as that in Example 6.22, meaning that the result is correct.

**Example 6.25.** If matrix $A$ has repeated eigenvalues, compute $\sin At$ and $\cos At$.

$$A = \begin{bmatrix} -7 & 2 & 0 & -1 \\ 1 & -4 & 2 & 1 \\ 2 & -1 & -6 & -1 \\ -1 & -1 & 0 & -4 \end{bmatrix}.$$

**Solutions.** With (6.4.7), the following statements can be used directly to compute the matrix sine and cosine functions:

```
>> A=[-7,2,0,-1; 1,-4,2,1; 2,-1,-6,-1; -1,-1,0,-4]; % matrix input
   syms t, A1=(expm(A*1j*t)-expm(-A*1j*t))/(2*1j);
   A1=simplify(A1), A2=(expm(A*1j*t)+expm(-A*1j*t))/2;
   A2=simplify(A2) % Euler's formulas evaluation
```

In fact, although simplifications are made in the function call, the results still contain exponential functions with complex arguments, in the new versions of MATLAB. The function rewrite() should be called to further simplify the results

```
>> simplify(rewrite(A1,'sin')), simplify(rewrite(A2,'sin'))
```

The results obtained are as follows, which are the same as those in the older versions of MATLAB:

$$\sin At = \begin{bmatrix} -2/9\sin 3t + (t^2 - 7/9)\sin 6t - 5/3t\cos 6t & -1/3\sin 3t + 1/3\sin 6t + t\cos 6t \\ -2/9\sin 3t + (t^2 + 2/9)\sin 6t + 1/3t\cos 6t & -1/3\sin 3t - 2/3\sin 6t + t\cos 6t \\ -2/9\sin 3t - (2t^2 - 2/9)\sin 6t + 4/3t\cos 6t & -1/3\sin 3t + 1/3\sin 6t - 2t\cos 6t \\ 4/9\sin 3t + (t^2 - 4/9)\sin 6t + 1/3t\cos 6t & 2/3\sin 3t - 2/3\sin 6t + t\cos 6t \end{bmatrix}$$

$$\begin{bmatrix} -2/9\sin 3t + (2/9 + t^2)\sin 6t - 2/3t\cos 6t & 1/9\sin 3t + (-1/9 + t^2)\sin 6t - 2/3t\cos 6t \\ -2/9\sin 3t + (2/9 + t^2)\sin 6t + 4/3t\cos 6t & 1/9\sin 3t - (1/9 - t^2)\sin 6t + 4/3t\cos 6t \\ -2/9\sin 3t - (7/9 + 2t^2)\sin 6t - 2/3t\cos 6t & 1/9\sin 3t - (1/9 + 2t^2)\sin 6t - 2/3t\cos 6t \\ 4/9\sin 3t - (4/9 - t^2)\sin 6t + 4/3t\cos 6t & -2/9\sin 3t - (7/9 - t^2)\sin 6t + 4/3t\cos 6t \end{bmatrix},$$

$$\cos At = \begin{bmatrix} 2/9\cos 3t - (t^2 + 7/9)\cos 6t - 5/3t\sin 6t & 1/3\cos 3t - 1/3\cos 6t + t\sin 6t \\ 2/9\cos 3t - (t^2 + 2/9)\cos 6t + 1/3t\sin 6t & 1/3\cos 3t + 2/3\cos 6t + t\sin 6t \\ 2/9\cos 3t + (2t^2 - 2/9)\cos 6t + 4/3t\sin 6t & 1/3\cos 3t - 1/3\cos 6t - 2t\sin 6t \\ -4/9\cos 3t - (t^2 - 4/9)\cos 6t + 1/3t\sin 6t & -2/3\cos 3t + 2/3\cos 6t + t\sin 6t \end{bmatrix}$$

$$\begin{bmatrix} 2/9\cos 3t - (2/9 + t^2)\cos 6t - 2/3t\sin 6t & -1/9\cos 3t + (1/9 - t^2)\cos 6t - 2/3t\sin 6t \\ 2/9\cos 3t - (2/9 + t^2)\cos 6t + 4/3t\sin 6t & -1/9\cos 3t + (1/9 - t^2)\cos 6t + 4/3t\sin 6t \\ 2/9\cos 3t + (7/9 + 2t^2)\cos 6t - 2/3t\sin 6t & -1/9\cos 3t + (1/9 + 2t^2)\cos 6t - 2/3t\sin 6t \\ -4/9\cos 3t + (4/9 - t^2)\cos 6t + 4/3t\sin 6t & 2/9\cos 3t + (7/9 - t^2)\cos 6t + 4/3t\sin 6t \end{bmatrix}.$$

**Example 6.26.** Consider the matrix in Example 6.11 and compute $\cos e^{At}$.

**Solutions.** The matrix can be input into MATLAB first, then the exponential can be evaluated. Then, the following statements can be used to compute the matrix cosine function:

```
>> A=[-3,-1,-1; 0,-3,-1; 1,2,0];
   syms t; A1=simplify(funm(expm(A*t),@cos))
```

The result is as follows:

$$
A_1 = \begin{bmatrix}
\cos\theta + t\theta\sin\theta & t\theta\sin\theta \\
t^2\theta^2\cos\theta + \sin\theta/(2\theta) & \cos\theta + t\theta\sin\theta + t^2\theta^2\cos\theta/2 + t^2\theta\sin\theta/2 \\
-t^2\theta^2\cos\theta/2 - t(t+2)\theta\sin\theta/2 & -2t\theta\sin\theta - t^2\theta^2\cos\theta/2 - t^2\theta\sin\theta/2
\end{bmatrix}
$$

$$
\begin{matrix}
t\theta\sin\theta \\
t\theta\sin\theta + t^2\theta^2\cos\theta/2 + t^2\theta\sin\theta/2 \\
\cos\theta - 2t\theta\sin\theta - t^2\theta^2\cos\theta/2 - t^2\theta\sin\theta/2
\end{matrix} \Bigg],
$$

where $\theta = e^{-2t}$.

## 6.5 Arbitrary matrix functions

Apart from the existing matrix exponentials, logarithmic and trigonometric functions, MATLAB can also be used to evaluate other nonlinear matrix functions. It should be pointed out that, although the new version of funm() function can be used to compute analytical matrix functions, there are still a few limitations.

In this section, Jordan transform-based matrix function evaluation algorithm is presented, followed by its MATLAB implementations[13, 29]. The algorithm and MATLAB code were introduced in 2004, when there were no other solvers for such kind of problems.

### 6.5.1 Nilpotent matrices

**Definition 6.10.** A commonly used nilpotent matrix is the matrix whose first super-diagonal elements are ones, while the other elements are zeros, with mathematical form

$$
H_n = \begin{bmatrix}
0 & 1 & 0 & \cdots & 0 \\
0 & 0 & 1 & \cdots & 0 \\
\vdots & \vdots & \vdots & \ddots & \vdots \\
0 & 0 & 0 & \cdots & 0
\end{bmatrix}. \tag{6.5.1}
$$

**Example 6.27.** For a given $4 \times 4$ nilpotent matrix $H$, observe its powers.

**Solutions.** A $4 \times 4$ nilpotent matrix can be generated first, then a loop structure can be used to compute the powers. Observe the changes in the position of the 1's.

```
>> H=diag([1 1 1],1)
   for i=2:4, H^i, end % observe the positions of 1's
```

In the following display, the first matrix is the nilpotent matrix, and the others are the powers of this matrix. The matrix $H^4$ and all subsequent ones are matrices of zeros.

$$H = \begin{bmatrix} 0 & 1 & 0 & 0 \\ 0 & 0 & 1 & 0 \\ 0 & 0 & 0 & 1 \\ 0 & 0 & 0 & 0 \end{bmatrix}, \quad H^2 = \begin{bmatrix} 0 & 0 & 1 & 0 \\ 0 & 0 & 0 & 1 \\ 0 & 0 & 0 & 0 \\ 0 & 0 & 0 & 0 \end{bmatrix},$$

$$H^3 = \begin{bmatrix} 0 & 0 & 0 & 1 \\ 0 & 0 & 0 & 0 \\ 0 & 0 & 0 & 0 \\ 0 & 0 & 0 & 0 \end{bmatrix}, \quad H^4 = \begin{bmatrix} 0 & 0 & 0 & 0 \\ 0 & 0 & 0 & 0 \\ 0 & 0 & 0 & 0 \\ 0 & 0 & 0 & 0 \end{bmatrix}.$$

**Theorem 6.9.** *If $H_m$ is an $m \times m$ nilpotent matrix, and if $k \geqslant m$, then $H_m^k \equiv \mathbf{0}$.*

### 6.5.2 Jordan transform-based matrix function evaluations

With the concept of nilpotent matrices, a Jordan matrix can be expressed as the sum of a diagonal matrix and a nilpotent matrix. With such an idea, the computation of an ordinary matrix function can be formulated.

Before further presenting the algorithm, the Taylor series expansion of $f(x)$ is presented first

$$f(\lambda + \Delta t) = f(\lambda) + f'(\lambda)\Delta t + \frac{1}{2!}f''(\lambda)\Delta t^2 + \cdots. \tag{6.5.2}$$

**Theorem 6.10.** *An $m_i \times m_i$ Jordanian block $J_i$ can be written as $J_i = \lambda_i I + H_{m_i}$, where $\lambda_i$ is the repeated eigenvalue of the Jordan matrix and $H_{m_i}$ is a nilpotent matrix. With the properties of nilpotent matrices, the matrix function $\psi(J_i)$ of the Jordan block $J_i$ can be evaluated from*

$$\psi(J_i) = \psi(\lambda_i)I_{m_i} + \psi'(\lambda_i)H_{m_i} + \cdots + \frac{\psi^{(m_i-1)}(\lambda_i)}{(m_i-1)!}H_{m_i}^{m_i-1}. \tag{6.5.3}$$

**Theorem 6.11.** *The equivalent form of the Jordan matrix in Theorem 6.10 can be expressed as*

$$\psi(J_i) = \begin{bmatrix} \psi(\lambda_i) & \psi'(\lambda_i)/1! & \cdots & \psi^{(m_i-1)}(\lambda_i)/(m_i-1)! \\ 0 & \psi(\lambda_i) & \cdots & \psi^{(m_i-2)}(\lambda_i)/(m_i-2)! \\ \vdots & \vdots & \ddots & \vdots \\ 0 & 0 & \cdots & \psi(\lambda_i) \end{bmatrix}. \tag{6.5.4}$$

**Theorem 6.12.** *If an arbitrary matrix A can be expressed using Jordan decomposition as*

$$
A = V \begin{bmatrix} J_1 & & & \\ & J_2 & & \\ & & \ddots & \\ & & & J_m \end{bmatrix} V^{-1}, \tag{6.5.5}
$$

*then an arbitrary function $\psi(A)$ can be evaluated from*

$$
\psi(A) = V \begin{bmatrix} \psi(J_1) & & & \\ & \psi(J_2) & & \\ & & \ddots & \\ & & & \psi(J_m) \end{bmatrix} V^{-1}, \tag{6.5.6}
$$

*where the functions of Jordan blocks can be evaluated as in Theorem 6.10.*

Based on the above algorithm, a MATLAB function `funmsym()` can be written, which can be used to compute the analytical value of any matrix function. The listing of the function is

```
function F=funmsym(A,fun,x)
[V,T]=jordan(sym(A)); vec=diag(T);
v1=[0,diag(T,1)',0]; v2=find(v1==0);  v_n=v2(2:end)-v2(1:end-1);
lam=vec(v2(1:end-1)); vec(v2(1:end-1));
m=length(lam); F=sym([]); % configure Jordan blocks and count numbers
for i=1:m                 % process Jordan blocks
    k=v2(i):v2(i)+v_n(i)-1; J1=T(k,k); fJ=funJ(J1,fun,x); F(k,k)=fJ;
end
F=V*F*inv(V); % compute matrix function with (6.5.5)
function fJ=funJ(J,fun,x), lam=J(1,1); % Jordan block processing
f1=fun; fJ=subs(fun,x,lam)*eye(size(J)); H=diag(diag(J,1),1); H1=H;
for i=2:length(J) % compute arbitrary matrix with nilpotent matrix
    f1=diff(f1,x); a1=subs(f1,x,lam); fJ=fJ+a1*H1; H1=H1*H/i;
end
```

The syntax of the function is $A_1$=`funmsym`$(A,\texttt{funx},x)$, where $x$ is a symbolic variable, while `funx` is the prototype function of $x$. For instance, if $e^A$ is expected, the prototype function `funx` should be expressed as `exp(x)`. In fact, `funx` can be used to describe functions of any complexity, for instance, `exp(x*t)` for computing $e^{At}$, where $t$ is another independent symbolic variable. Also the prototype function can be specified as `exp(x*cos(x*t))` for computing the composite matrix function $\psi(A) = e^{A\cos At}$.

**Example 6.28.** Solve the matrix computing problem again in Example 6.15.

**Solutions.** Since matrix function $e^{e^{At}t}$ is expected, the prototype function should be specified as $\exp(\exp(x*t)*t)$. Therefore, the following statements can be used, and the results are exactly the same as those obtained in Example 6.15:

```
>> syms t; A=[-3,-1,-1; 0,-3,-1; 1,2,0]; A=sym(A);
   syms x; A2=funmsym(A,exp(exp(x*t)*t),x)
```

**Example 6.29.** Compute $\psi(A) = e^{A \cos At}$, if

$$A = \begin{bmatrix} -7 & 2 & 0 & -1 \\ 1 & -4 & 2 & 1 \\ 2 & -1 & -6 & -1 \\ -1 & -1 & 0 & -4 \end{bmatrix}.$$

**Solutions.** For the matrix function $\psi(A) = e^{A \cos At}$, the prototype function should be described as $f=\exp(x*\cos(x*t))$. The following statements should be used to compute the composite matrix function:

```
>> A=[-7,2,0,-1; 1,-4,2,1; 2,-1,-6,-1; -1,-1,0,-4];    % input matrix
   syms x t; A=sym(A);
   A1=funmsym(A,exp(x*cos(x*t)),x) % matrix function
   A2=expm(A*funm(A*t,@cos)) % the results with new MATLAB functions
```

The result obtained is rather complicated, therefore, only the upper left corner element is displayed as follows:

$$\psi_{1,1}(A) = 2/9e^{-3\cos 3t} + (2t\sin 6t + 6t^2\cos 6t)e^{-6\cos 6t} + (\cos 6t - 6t\sin 6t)^2 e^{-6\cos 6t}$$
$$- 5/3(\cos 6t - 6t\sin 6t)e^{-6\cos 6t} + 7/9e^{-6\cos 6t}.$$

It can be seen that the $e^{-6\cos 6t}$ factor appears in the coefficients of many terms of $\psi_{1,1}(t)$, so it can be collected with the following statements to simplify the result:

```
>> collect(A1(1,1),exp(-6*cos(6*t)))   % collect the terms
```

and the simplified result is

$$\psi_{1,1}(A) = [12t\sin 6t + 6t^2\cos 6t + (\cos 6t - 6t\sin 6t)^2 - 5\cos 6t/3 + 7/9]e^{-6\cos 6t}$$
$$+ 2e^{-3\cos 3t}/9.$$

Further, letting $t = 1$, an accurate numerical value of $e^{A \cos A}$ can be calculated with the following statement:

```
>> subs(A1,t,1)
```

The result obtained is

$$e^{A \cos A} = \begin{bmatrix} 4.3583 & 6.5044 & 4.3635 & -2.1326 \\ 4.3718 & 6.5076 & 4.3801 & -2.116 \\ 4.2653 & 6.4795 & 4.2518 & -2.2474 \\ -8.6205 & -12.984 & -8.6122 & 4.3832 \end{bmatrix}.$$

The result is more accurate than that in `expm(A*funm(A,'cos'))`.

**Example 6.30.** Consider again the matrix in Example 6.16. Compute $e^A$ and also the logarithmic function of the result. See whether the original matrix can be restored.

**Solutions.** Analytical value of the exponential function $e^A$ of the matrix can be computed, from which the logarithmic function can be evaluated, where a symbolic matrix $A_2$ is returned. It can be that $A_2$ and $A$ are identical, meaning the results here are reliable.

```
>> A=[-3,-1,-1; 0,-3,-1; 1,2,0]; A=sym(A);
   syms x; A1=expm(A);   A2=simplify(funmsym(A1,log(x),x))
```

**Example 6.31.** Compute the square root of the matrix in Example 6.29.

**Solutions.** Since there is a triple eigenvalue of the matrix, all three numerical methods in MATLAB fail. With the new function `funmsym()`, correct results can be obtained, indicating that the function is effective. The other two analytical solutions, $A_4$ and $A_5$, can also be obtained:

```
>> A=[-7,2,0,-1; 1,-4,2,1; 2,-1,-6,-1; -1,-1,0,-4]; % matrix input
   A1=sqrtm(A), A1^2, A2=A^(1/2), A2^2,
   A3a=funm(A,@sqrt)    % the three unsuccessful methods
   syms x; A4=funmsym(sym(A),sqrt(x),x),
   simplify(A4^2), A5=-A4 % the other two solutions
```

The analytical solution of the problem is

$$A_4 = j \begin{bmatrix} 2\sqrt{3}/9 + 131\sqrt{6}/144 & \sqrt{3}/3 - 5\sqrt{6}/12 & 2\sqrt{3}/9 - 25\sqrt{6}/144 & -\sqrt{3}/9 + 23\sqrt{6}/144 \\ 2\sqrt{3}/9 - 37\sqrt{6}/144 & \sqrt{3}/3 + 7\sqrt{6}/12 & 2\sqrt{3}/9 - 49\sqrt{6}/144 & -\sqrt{3}/9 - \sqrt{6}/144 \\ 2\sqrt{3}/9 - 23\sqrt{6}/72 & \sqrt{3}/3 - \sqrt{6}/6 & 2\sqrt{3}/9 + 61\sqrt{6}/72 & \sqrt{3}/9 + 13\sqrt{6}/72 \\ -4\sqrt{3}/9 + 59\sqrt{6}/144 & -2\sqrt{3}/3 + 7\sqrt{6}/ & -4\sqrt{3}/9 + 47\sqrt{6}/144 & 2\sqrt{3}/9 + 95\sqrt{6}/12\,144 \end{bmatrix}.$$

### 6.5.3 User-defined matrix functions

In real applications, matrix functions are not restricted to the transcendental functions presented earlier, user-defined functions should be allowed, such that matrix functions of any complexity should be supported, for instance, matrix Mittag-Leffler functions, which are extensions of the exponential functions. In this section, Mittag-Leffler function is given here to represent user-defined matrix functions.

**Definition 6.11.** A univariate Mittag-Leffler function is defined as

$$E_\alpha(x) = \sum_{k=1}^{\infty} \frac{1}{\Gamma(\alpha k + 1)} x^k, \tag{6.5.7}$$

where $\Gamma(\cdot)$ is the Gamma function.

Mittag-Leffler function was proposed in 1903 by Swedish mathematician Magnus Gustaf Mittag-Leffler (1846–1927). The function is an important special function in fractional calculus.

**Example 6.32.** For the given matrix $A$, compute Mittag-Leffler matrix function $\Phi(t) = E_\alpha(At^\alpha)$, where $E_\alpha(\cdot)$ is the univariate Mittag-Leffler matrix function[28] and

$$A(t) = \begin{bmatrix} -2 & 0 & -1 & 0 \\ -1 & -3 & 1 & 0 \\ 2 & 1 & 1 & 1 \\ 0 & 1 & -2 & -2 \end{bmatrix}.$$

**Solutions.** Defining a symbolic function $E(x)$ for the Mittag-Leffler function $E_\alpha(x)$, the matrix function of $E_\alpha(At^\alpha)$ can be expressed as a template function $E(x*t^\alpha)$, such that funmsym() function can be used to compute the Mittag-Leffler function matrix:

```
>> syms t x a E(x)              % declaring symbolic variables and function
   A=[-2,0,-1,0; -1,-3,1,0; 2,1,1,1; 0,1,-2,-2];   % input matrix
   Phi=simplify(funmsym(A,E(x*t^a),x))  % Compute matrix function
```

The matrix Mittag-Leffler function obtained is

$$\Phi(t) = \begin{bmatrix} E_\alpha(-t^\alpha) - t^{2\alpha} E_\alpha''(-t^\alpha)/2 - t^\alpha E_\alpha'(-t^\alpha) \\ E_\alpha(-3t^\alpha) - E_\alpha(-t^\alpha) + t^{2\alpha}E_\alpha''(-t^\alpha)/2 + t^\alpha E_\alpha'(-t^\alpha) \\ t^{2\alpha}E_\alpha''(-t^\alpha)/2 + 2t^\alpha E_\alpha'(-t^\alpha) \\ E_\alpha(-t^\alpha) - E_\alpha(-3t^\alpha) - t^{2\alpha}E_\alpha''(-t^\alpha)/2 - 2t^\alpha E_\alpha'(-t^\alpha) \end{bmatrix}$$

$$\begin{matrix} -t^{2\alpha}E_\alpha''(-t^\alpha)/2 \\ E_\alpha(-3t^\alpha) + t^{2\alpha}E_\alpha''(-t^\alpha)/2 \\ t^{2\alpha}E_\alpha''(-t^\alpha)/2 + t^\alpha E_\alpha'(-t^\alpha) \\ E_\alpha(-t^\alpha) - E_\alpha(-3t^\alpha) - t^{2\alpha}E_\alpha''(-t^\alpha)/2 - t^\alpha E_\alpha'(-t^\alpha) \end{matrix}$$

$$\begin{matrix} -t^{2\alpha}E_\alpha''(-t^\alpha)/2 - t^\alpha E_\alpha'(-t^\alpha) & & -t^{2\alpha}E_\alpha''(-t^\alpha)/2 \\ t^{2\alpha}E_\alpha''(-t^\alpha)/2 + t^\alpha E_\alpha'(-t^\alpha) & & t^{2\alpha}E_\alpha''(-t^\alpha)/2 \\ E_\alpha(-t^\alpha) + t^{2\alpha}E_\alpha''(-t^\alpha)/2 + 2t^\alpha E_\alpha'(-t^\alpha) & & t^{2\alpha}E_\alpha''(-t^\alpha)/2 + t^\alpha E_\alpha'(-t^\alpha) \\ -t^{2\alpha}E_\alpha''(-t^\alpha)/2 - 2t^\alpha E_\alpha'(-t^\alpha) & & E_\alpha(-t^\alpha) - t^{2\alpha}E_\alpha''(-t^\alpha)/2 - t^\alpha E_\alpha'(-t^\alpha) \end{matrix} \Bigg].$$

where $E'_\alpha(\cdot)$ and $E''_\alpha(\cdot)$ are respectively the first- and second-order derivatives of Mittag-Leffler function $E_\alpha(\cdot)$ with respect to $t$.

If function funm() is used, the following statements can be employed to solve the problem directly. The result obtained is exactly the same as that obtained earlier:

```
>> P1=simplify(funm(A*t^a,E))
```

## 6.6 Matrix powers

The computation of the $k$th power of a square matrix $A$, i. e., $A^k$, is considered here, where $k$ is an integer. If $k$ is not an integer, the evaluation of $A^k$ is not simple, since an infinite sum is involved. An example is given in this section, followed by a universal way of manipulating matrix power problems.

### 6.6.1 Matrix powers via Jordan transform

**Theorem 6.13.** *Assume that matrix $A$ can be transformed to a Jordan matrix, as $A = VJV^{-1}$. Matrix power can then be written as $A^k = VJ^kV^{-1}$, from which $J^k$ can be computed.*

**Theorem 6.14.** *It was indicated that $J = \lambda I + H_m$, where $H_m$ is an $m \times m$ nilpotent matrix and, when $k \geqslant m$, one has $H^k_m \equiv 0$. It is known from binomial expansion that*

$$J^k = \lambda^k I + k\lambda^{k-1} H_m + \frac{k(k-1)}{2!}\lambda^{k-2}H^2_m + \cdots. \tag{6.6.1}$$

Since $H^m_m$ and its subsequent powers are zero matrices, the infinite series can be computed directly with $m$ terms. Therefore, the matrix $J^k$ can be computed analytically as

$$J^k = \lambda^k I + k\lambda^{k-1} H_m + \frac{k(k-1)}{2!}\lambda^{k-2}H^2_m + \cdots + \frac{k!}{(m-1)!(k-m)!}\lambda^{k-1}H^{m-1}_m. \tag{6.6.2}$$

**Example 6.33.** Consider matrix $A$ in Example 6.11, given below. Compute $A^k$, where $k$ can be any integer, for

$$A = \begin{bmatrix} -3 & -1 & -1 \\ 0 & -3 & -1 \\ 1 & 2 & 0 \end{bmatrix}.$$

**Solutions.** Jordan transform should be carried out first, with

```
>> A=sym([-3,-1,-1; 0,-3,-1; 1,2,0]);
   syms k, [V J]=jordan(sym(A))
```

It can be seen that $J$ is a $3 \times 3$ Jordan matrix, whose eigenvalues are all equal to $\lambda = -2$. Now $H = J - \lambda I$ can be used to extract the nilpotent matrix. Therefore, the following commands can be used to compute the power:

```
>> A0=-2*eye(3); H=J-A0; % extract nilpotent matrix
   J1=A0^k+k*A0^(k-1)*H+k*(k-1)/2*A0^(k-2)*H^2; % add three terms
   F=simplify(V*J1*inv(V))
```

The $k$th power of matrix $A$ can be obtained as follows, where integer $k$ can be both positive and negative:

$$F = \begin{bmatrix} (-2)^k(k+2)/2 & (-2)^k k/2 & (-2)^k k/2 \\ -(-2)^{(k-2)}k(k-1)/2 & (-2)^k(-k^2+5k+8)/8 & -(-2)^k k(k-5)/8 \\ (-2)^k k(k-5)/8 & (-2)^k k(k-9)/8 & (-2)^k(k^2-9k+8)/8 \end{bmatrix}.$$

**Example 6.34.** What will happen in Example 6.33, if $k$ is selected as a rational number, rather than integer?

**Solutions.** In fact, $k$ can be a rational number for the results in Example 6.33. Letting $k = 1/121$, the power can also be found and validated. The case where $k$ is a different rational number cannot be validated here; this will be demonstrated later.

```
>> A=sym([-3,-1,-1; 0,-3,-1; 1,2,0]); syms k
   [V J]=jordan(A); A0=-2*eye(3); H=J-A0;
   J1=A0^k+k*A0^(k-1)*H+k*(k-1)/2*A0^(k-2)*H^2;
   F=simplify(V*J1*inv(V))
   m=sym(121); F1=subs(F,k,1/m), F1^m
```

The result obtained is given as follows, and it is noted that it cannot be obtained by the direct command of $A^{\wedge}(1/121)$.

$$F_1 = \begin{bmatrix} 243\sqrt[121]{-2}/242 & \sqrt[121]{-2}/242 & \sqrt[121]{-2}/242 \\ 15\sqrt[121]{-2}/14\,641 & 29\,433\sqrt[121]{-2}/29\,282 & 151\sqrt[121]{-2}/29\,282 \\ -151\sqrt[121]{-2}/29\,282 & -136\sqrt[121]{-2}/14\,641 & 14\,505\sqrt[121]{-2}/14\,641 \end{bmatrix}.$$

### 6.6.2 A matrix power solver

Consider again the ideas in function `funmsym()`. A matrix power function can also be written in a similar way, where the kernel subfunction `funJ()` can be replaced by another one, `powJ()`.

```
function F=mpowersym(A,k)
A=sym(A); [V,T]=jordan(A); vec=diag(T); v1=[0,diag(T,1)',0];
```

```
v2=find(v1==0); lam=vec(v2(1:end-1)); m=length(lam);
for i=1:m, % loop structure for each Jordan block
    k0=v2(i):v2(i+1)-1; J1=T(k0,k0); F(k0,k0)=powJ(J1,k);
end
F=simplify(V*F*inv(V));    % simplify the matrix power
function fJ=powJ(J,k)       % compute the kth power of Jordan matrix
lam=J(1,1); I=eye(size(J)); H=J-lam*I; fJ=lam^k*I; H1=k*H;
for i=2:length(J), fJ=fJ+lam^(k+1-i)*I*H1; H1=H1*H*(k+1-i)/i; end
```

**Example 6.35.** Consider matrix $A$ in Example 6.29, compute $F = A^k$ if

$$A = \begin{bmatrix} -7 & 2 & 0 & -1 \\ 1 & -4 & 2 & 1 \\ 2 & -1 & -6 & -1 \\ -1 & -1 & 0 & -4 \end{bmatrix}.$$

**Solutions.** The following commands can be used to compute the matrix power:

```
>> A=[-7,2,0,-1; 1,-4,2,1; 2,-1,-6,-1; -1,-1,0,-4]; % input A
   syms k, A=sym(A); F=mpowersym(A,k)    % compute the power A^k
   F=collect(F,(-6)^k)                    % combine terms of (-6)^k
```

The simplified matrix power can be obtained as

$$F = \begin{bmatrix} (k^2/36 + k/4 + 7/9)(-6)^k + 2(-3)^k/9 & (-k/6 - 1/3)(-6)^k + (-3)^k/3 \\ (k^2/36 - k/12 - 2/9)(-6)^k 2 + (-3)^k/9 & (2/3 - k/6)(-6)^k + (-3)^k/3 \\ (-k^2/18 - k/6 - 2/9)(-6)^k + 2(-3)^k/9 & (k/3 - 1/3)(-6)^k + (-3)^k/3 \\ (k^2/36 - k/12 + 4/9)(-6)^k - 4(-3)^k/9 & (2/3 - k/6)(-6)^k - 2(-3)^k/3 \end{bmatrix}$$

$$\begin{bmatrix} (k^2/36 + k/12 - 2/9)(-6)^k + 2(-3)^k/9 & (k^2/36 + k/12 + 1/9)(-6)^k - (-3)^k/9 \\ (k^2/36 - k/4 - 2/9)(-6)^k + 2(-3)^k/9 & (k^2/36 - k/4 + 1/9)(-6)^k - (-3)^k/9 \\ (-k^2/18 + k/6 + 7/9)(-6)^k + 2(-3)^k/9 & (-k^2/18 + k/6 + 1/9)(-6)^k - (-3)^k/9 \\ (k^2/36 - k/4 + 4/9)(-6)^k + 4(-3)^k/9 & (k^2/36 - k/4 + 7/9)(-6)^k + 2(-3)^k/9 \end{bmatrix}.$$

Two different methods can be used to compute $A^{12345}$ and the square root of $A$, and they are identical, indicating that the analytical form obtained is reliable:

```
>> simplify(A^12345-subs(F,k,12345)) % validation of a matrix power
   syms x; A3=funmsym(sym(A),sqrt(x),x)
   simplify(A3-subs(F,k,1/2)) % matrix square root test
```

**Example 6.36.** Validate that the $k$th power also works for rational values of $k$.

**Solutions.** For instance, if $r = 1/3$, the following commands can be issued and the cubic root of the matrix can be found and validated.

```
>> A=sym([-3,-1,-1; 0,-3,-1; 1,2,0]);
   syms k, F=mpowersym(A,k)
   r=sym(1/3); F1=subs(F,k,r),
   F2=mpowersym(F1,k); F3=subs(F2,k,1/r)
```

**Example 6.37.** In fact, funmsym() function can also be used to evaluate $A^k$. Solve again the matrix power problem in Example 6.35, and compare the results.

**Solutions.** The following statements can be used to compute the matrix power using two methods. It can be seen that the two results are identical:

```
>> A=[-7,2,0,-1; 1,-4,2,1; 2,-1,-6,-1; -1,-1,0,-4];
   syms k, A=sym(A); F=mpowersym(A,k)
   syms x, F1=funmsym(A,x^k,x), simplify(F-F1) % direct computation
```

### 6.6.3 Matrix powers via the z transform

Another method for computing matrix powers $A^k$ is provided in [32], where the inverse $z$ transform is involved. The main result is summarized below.

**Theorem 6.15.** *For a square matrix $A$, its kth power can be evaluated directly from*

$$A^k = \mathscr{Z}^{-1}[z(zI - A)^{-1}], \tag{6.6.3}$$

*where $\mathscr{Z}^{-1}[\cdot]$ is the inverse z transform of the function.*

Inverse $z$ transforms can be evaluated symbolically with function iztrans(). An example is given below to compute matrix power with such an algorithm.

**Example 6.38.** Solve the matrix power problem again in Example 6.35 with the $z$ transform method.

**Solutions.** If the $z$ transform method is used, the following MATLAB statements can be employed for direct computation:

```
>> A=[-7,2,0,-1; 1,-4,2,1; 2,-1,-6,-1; -1,-1,0,-4];
   syms z k; A1=iztrans(z*inv(z*eye(4)-A),z,k);
   simplify(A1)
```

The result is given below, and it can be seen that many terms like nchoosek(k-1,2) are returned. Further automatic computation is not possible. In fact, the term is $C_{k-1}^2$. Therefore, further simplifications are needed:

$$A_1 = \begin{bmatrix} (-6)^k C_{k-1}^2/18 + (-6)^k k/3 + 2(-3)^k/9 + 13(-6)^k/18 \\ (-6)^k C_{k-1}^2/18 + 2(-3)^k/9 - 5(-6)^k/18 \\ 2(-3)^k/9 - (-6)^k k/3 - (-6)^k C_{k-1}^2/9 - (-6)^k/9 \\ (-6)^k C_{k-1}^2/18 - 4(-3)^k/9 + 7(-6)^k/18 \end{bmatrix}$$

$$\begin{matrix} (-3)^k/3 - (-6)^k k/6 - (-6)^k/3 \\ (-3)^k/3 - (-6)^k k/6 + 2(-6)^k/3 \\ (-3)^k/3 + (-6)^k(k-1)/3 \\ 2(-6)^k/3 - 2(-3)^k/3 - (-6)^k k/6 \end{matrix}$$

$$\begin{matrix} (-6)^k C_{k-1}^2/18 + (-6)^k k/6 + 2(-3)^k/9 - 5(-6)^k/18 \\ (-6)^k C_{k-1}^2/18 - (-6)^k k/6 + 2(-3)^k/9 - 5(-6)^k/18 \\ 2(-3)^k/9 - (-6)^k C_{k-1}^2/9 + 8(-6)^k/9 \\ (-6)^k C_{k-1}^2/18 - (-6)^k k/6 - 4(-3)^k)/9 + 7(-6)^k)/18 \end{matrix}$$

$$\begin{matrix} (-6)^k C_{k-1}^2/18 + (-6)^k k/6 - (-3)^k/9 + (-6)^k/18 \\ (-6)^k C_{k-1}^2/18 - (-6)^k k/6 - (-3)^k/9 + (-6)^k/18 \\ 2(-6)^k/9 - (-3)^k/9 - (-6)^k C_{k-1}^2/9 \\ (-6)^k C_{k-1}^2/18 - (-6)^k k/6 + 2(-3)^k/9 + 13(-6)^k/18 \end{matrix}$$

One may substitute $C_{k-1}^2$ with $(k-1)(k-2)/2$, and with the following statements, the result can be simplified, and it can be seen that the result is identical to those obtained in Example 6.35:

```
>> F2=simplify(subs(F1,nchoosek(k-1,2),(k-1)*(k-2)/2))
```

Although several methods are provided for computing $A^k$, they are, in fact, different. We suggest using the unified function funmsym() directly.

### 6.6.4 Computing $k^A$

There is no function provided in MATLAB Symbolic Math Toolbox for computing $k^A$. With the use of the function funmsym(), and the prototype expression specified as $k$^$x$, the matrix function $k^A$ can be obtained immediately. Examples are given below to demonstrate the computation.

**Example 6.39.** For matrix $A$ in Example 6.35, compute $k^A$.

**Solutions.** Matrix power $k^A$ can be evaluated directly by

```
>> A=[-7,2,0,-1; 1,-4,2,1; 2,-1,-6,-1; -1,-1,0,-4];
   syms x k; A1=funmsym(A,k^x,x); simplify(A1)
```

and the result is

$$A_1 = \frac{1}{9k^6} \begin{bmatrix} 9\ln^2 k - 15\ln k + 2k^3 + 7 & 9\ln k + 3k^3 - 3 \\ 3\ln k + 9\ln^2 k + 2k^3 - 2 & 9\ln k + 3k^3 + 6 \\ 12\ln k - 18\ln^2 k + 2k^3 - 2 & -18\ln k + 3k^3 - 3 \\ 3\ln k + 9\ln^2 k - 4k^3 + 4 & 9\ln k - 6k^3 + 6 \end{bmatrix}$$

$$\begin{bmatrix} -6\ln k + 9\ln^2 k + 2k^3 - 2 & -6\ln k + 9\ln^2 k - k^3 + 1 \\ 12\ln k + 9\ln^2 k + 2k^3 - 2 & 12\ln k + 9\ln^2 k - k^3 + 1 \\ -6\ln k - 18\ln^2 k + 2k^3 + 7 & -6\ln k - 18\ln^2 k - k^3 + 1 \\ 12\ln k + 9\ln^2 k - 4k^3 + 4 & 12\ln k + 9\ln^2 k + 2k^3 + 7 \end{bmatrix}.$$

## 6.7 Problems

6.1 Generate a standard uniform distributed random array having size $3 \times 10 \times 80 \times 5$. Find the mean value of all the data. Also find the position of the maximum data sample.

6.2 Compute element-by-element sine values for the four-dimensional array in Problem 6.1, then compute arcsine of each value of the result and see whether the original array can be restored. Why?

6.3 Some matrix functions can be evaluated from polynomial functions. For instance, Taylor series is the most widely used one. If the independent variable $x$ is replaced by matrix $A$, nonlinear matrix functions can be evaluated. Taylor series expansions of $\ln A$, $\cos A$, and $\arcsin A$ are presented in Theorems 6.4, 6.6, and 6.7. Write down MATLAB functions to implement these formulas, and compare the results with those from functions funm() and funmsym().

6.4 For an autonomous linear differential equation $x'(t) = Ax(t)$, it is known that the analytical solution can be expressed as $x(t) = e^{At}x(0)$. Solve analytically the following autonomous differential equation:

$$x'(t) = \begin{bmatrix} -3 & 0 & 0 & 1 \\ -1 & -1 & 1 & -1 \\ 1 & 0 & -2 & 1 \\ 0 & 0 & 0 & -4 \end{bmatrix} x(t), \quad x(0) = \begin{bmatrix} -1 \\ 0 \\ 3 \\ 1 \end{bmatrix}.$$

6.5 For the given matrix $A$, compute the logarithmic matrices $\ln A$ and $\ln At$, and validate the results with the reliable function expm() if

$$A = \begin{bmatrix} -1 & -1/2 & 1/2 & -1 \\ -2 & -5/2 & -1/2 & 1 \\ 1 & -3/2 & -5/2 & -1 \\ 3 & -1/2 & -1/2 & -4 \end{bmatrix}.$$

6.6 Compute the numerical and analytical values of the square root matrix, and validate the results for

$$A = \begin{bmatrix} -30 & 0 & 0 & 0 \\ 28 & -2 & 1 & 1 \\ -27 & -28 & -31 & -1 \\ 27 & 28 & 29 & -1 \end{bmatrix}.$$

6.7 Compute the matrix trigonometric functions such as $\sin At$, $\cos At$, $\tan At$ and $\cot At$ for

$$A_1 = \begin{bmatrix} -15/4 & 3/4 & -1/4 & 0 \\ 3/4 & -15/4 & 1/4 & 0 \\ -1/2 & 1/2 & -9/2 & 0 \\ 7/2 & -7/2 & 1/2 & -1 \end{bmatrix}, \quad A_2 = \begin{bmatrix} -1 & 0 & 0 & 0 \\ 0 & -1 & 1 & 0 \\ 2 & 0 & -2 & 1 \\ -1 & 0 & 0 & -2 \end{bmatrix}.$$

6.8 For a block Jordan matrix $A$ with

$$A = \begin{bmatrix} A_1 & & \\ & A_2 & \\ & & A_3 \end{bmatrix}, \quad \text{where}$$

$$A_1 = \begin{bmatrix} -3 & 1 & 0 \\ 0 & -3 & 1 \\ 0 & 0 & -3 \end{bmatrix}, \quad A_2 = \begin{bmatrix} -5 & 1 \\ 0 & -5 \end{bmatrix}, \quad A_3 = \begin{bmatrix} -1 & 1 & 0 & 0 \\ 0 & -1 & 1 & 0 \\ 0 & 0 & -1 & 1 \\ 0 & 0 & 0 & -1 \end{bmatrix},$$

compute the analytical expression of the functions $e^{At}$ and $\sin(2At + \pi/3)$.

6.9 Compute $e^{A^2 t}A^2 + \sin(A^3 t)At + e^{\sin At}$ for

$$A = \begin{bmatrix} -3 & -1 & 1 & 0 \\ -28 & -57 & 27 & -1 \\ -29 & -56 & 26 & -1 \\ 1 & 29 & -29 & -30 \end{bmatrix}.$$

6.10 Compute $e^{At}$, $\sin At$ and $e^{At}\sin(A^2 e^{At}t)$ if

$$A = \begin{bmatrix} -4.5 & 0 & 0.5 & -1.5 \\ -0.5 & -4 & 0.5 & -0.5 \\ 1.5 & 1 & -2.5 & 1.5 \\ 0 & -1 & -1 & -3 \end{bmatrix}.$$

6.11 For the matrix $A$ in Example 6.10, compute $A^k$.

6.12 For the matrix $A$ in Example 6.5, compute $k^A$ and $5^A$, and validate the results.

# 7 Linear algebra applications

Linear algebra is the most widely used mathematical modeling tool in science and technology. In this chapter, linear algebra applications are introduced. In Section 7.1, modeling and solutions of linear equations in certain applications are presented. Examples are used to show the applications in electric network analysis, structural analysis, and chemical reaction equation balancing. In Section 7.2, linear algebra applications in linear control system analysis are demonstrated to investigate system stability, controllability and observability. An introduction to solving some linear state space equations is also presented. In Section 7.3, linear algebra applications in digital image processing are presented. In Section 7.4, applications in graph theory problems are illustrated. Graph representations and incident matrices are presented, and the shortest path problem and complicated block diagram simplification are illustrated. In Section 7.5, analytical and numerical solutions of difference equations are introduced, simulation solutions of nonlinear difference equations are presented and the concept and solutions of Markov chain models are proposed. In Section 7.6, linear algebra applications in data processing are presented. Linear regression and least squares curve fitting problems are described, followed by Bézier curve plotting and principal component analysis method with MATLAB. Applications in dimension reduction are also addressed.

## 7.1 Linear equations applications

The development of linear algebra studies originated from linear algebraic equations and their solutions. In this section, we demonstrate linear equation modeling in some specific scientific and technology problems and solution processes. The applications in electric network and static structure analysis are studied first to show how to write equations and how to solve them. Then in chemical equation balancing problems, examples are used to show how to first establish linear equations, and then how to balance chemical equations.

### 7.1.1 Electric network analysis

In ordinary electric circuits, the commonly used elements are resistor $R$, inductor $L$, and capacitor $C$. From the well-known Ohm's law, it is known that the voltage $u$ across the resistor satisfies $u = Ri$, where $R$ is the resistance and $i$ is the current. For inductors and capacitors, the relationship between the voltage and current is no longer static. The following differential equations are used to describe the relationships:

$$u(t) = L\frac{di(t)}{dt}, \quad i(t) = C\frac{du(t)}{dt}. \tag{7.1.1}$$

https://doi.org/10.1515/9783110666991-007

The solutions of the differential equations are complicated. For simplicity, Laplace transform can be used to convert the differential equations into linear algebraic ones:

$$U(s) = sLI(s) - LI_0, \quad I(s) = sCU(s) - CU_0, \tag{7.1.2}$$

where $I_0$ is the initial current in the inductor and $U_0$ is the initial voltage of the capacitor. In general static circuit analysis, the initial values can be set to 0.

When the impedance $Z(s)$ is used, the Ohm's law under the Laplace transform framework can be rewritten as

$$U(s) = Z(s)I(s). \tag{7.1.3}$$

For the commonly used elements, the impedance of resistor is $R$, and the impedance of inductors and capacitors are respectively $sL$ and $1/(sC)$. The reciprocal of the impedance is also known as the admittance $Y(s)$, so that another version of Ohm's law can be written as

$$I(s) = Y(s)U(s). \tag{7.1.4}$$

Based on the important Kirchhoff's current and voltage law, it is not difficult to derive the current equation in the loops, and voltage equation on the nodes for a complicated electric network. Now for the examples, the corresponding equations can be written, and the solution and validation of the equations are also made.

**Theorem 7.1** (Loop current equation). *For each loop in a circuit, if the current is $i_i$, the following linear equations can be established:*

$$\mathbf{Z}i = v, \tag{7.1.5}$$

*where the impedance matrix $\mathbf{Z}$ can be constructed as*
*(1) The diagonal element $z_{ii}$ equals to sum of all the impedances connected to the ith node.*
*(2) The off-diagonal element $z_{ij}$ equals the sum of the impedances between the ith and jth nodes, times −1.*

*The element $v_i$ in vector $v$ is the sum of external voltages connected to the ith loop.*

**Theorem 7.2** (Nodal voltage equations). *For each node in a circuit, if its potential is $v_i$, the following linear equations can be established:*

$$\mathbf{Y}v = j, \tag{7.1.6}$$

*where the admittance matrix $\mathbf{Y}$ can be written as follows:*
*(1) The diagonal element $y_{ii}$ equals to the sum of all the admittances connected to the ith node.*

(2) *The off-diagonal element $y_{ij}$ equals to the sum of the admittances between the ith and jth nodes, times –1.*

*The element $j_i$ in vector $\mathbf{j}$ is the sum of all the external currents flowing into the ith node.*

**Example 7.1.** Assume that the circuit of a resistive network is given in Figure 7.1[24]. Write down the circuit equation for the currents of each loop.

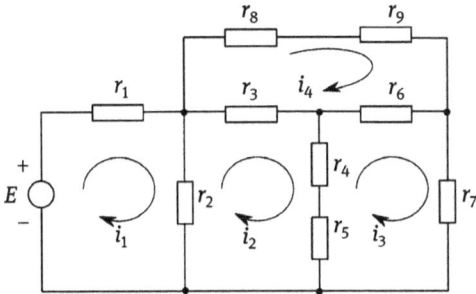

**Figure 7.1:** A resistive network.

**Solutions.** Let us consider the first loop. The voltage of the source $E$ equals to the voltage drops on $r_1$ and $r_2$. If the current running into $r_1$ is $i_1$, and that running into $r_2$ is $i_1 - i_2$, the following loop currents equation can be written as

$$E = r_1 i_1 + r_2(i_1 - i_2) = (r_1 + r_2)i_1 - r_2 i_2.$$

In the second loop, if the clockwise direction is considered as positive, the current running into $r_2$ is $i_2 - i_1$, that running into $r_3$ is $i_2 - i_4$, while that running into $r_4$ and $r_5$ is $i_2 - i_3$. The following linear equation can be written:

$$r_2(i_2 - i_1) + r_3(i_2 - i_4) + (r_4 + r_5)(i_2 - i_3) = 0.$$

Further processing the result, it is found that

$$-r_2 i_1 + (r_2 + r_3 + r_4 + r_5)i_2 - (r_4 + r_5)i_3 - r_3 i_4 = 0.$$

The other two loop equations can be written as:

$$(r_4 + r_5)(i_3 - i_2) + r_6(i_3 - i_4) + r_7 i_3 = -(r_4 + r_5)i_2 + (r_4 + r_5 + r_6 + r_7)i_3 - r_6 i_4 = 0,$$

$$r_3(i_2 - i_4) + (r_8 + r_9)i_4 + r_6(i_4 - i_3) = -r_3 i_2 - r_6 i_3 + (r_3 + r_6 + r_8 + r_9)i_4 = 0.$$

The matrix form of the loop equations can be obtained as

$$\begin{bmatrix} r_1 + r_2 & -r_2 & 0 & 0 \\ -r_2 & r_2 + r_3 + r_4 + r_5 & -r_4 - r_5 & -r_3 \\ 0 & -r_4 - r_5 & r_4 + r_5 + r_6 + r_7 & -r_6 \\ 0 & -r_3 & -r_6 & r_3 + r_6 + r_8 + r_9 \end{bmatrix} \begin{bmatrix} i_1 \\ i_2 \\ i_3 \\ i_4 \end{bmatrix} = \begin{bmatrix} E \\ 0 \\ 0 \\ 0 \end{bmatrix}.$$

It can be seen from the matrix equation that the result is the same as that in Theorem 7.1. After inputting the matrices into MATLAB environment, the linear equation can be solved directly to find the current signals. Since the results are too complicated, they are not listed here. It can be seen through validation that the results indeed satisfy the original equations.

```
>> syms r1 r2 r3 r4 r5 r6 r7 r8 r9 E
   Z=[r1+r2, -r2, 0, 0;
      -r2, r2+r3+r4+r5, -r4-r5, -r3;
      0, -r4-r5, r4+r5+r6+r7, -r6;
      0, -r3, -r6, r3+r6+r8+r9];
   v=[E; 0; 0; 0]; i=simplify(inv(Z)*v), simplify(Z*i-v)
```

**Example 7.2.** For a resistor–capacitor–inductor circuit shown in Figure 7.2[24], write down the nodal equations of the circuit. Find the Laplacian transform expressions of the voltages.

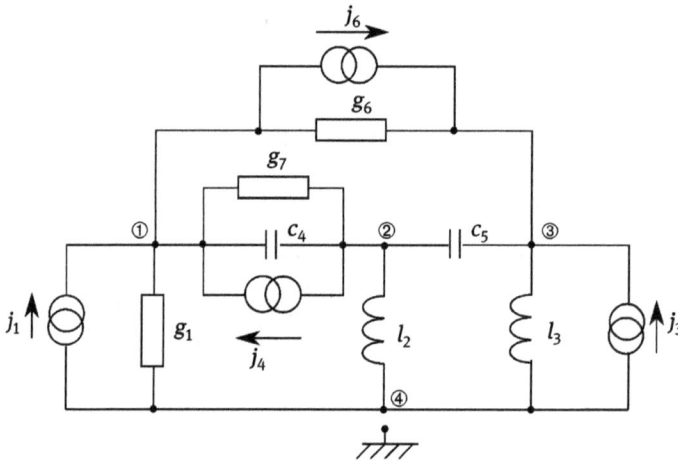

**Figure 7.2:** Resistor–capacitor–inductor circuit.

**Solutions.** Let us observe first node ①. The effective current running into this node is $j_1 + j_4 - j_6$. Besides, from the nodes directly connected to node ①, the current equations can be set up as

$$g_1(v_1 - v_4) + g_6(v_1 - v_3) + (g_7 + sc_4)(v_1 - v_2) = j_1 + j_4 - j_6,$$

where $g_1 = 1/r_1$ is the reciprocal of resistance $r_1$, known as conductance. Besides, by convention, the left-hand side of the equation is the current running into the node, while on the right we have the current running out of the node. The following equations are also written according to the convention.

Manipulating the equation, it can be rewritten as

$$(g_1 + g_6 + g_7 + sc_4)v_1 - (g_7 + sc_4)v_2 - g_6v_3 - g_1v_4 = j_1 + j_4 - j_6.$$

For node ②, the current equation can also be built up as

$$(g_7 + sc_4)(v_2 - v_1) + \frac{1}{sl_2}(v_2 - v_4) + sc_5(v_2 - v_3) = -j_4,$$

which can be rewritten as

$$-(g_7 + sc_4)v_1 + \left(g_7 + sc_4 + \frac{1}{sl_2} + sc_5\right)v_2 - sc_5v_3 - \frac{1}{sl_2}v_4 = -j_4.$$

Similarly, for nodes ③ and ④, the following equations can be written:

$$-g_6v_1 - sc_5v_2 + \left(\frac{1}{sl_3} + sc_5 + g_6\right)v_3 - \frac{1}{sl_3}v_4 = j_3 + j_6,$$

$$-g_1v_1 - \frac{1}{sl_2}v_2 - \frac{1}{sl_3}v_3 + \left(g_1 + \frac{1}{sl_2} + \frac{1}{sl_3}\right)v_4 = -j_1 - j_3.$$

It is not hard to express the equations in matrix form as

$$Y\begin{bmatrix} v_1 \\ v_2 \\ v_3 \\ v_4 \end{bmatrix} = \begin{bmatrix} j_1 + j_4 - j_6 \\ -j_4 \\ j_3 + j_6 \\ -j_1 - j_3 \end{bmatrix},$$

where

$$Y = \begin{bmatrix} g_1 + g_6 + g_7 + sc_4 & -(g_7 + sc_4) & -g_6 & -g_1 \\ -(g_7 + sc_4) & g_7 + sc_4 + 1/(sl_2) + sc_5 & -sc_5 & -1/(sl_2) \\ -g_6 & -sc_5 & 1/(sl_3) + sc_5 + g_6 & -1/(sl_3) \\ -g_1 & -1/(sl_2) & -1/(sl_3) & g_1 + 1/(sl_2) + 1/(sl_3) \end{bmatrix}.$$

Through low-level formulation, the admittance matrix obtained is the same as that described in Theorem 7.2. The subsequent circuit modeling can be made directly from Theorem 7.2, rather than low-level formulation.

With the mathematical form, it is not hard to input the matrix into MATLAB, then solve the matrix equations directly, so that the potentials $v_i$ can be found. Unfortunately, if the equations are analyzed, it is found that $Y$ is a singular matrix, with rank of 3 rather than 4. The equations have infinitely many solutions.

```
>> syms g1 g6 g7 c4 c5 12 13 s j1 j3 j4 j6
   Y=[g1+g6+g7+s*c4, -(g7+s*c4), -g6, -g1;
      -(g7+s*c4), g7+s*c4+1/(s*12)+s*c5, -s*c5, -1/(s*12);
      -g6, -s*c5, 1/(s*13)+s*c5+g6, -1/(s*13);
      -g1, -1/(s*12), -1/(s*13), g1+1/(s*12)+1/(s*13)];
   B=[j1+j4-j6; -j4; j3+j6; -j1-j3]; rank(Y)
```

To make things simpler, letting $v_4 = 0$, the nodal voltage equations can be simplified as

$$\begin{bmatrix} g_1 + g_6 + g_7 + sc_4 & -(g_7 + sc_4) & -g_4 \\ -(g_7 + sc_4) & g_7 + sc_4 + 1/(sl_2) + sc_5 & -sc_5 \\ -g_6 & -sc_5 & 1/(sl_3) + sc_5 + g_6 \end{bmatrix} \begin{bmatrix} v_1 \\ v_2 \\ v_3 \end{bmatrix} = \begin{bmatrix} j_1 + j_4 - j_6 \\ -j_4 \\ j_3 + j_6 \end{bmatrix}.$$

With the following statements, the original equations can be solved directly. It can be validated by substituting the result back into the equations that the error after simplification is zero, indicating that the solution obtained satisfies the original equations.

```
>> Y=[g1+g6+g7+s*c4, -(g7+s*c4), -g6;
     -(g7+s*c4), g7+s*c4+1/(s*12)+s*c5, -s*c5;
     -g6, -s*c5, 1/(s*13)+s*c5+g6];
  B=[j1+j4-j6; -j4; j3+j6];
  v=simplify(inv(Y)*B), simplify(Y*v-B)
```

Since the result obtained is a Laplacian transform expression, the analytical time-domain solution cannot be found unless specific values of resistance, capacitance, and inductance are given.

**Example 7.3.** Consider the circuit in Figure 7.3. There are operational amplifiers in the network[24]. If the gains of the operational amplifiers are all $+\infty$, it means that the potentials at the two input terminals are the same, such that the current running into the amplifier is zero. Write down the voltage model of the circuit, and find the voltage at some nodes.

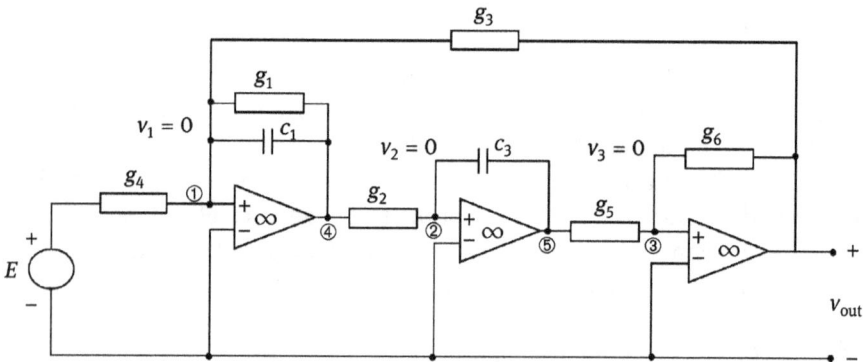

**Figure 7.3:** Operational amplifier network.

**Solutions.** Let us see node ① first. Since one of the input terminals is connected to zero potential, according to the rules that the two input terminals have the same potential, we have $v_1 = 0$, and the current running into the amplifier is zero. The following current equation can be established:

$$(g_4 + sc_1 + g_1 + g_3)v_1 - (g_1 + sc_1)v_4 - g_3v_{out} \xrightarrow{v_1 = 0} -(g_1 + sc_1)v_4 - g_3v_{out} = Eg_4.$$

Since the current running into the second amplifier is zero, and $v_2 = 0$, it is found for node ② that

$$(g_2 + sc_2)v_2 - g_2v_4 - sc_2v_5 \xrightarrow{v_2 = 0} -g_2v_4 - sc_2v_5 = 0.$$

From node ③, $v_3 = 0$ and the current running into the amplifier is zero, so it is found that

$$(g_5 + g_6)v_3 - g_5v_5 - g_6v_{out} \xrightarrow{v_3 = 0} -g_5v_5 - g_6v_{out} = 0.$$

Summarizing the above, the matrix equation can be established as

$$\begin{bmatrix} -(g_1 + sc_1) & 0 & -g_3 \\ -g_2 & -sc_2 & 0 \\ 0 & -g_5 & -g_6 \end{bmatrix} \begin{bmatrix} v_4 \\ v_5 \\ v_{out} \end{bmatrix} = \begin{bmatrix} g_4 E \\ 0 \\ 0 \end{bmatrix}.$$

The following matrix equation can be sent directly to MATLAB environment, such that the Laplace transform expressions of the voltages can be found, from which the time-domain solutions of the voltages can be obtained through the inverse Laplace transform.

```
>> syms E s g1 g2 g3 g4 g5 g6 c1 c2
   Y=[-(g1+s*c1), 0, -g3; -g2, -s*c2, 0; 0, -g5, -g6];
   v=simplify(inv(Y)*[g4*E; 0; 0]), v1=ilaplace(v)
```

The Laplace transform expressions of the voltages are

$$v = \frac{E}{c_1c_2g_6s^2 + c_2g_1g_6s + g_2g_3g_5} \begin{bmatrix} -c_2g_4g_6s \\ g_2g_4g_6 \\ -g_2g_4g_5 \end{bmatrix}.$$

Through inverse Laplace transform, the analytical solution of time-domain expression of the voltages can be found, and simplified manually as

$$v_4 = -\frac{Eg_4}{c_1} e^{-g_1t/(2c_1)} \left( \cosh \psi t - \frac{g_1\sqrt{c_2g_6}}{\delta} \sinh \psi t \right),$$

$$v_5 = \frac{2Eg_2g_4\sqrt{g_6}}{\sqrt{c_2}\delta} e^{-g_1t/(2c_1)} \sinh \psi t,$$

$$v_{out} = -\frac{2Eg_2g_4g_5}{\sqrt{c_2g_6}\delta} e^{-g_1t/(2c_1)} \sinh \psi t,$$

where

$$\delta = \sqrt{c_2g_1^2g_6 - 4c_1g_2g_3g_5}, \quad \psi = \frac{\delta}{2c_1\sqrt{c_2g_6}}.$$

### 7.1.2 Analysis of truss structures

In Chapter 1, a truss system has been modeled and solved. Complicated mechanics problems can be modeled with linear equations, and the solutions can be found directly with MATLAB. The analytical and numerical solutions can usually be found.

For ordinary truss system balancing problems, the summed forces in the vertical and horizontal directions are both zeros. Based on the rule, the corresponding linear equations can be formulated. MATLAB can be used to find directly the analytical and numerical solutions.

The method demonstrated in the example is only carried out on static forces. In real applications, more complicated and practical problems must be considered. For instance, the truss distortions should be considered. These problems cannot be described easily and accurately with linear algebraic equations. Special tools such as finite element method should be introduced to solve real problems. The topics are not covered in this book.

### 7.1.3 Balancing chemical reaction equations

Chemical reaction equation balancing problems can be implemented by modeling with linear algebraic equations and solving them. The general method used is that undetermined constants can be introduced to all the terms on both sides of the chemical reaction equation, and based on the rules that the same elements on both sides are equal, linear algebraic equations can be set up and solved, such that the chemical reaction equations can be balanced. Examples are given in this section to show how to use linear algebra and computer tools to balance chemical reaction equations.

**Example 7.4.** Balance the following chemical reaction equation:

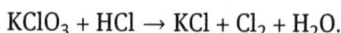

$$KClO_3 + HCl \rightarrow KCl + Cl_2 + H_2O.$$

**Solutions.** Balancing a chemical reaction equation means finding coefficients for each term in the reaction equation, such that the elements on both sides are equalized. Assuming that nothing is known for the above equation, the following reaction equation can be written:

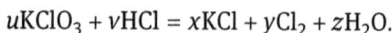

$$uKClO_3 + vHCl = xKCl + yCl_2 + zH_2O.$$

If the algebraic equations for the undetermined constants $u$, $v$, $x$, $y$, and $z$ can be established, linear equation solution method can be applied in finding the undetermined constants. Let us consider K elements on both sides of the equation. The first equation can be found as $u = x$. Now observe the Cl element. It is seen that $u+v = x+2y$. For element H, it is not hard to find that $v = 2z$. For element O, it is found that $3u = z$. Summarizing the above, the following system of linear equations can be established:

$$\begin{cases} K & \rightarrow & u - x = 0, \\ Cl & \rightarrow & u + v - x - 2y = 0, \\ H & \rightarrow & v - 2z = 0, \\ O & \rightarrow & 3u - z = 0. \end{cases}$$

It can be seen that there are 5 unknowns and 4 equations. The equations then have infinitely many solutions. This is a normal phenomenon. The least integer solutions are expected for

$$\begin{bmatrix} 1 & 0 & -1 & 0 & 0 \\ 1 & 1 & -1 & -2 & 0 \\ 0 & 1 & 0 & 0 & -2 \\ 3 & 0 & 0 & 0 & -1 \end{bmatrix} \begin{bmatrix} u \\ v \\ x \\ y \\ z \end{bmatrix} = \mathbf{0}.$$

Matrix $A$ can be entered into MATLAB environment, and through simplification, reduced row echelon form can be found, such that all the independent variables can be converted to the multiples of $z$:

```
>> A=[1,0,-1,0,0; 1,1,-1,-2,0; 0,1,0,0,-2; 3,0,0,0,-1]
   A=sym(A); C=rref(A)
```

The result obtained is

$$C = \begin{bmatrix} 1 & 0 & 0 & 0 & -1/3 \\ 0 & 1 & 0 & 0 & -2 \\ 0 & 0 & 1 & 0 & -1/3 \\ 0 & 0 & 0 & 1 & -1 \end{bmatrix}.$$

It is obvious that $u = z/3$, $v = 2z$, $x = z/3$, and $y = z$. When substituted back to the equation, they yield

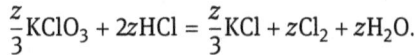

$$\frac{z}{3} KClO_3 + 2z HCl = \frac{z}{3} KCl + z Cl_2 + z H_2O.$$

It can be seen that by canceling the common term $z$ on both sides, and multiplying by 3 to cancel the common denominator, the chemical reaction equation can be balanced as

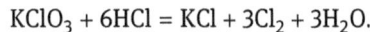

$$KClO_3 + 6HCl = KCl + 3Cl_2 + 3H_2O.$$

**Example 7.5.** Balance the following chemical reaction equation:

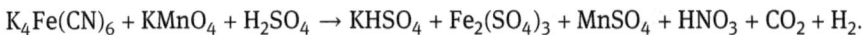

$$K_4Fe(CN)_6 + KMnO_4 + H_2SO_4 \rightarrow KHSO_4 + Fe_2(SO_4)_3 + MnSO_4 + HNO_3 + CO_2 + H_2.$$

**Solutions.** This is a chemical reaction equation which is hard to balance with manual methods. Computer tools should be introduced to complete the task. Since there are

9 terms in the equation altogether, 9 undetermined constants $x_1, x_2, \ldots, x_9$ should be introduced such that

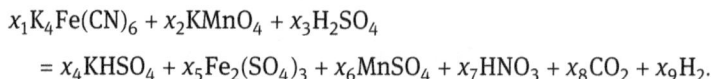

$$x_1 K_4 Fe(CN)_6 + x_2 KMnO_4 + x_3 H_2 SO_4$$
$$= x_4 KHSO_4 + x_5 Fe_2(SO_4)_3 + x_6 MnSO_4 + x_7 HNO_3 + x_8 CO_2 + x_9 H_2.$$

There are 8 elements in the chemical reaction equation, and the following linear equations can be written:

$$\begin{cases} K & \to & 4x_1 + x_2 - x_4 = 0, \\ Fe & \to & x_1 - 2x_5 = 0, \\ C & \to & 6x_1 - x_8 = 0, \\ N & \to & 6x_1 - x_7 = 0, \\ Mn & \to & x_2 - x_6 = 0, \\ S & \to & x_3 - x_4 - 3x_5 - x_6 = 0, \\ O & \to & 4x_2 + 4x_3 - 4x_4 - 12x_5 - 4x_6 - 3x_7 - 2x_8 = 0, \\ H & \to & 2x_3 - x_4 - x_7 - 2x_9 = 0, \end{cases}$$

from which the matrix form can be written as

$$\begin{bmatrix} 4 & 1 & 0 & -1 & 0 & 0 & 0 & 0 & 0 \\ 1 & 0 & 0 & 0 & -2 & 0 & 0 & 0 & 0 \\ 6 & 0 & 0 & 0 & 0 & 0 & 0 & -1 & 0 \\ 6 & 0 & 0 & 0 & 0 & 0 & -1 & 0 & 0 \\ 0 & 1 & 0 & 0 & 0 & -1 & 0 & 0 & 0 \\ 0 & 0 & 1 & -1 & -3 & -1 & 0 & 0 & 0 \\ 0 & 4 & 4 & -4 & -12 & -4 & -3 & -2 & 0 \\ 0 & 0 & 2 & -1 & 0 & 0 & -1 & 0 & -2 \end{bmatrix} \begin{bmatrix} x_1 \\ x_2 \\ x_3 \\ x_4 \\ x_5 \\ x_6 \\ x_7 \\ x_8 \\ x_9 \end{bmatrix} = \mathbf{0}.$$

Inputting the **A** matrix into MATLAB, the reduced row echelon form can be found easily with the following statements:

```
>> A=[4,1,0,-1,0,0,0,0,0; 1,0,0,0,-2,0,0,0,0;
      6,0,0,0,0,0,0,-1,0; 6,0,0,0,0,0,0,-1,0,0;
      0,1,0,0,0,-1,0,0,0; 0,0,1,-1,-3,-1,0,0,0;
      0,4,4,-4,-12,-4,-3,-2,0; 0,0,2,-1,0,0,-1,0,-2];
   C=rref(sym(A))
```

The result obtained is

$$
C = \begin{bmatrix}
1 & 0 & 0 & 0 & 0 & 0 & 0 & 0 & -4/47 \\
0 & 1 & 0 & 0 & 0 & 0 & 0 & 0 & -30/47 \\
0 & 0 & 1 & 0 & 0 & 0 & 0 & 0 & -82/47 \\
0 & 0 & 0 & 1 & 0 & 0 & 0 & 0 & -46/47 \\
0 & 0 & 0 & 0 & 1 & 0 & 0 & 0 & -2/47 \\
0 & 0 & 0 & 0 & 0 & 1 & 0 & 0 & -30/47 \\
0 & 0 & 0 & 0 & 0 & 0 & 1 & 0 & -24/47 \\
0 & 0 & 0 & 0 & 0 & 0 & 0 & 1 & -24/47
\end{bmatrix}.
$$

It can be seen from the results that the undetermined constants can all be written as multiples of $x_9$:

$$
x_1 = \frac{4x_9}{47}, \quad x_2 = \frac{30x_9}{47}, \quad x_3 = \frac{82x_9}{47}, \quad x_4 = \frac{46x_9}{47}, \quad x_5 = \frac{2x_9}{47},
$$
$$
x_6 = \frac{30x_9}{47}, \quad x_7 = \frac{24x_9}{47}, \quad x_8 = \frac{24x_9}{47}.
$$

The original chemical reaction equation can be written as

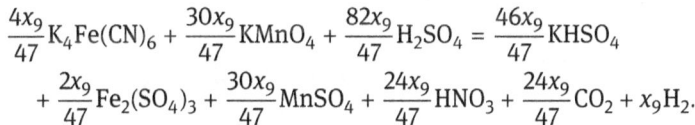

$$
\frac{4x_9}{47} K_4 Fe(CN)_6 + \frac{30x_9}{47} KMnO_4 + \frac{82x_9}{47} H_2SO_4 = \frac{46x_9}{47} KHSO_4
$$
$$
+ \frac{2x_9}{47} Fe_2(SO_4)_3 + \frac{30x_9}{47} MnSO_4 + \frac{24x_9}{47} HNO_3 + \frac{24x_9}{47} CO_2 + x_9 H_2.
$$

Multiplying by 47 and then dividing by $x_9$ both sides, the balanced chemical reaction equation can be written as

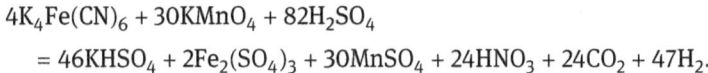

$$
4K_4 Fe(CN)_6 + 30KMnO_4 + 82H_2SO_4
$$
$$
= 46KHSO_4 + 2Fe_2(SO_4)_3 + 30MnSO_4 + 24HNO_3 + 24CO_2 + 47H_2.
$$

## 7.2 Applications in linear control systems

From a certain viewpoint, the problems of analysis and design of linear control systems are, in fact, linear algebra problems. In this section, conversion from a state space model to transfer functions is introduced, then the analysis of system stability, controllability and observability is presented. Analytical solutions to some linear differential equations are provided with matrix computation methods.

### 7.2.1 Control system model conversions

There are two categories of methods for describing linear time-invariant control systems: one is the state space model and the other is the transfer function model. In control systems theory, the transfer function model is regarded as an external model since the relationships of input and output signals are modeled. State space models, on the other hand, are referred to as internal, since not only the relationships between input and output signals, but also the internal signals such as states are studied. In this section, the conversion from a state space model to a transfer function model is demonstrated.

**Definition 7.1.** The state space model of the linear control system is

$$\begin{cases} x'(t) = Ax(t) + Bu(t), \\ y(t) = Cx(t) + Du(t), \end{cases} \tag{7.2.1}$$

where $x(t)$ is the state vector, $u(t)$ is the input signal, while $y(t)$ is the output signal of the system. The state space model is sometimes briefly denoted as $(A, B, C, D)$.

**Definition 7.2.** The state space model of a discrete system is described as

$$\begin{cases} x(t + 1) = Ax(t) + Bu(t), \\ y(t) = Cx(t) + Du(t), \end{cases} \tag{7.2.2}$$

where $x(t)$ is the state vector, $u(t)$ is the input signal, while $y(t)$ is the output signal of the system.

**Theorem 7.3.** *If the state space model of the continuous system is described as* $(A, B, C, D)$, *the equivalent transfer function model can be written as*

$$G(s) = C(sI - A)^{-1}B + D. \tag{7.2.3}$$

**Example 7.6.** A continuous multivariate state space model is given below. Find its equivalent transfer function matrix model, if

$$x'(t) = \begin{bmatrix} -9 & -3 & -8 & -4 & -9/2 \\ -8 & -4 & -8 & -2 & -4 \\ 0 & 1 & -1 & 2 & 1 \\ -7 & -2 & -7 & -4 & -7/2 \\ 22 & 6 & 22 & 6 & 9 \end{bmatrix} x(t) + \begin{bmatrix} -4 & 5/2 \\ 2 & 0 \\ 4 & -4 \\ -2 & 1/2 \\ 0 & 3 \end{bmatrix} u(t),$$

$$y(t) = \begin{bmatrix} 2 & 3 & 3 & 4 & 3 \\ 0 & -1 & 0 & 0 & 0 \end{bmatrix} x(t).$$

**Solutions.** The system has two inputs and two outputs. The following statements can be used to directly compute the transfer function matrix:

```
>> A=[-9,-3,-8,-4,-9/2; -8,-4,-8,-2,-4; 0,1,-1,2,1;
      -7,-2,-7,-4,-7/2; 22,6,22,6,9];
   B=[-4,5/2; 2,0; 4,-4; -2,1/2; 0,3];
   C=[2,3,3,4,3; 0,-1,0,0,0];
   A=sym(A); B=sym(B); C=sym(C); syms s
   G1=C*inv(s*eye(5)-A)*B; G1=simplify(G)
```

The transfer function matrix is

$$G_1(s) = \begin{bmatrix} 2/(s+1) & 4/(s+2) \\ -2/(s+2) & 1/(s^2+4s+3) \end{bmatrix}.$$

A powerful Control System Toolbox is provided in MATLAB. Some of the low-level operations can be implemented with numerical methods. For instance, the state space model can be expressed by $G$=ss $(A,B,C,D)$ command. If the transfer function model is expected, one may use the command $G_1$=tf $(G)$ to find it.

**Example 7.7.** Solve again the model conversion problem in Example 7.6.

**Solutions.** The following commands can be used to compute the transfer function matrix. The result obtained is exactly the same as that obtained in Example 7.6. The data structure used here is different, since the transfer function object is used here.

```
>> A=[-9,-3,-8,-4,-4.5; -8,-4,-8,-2,-4; 0,1,-1,2,1;
      -7,-2,-7,-4,-3.5; 22,6,22,6,9];
   B=[-4,2.5; 2,0; 4,-4; -2,0.5; 0,3];
   C=[2,3,3,4,3; 0,-1,0,0,0]; G=ss(A,B,C,0);
   G1=tf(G); G1=minreal(G1,1e-7)
```

### 7.2.2 Qualitative analysis of linear systems

System stability is the most important property of control systems. Besides, controllability and observability of the systems can also be judged. The solutions of these problems rely on the linear algebra support. If equipped with powerful tools such as MATLAB, the seemingly complicated qualitative analysis of linear systems can be solved easily.

**Theorem 7.4.** *For a linear time-invariant continuous system, if all the real parts of the system matrix $A$ are smaller than zero, the system is stable. If the system is a discrete one, then if all the eigenvalues of $A$ lie inside the unit circle, the system is stable.*

**Example 7.8.** The coefficient matrix $A$ of a linear continuous system is given below. Test whether the system is stable or not. If the given matrix is of a discrete system, check whether the system is stable:

$$A = \begin{bmatrix} -2 & 1 & -2 & 1 \\ 0 & -2 & 0 & 2 \\ 2 & 1 & -2 & -1 \\ 0 & 2 & -1 & -2 \end{bmatrix}.$$

**Solutions.** In classical control, it is rather complicated to assess the stability of linear systems. Indirect methods such as Routh criterion should be used. Equipped with powerful tools such as MATLAB, direct assessment of continuous systems stability is very simple and straightforward. The eigenvalues of matrix $A$ can be evaluated directly, based on the positions of the eigenvalues of the coefficient matrix.

```
>> A=[-2,1,-2,1; 0,-2,0,2; 2,1,-2,-1; 0,2,-1,-2];
   d=eig(A)
```

The eigenvalues are located at $-4.2616$, $-1.7127 \pm 1.7501j$, and $-0.3131$. All of them have negative real parts. Therefore the system is stable. If the system is discrete, it is obvious that the absolute values of the first three eigenvalues are larger than 1, located outside the unit circle, indicating that the discrete system is unstable.

Generally speaking, controllability is the property to indicate whether the states can be arbitrarily controlled by input signals. The observability is the property to indicate whether the states of the system can be reconstructed from the measured input and output signals. Here, the testing of controllability and observability of linear systems is presented with examples.

**Theorem 7.5.** *The system $(A, B)$ is fully controllable if the following matrix has full-rank:*

$$T = [B, AB, A^2B, \ldots, A^{n-1}B].$$ (7.2.4)

**Theorem 7.6.** *The system $(A, C)$ is fully observable if the following testing matrix has full-rank:*

$$T = \begin{bmatrix} C \\ CA \\ \vdots \\ CA^{n-1} \end{bmatrix}.$$ (7.2.5)

**Theorem 7.7.** *The system $(A, B, C, D)$ is fully output controllable if the following matrix is of full-rank:*

$$T = [CB, CAB, CA^2B, \ldots, CA^{n-1}B, D].$$ (7.2.6)

Generally speaking, the output controllability has nothing to do with the state controllability. It is possible that a system is not fully controllable, but is output controllable.

The above theorem applies to the continuous state space model, and it applies also to discrete systems. The rank of the matrix can be evaluated directly with function `rank()`.

**Example 7.9.** Assess now the controllability of the system in Example 7.6.

**Solutions.** The matrices of the system can be entered into MATLAB workspace. Then the testing matrix $T$ can be constructed. The rank of the matrix can be found, which is 5, indicating that the system is fully controllable.

```
>> A=[-9,-3,-8,-4,-9/2; -8,-4,-8,-2,-4; 0,1,-1,2,1;
      -7,-2,-7,-4,-7/2; 22,6,22,6,9];
   B=[-4,5/2; 2,0; 4,-4; -2,1/2; 0,3];
   T=[B A*B A^2*B A^3*B A^4*B], rank(T)
```

The constructed testing matrix $T$ is a $5 \times 10$ matrix given below. It is not an easy job to find its rank without computers. With powerful facilities of MATLAB, the rank can be found immediately

$$T = \begin{bmatrix} -4 & 2.5 & 6 & -6 & -10 & 14.5 & 18 & -36 & -34 & 92.5 \\ 2 & 0 & -4 & -1 & 8 & 4 & -16 & -13 & 32 & 40 \\ 4 & -4 & -6 & 8 & 10 & -16 & -18 & 32 & 34 & -64 \\ -2 & 0.5 & 4 & -2 & -8 & 6.5 & 16 & -20 & -32 & 60.5 \\ 0 & 3 & 0 & -3 & 0 & -1 & 0 & 21 & 0 & -97 \end{bmatrix}.$$

### 7.2.3 Transmission zeros in multivariate systems

The poles of a multivariate system can easily be found. For instance, the common denominator can be extracted, from which the roots of the denominator polynomials can be found immediately. They are the poles of the system. The transmission zeros are rather complicated to find. Here an example is given to demonstrate the solution methods.

**Theorem 7.8.** *For the state space model* $(A, B, C, D)$*, the transmission zeros can be found from*

$$\det\left(\begin{bmatrix} A - \lambda I & B \\ C & D \end{bmatrix}\right) = 0. \tag{7.2.7}$$

**Example 7.10.** Find the transmission zeros of the system in Example 7.6.

**Solutions.** In Control System Toolbox in MATLAB, function tzero() is provided to find the transmission zeros of a multivariate system. Besides, from Theorem 7.8, the polynomial of the zeros is $-8s^3 - 42s^2 - 64s - 32$, from which more accurate transmission zeros can be found.

```
>> A=[-9,-3,-8,-4,-9/2; -8,-4,-8,-2,-4; 0,1,-1,2,1;
      -7,-2,-7,-4,-7/2; 22,6,22,6,9];
   B=[-4,5/2; 2,0; 4,-4; -2,1/2; 0,3]; D=zeros(2);
   C=[2,3,3,4,3; 0,-1,0,0,0]; G=ss(A,B,C,0); z=tzero(G)
   syms s; vpasolve(det([sym(A)-s*eye(5),B; C,D]))
```

With the high precision polynomial equation solver vpasolve(), the transmission zeros can be found as below. The results are same as those obtained with function tzero(), with higher accuracy.

$$-3.0665928333206257352, -1.0917035833396871324 \pm 0.335503377871855j.$$

### 7.2.4 Direct solutions of linear differential equations

There are various ways for solving linear differential equations. An introduction to the matrix exponential solution method is proposed, then matrix Sylvester differential equation solutions are studied.

**Theorem 7.9.** *For a differential equation* $x'(t) = Ax(t)$*, with* $x(0) = x_0$*, the analytical solution is* $x(t) = e^{At}x_0$*.*

**Example 7.11.** Solve the following linear differential equation:

$$x'(t) = \begin{bmatrix} -1 & -2 & 0 & -1 \\ -1 & -3 & -1 & -2 \\ -1 & 1 & -2 & 0 \\ 1 & 2 & 1 & 1 \end{bmatrix} x(t), \quad x(0) = \begin{bmatrix} 0 \\ 1 \\ 1 \\ 0 \end{bmatrix}.$$

**Solutions.** From the knowledge presented before, it is known that $e^{At}$ can be evaluated with expm() function directly. Therefore, the following commands can be used to solve the differential equation:

```
>> A=[-1,-2,0,-1; -1,-3,-1,-2; -1,1,-2,0; 1,2,1,1];
   A=sym(A); x0=sym([0; 1; 1; 0]); syms t
   x=simplify(expm(A*t)*x0)
   simplify(diff(x)-A*x), subs(x,t,0)-x0
```

The solutions obtained are shown as follows. Substituting the results back into the original equation, it can be found that the equation and initial conditions are both satisfied. The error vectors are all zero while

$$x(t) = \begin{bmatrix} e^{-2t}(t^2e^t - 4e^t + 4)/2 \\ -e^{-2t}(e^t + te^t - 2) \\ -e^{-t}(t^2 - 2)/2 \\ e^{-2t}(2e^t + te^t - 2) \end{bmatrix}.$$

For the ordinary state space model of a control system, the states are often expressed as a state vector $x(t)$, while in certain applications, differential equations with state matrices are involved. For instance, in this section, matrix Sylvester differential equations are studied.

**Definition 7.3.** The general form of a matrix Sylvester differential equation is[15]

$$X'(t) = AX(t) + X(t)B, \quad X(0) = C,$$ (7.2.8)

where $A \in \mathcal{R}^{n \times n}$, $B \in \mathcal{R}^{m \times m}$, $X, C \in \mathcal{R}^{n \times m}$.

**Theorem 7.10.** *A matrix Sylvester differential equation has the solution*[15] $X(t) = e^{At} C e^{Bt}$.

**Example 7.12.** Solve the following matrix differential equation:

$$X'(t) = \begin{bmatrix} -1 & -2 & 0 & -1 \\ -1 & -3 & -1 & -2 \\ -1 & 1 & -2 & 0 \\ 1 & 2 & 1 & 1 \end{bmatrix} X(t) + X(t) \begin{bmatrix} -2 & 1 \\ 0 & -2 \end{bmatrix}, \quad X(0) = \begin{bmatrix} 0 & -1 \\ 1 & 1 \\ 1 & 0 \\ 0 & 1 \end{bmatrix}.$$

**Solutions.** The matrices should be entered into MATLAB first, then from Theorem 7.10, the solution of the matrix differential equation can be found directly:

```
>> A=[-1,-2,0,-1; -1,-3,-1,-2; -1,1,-2,0; 1,2,1,1];
   B=[-2,1; 0,-2]; X0=[0,-1; 1,1; 1,0; 0,1];
   A=sym(A); B=sym(B); X0=sym(X0); syms t
   X=simplify(expm(A*t)*X0*expm(B*t))
   simplify(diff(X)-A*X-X*B), subs(X,t,0)-X0
```

The following solution can be found. Substituting the solution back to the original equation, it is found that the differential equation and initial values are all satisfied. The error matrices are all zero while

$$X(t) = \begin{bmatrix} (t^2/2 - 2)e^{-3t} + 2e^{-4t} & (2t+1)e^{-4t} + (-2+t^2+t^3-4t)e^{-3t} \\ 2e^{-4t} - (t+1)e^{-3t} & (2t+1)e^{-4t} - (t^2+3t)e^{-3t} \\ -e^{-3t}(t^2-2)/2 & -te^{-3t}(t^2+2t-6)/2 \\ (2+t)e^{-3t} - 2e^{-4t} & (2+t^2+4t)e^{-3t} - (2t+1)e^{-4t} \end{bmatrix}.$$

## 7.3 Introduction to digital image processing

Digital images can be directly expressed by matrices. Some methods in matrix analysis can be tried to process images. In this section, reading and writing methods of digital images are introduced first, based on the Image Processing Toolbox. Then a singular value decomposition-based image compression method is presented. Besides, variation and rotation of images are discussed, together with some of the practical functions in the toolbox.

### 7.3.1 Image input and display

Homochromous grayscale images can be expressed as a pixel matrix in MATLAB. The data structure is an unsigned 8 bit integer matrix, with the range $[0, 255]$. Value 0 indicates black in color, and 255 stands for white. Colored images have many different representation methods. Normally, the three-dimensional array of 8 bit unsigned data type is used. The third dimension describes the color components. This method will be demonstrated in examples.

Function `imread()` is provided in Image Processing Toolbox in MATLAB, allowing the users to load the images in files into MATLAB environment. Besides, function `imshow()` can be used to show the image, and function `imtool()` is allowed to make simple manipulations of images in a graphical user interface. The use of these functions will be demonstrated by examples, for establishing a good foundation for further studies. Equipped with the tools, the readers are ready to deal with various image files.

**Example 7.13.** Lena image is the most widely used testing image in the field of digital signal processing. The original image is a color image of size $512 \times 512$. The original file is given in lena512color.tif. Load the data file into MATLAB workspace, and display its red, green, and blue components.

**Solutions.** The image can be loaded into MATLAB as a three-dimensional array. The three color components can be extracted, and the image components are shown in Figure 7.4.

```
>> W=imread('lena512color.tif'); imshow(W(:,:,1))
   figure, imshow(W(:,:,2)), figure, imshow(W(:,:,3))
```

**Example 7.14.** Use function `imtool()` to perform simple image manipulation.

**Solutions.** An image editing and processing interface `imtool()` is provided in the MATLAB Image Processing Toolbox. It can be used to manipulate directly the image in

(a) red component          (b) green component          (c) blue component

**Figure 7.4:** Grayscale components of Lena color image.

MATLAB workspace, as shown in Figure 7.5. Apart from displaying the image, other information can also be displayed, so as to show the information of the pixels of interest. Other buttons can be used to zoom and cut the images.

```
>> W=imread('lena512color.tif'); imtool(W)
```

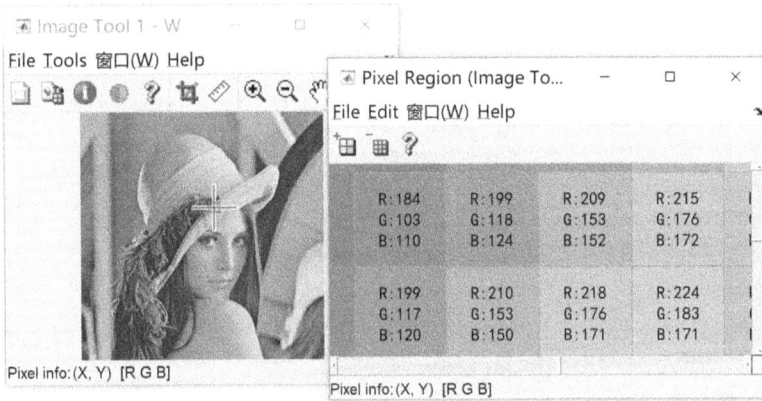

**Figure 7.5:** Editing interface `imtool()`.

## 7.3.2 Singular value decomposition applications

In Section 4.6, singular value decomposition of matrices was introduced. With this method, the diagonal elements of the singular values are arranged in descending order automatically. Normally speaking, larger singular values correspond to richer information; while smaller singular values correspond to poorer information. If the change of the diagonal elements is significant, the smaller singular values may not be important in the problem. Eliminating them may not affect the information of the image.

Assuming that the size of the image is $n \times m$, it is safe to assume that $n \geq m$. The storage space for the image is $n \times m$ bytes. If singular value decomposition is required, three matrices $U$, $S$, and $V$ are expected, where $U$ is an $n \times n$ matrix, $S$ is an $n \times m$ matrix, and $V$ is an $m \times m$ matrix. They are all stored in the double precision format. The storage space needed is far larger than that required for the original image.

Selecting the first $r$ singular values to replace all the singular values, the number of variables to be stored becomes $nr + r^2 + mr$. Note that since the double precision format is used, the actual storage space is $8(nr + r^2 + mr)$ bytes. The value and efficiency in singular value decomposition-based image compression will be studied later through examples.

**Example 7.15.** For the homochromous grayscale Lena image, use the singular value technique to compress it and see the results. The original image is a color one, so function rgb2gray() can be called to convert it into grayscale. See whether singular value decomposition-based compression is feasible or not.

**Solutions.** The data type of the image variable is uint8, while in the singular value decomposition, double precision data types are used. In the manipulation process, the data will be converted repeatedly between these data types. For different values of $r$, the compression results are obtained, as shown in Figure 7.6. For $r = 120$ or larger, details of the original image may be retained.

```
>> W=imread('lena512color.tif'); W=rgb2gray(W);
   imshow(W); [U S V]=svd(double(W));
   r=1:50;  W1=U*S(:,r)*V(:,r)'; figure; imshow(uint8(W1));
   r=1:80;  W1=U*S(:,r)*V(:,r)'; figure; imshow(uint8(W1));
   r=1:100; W1=U*S(:,r)*V(:,r)'; figure; imshow(uint8(W1));
   r=1:120; W1=U*S(:,r)*V(:,r)'; figure; imshow(uint8(W1));
   r=1:150; W1=U*S(:,r)*V(:,r)'; figure; imshow(uint8(W1));
```

It can be seen from the storage space viewpoint that, for the grayscale image, the storage needed is 262 144 bytes. Even if $r = 50$ with poor quality is selected, since the dou-

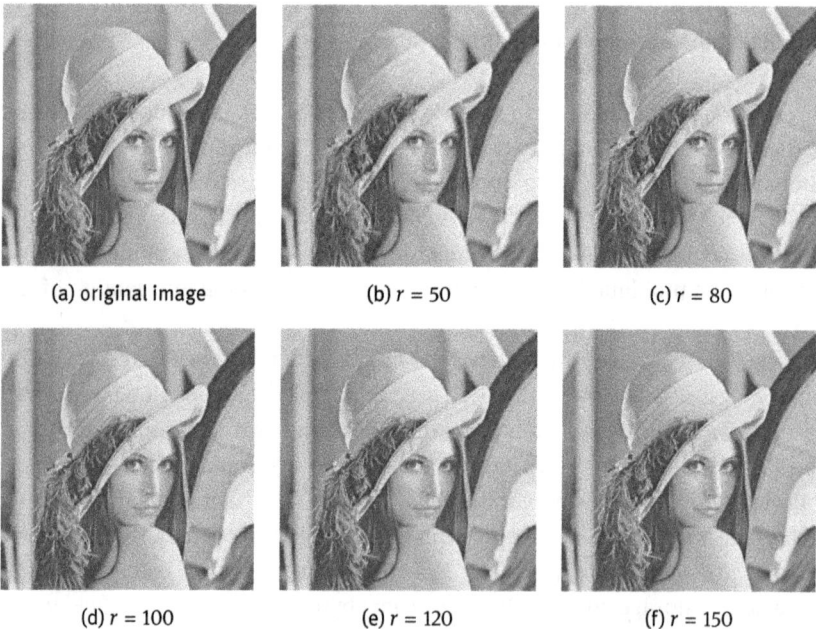

| (a) original image | (b) $r = 50$ | (c) $r = 80$ |

| (d) $r = 100$ | (e) $r = 120$ | (f) $r = 150$ |

**Figure 7.6:** Various compression results of the Lena image.

ble precision structure is needed, each number occupies 8 bytes, so the total storage space is about 429 600 bytes, 64 % more than the original image. Therefore, the singular value-based compression needs more storage space, while not all the detailed information may be retained. The method here cannot be used to really compress the images.

```
>> 512*512, r=50; (512*r+r*r+r*512)*8, 3*8*512*512
```

If $r = 512$, that is, all the details are to be retained, the storage space is 24 times the original image. Even though single precision data type is used, it still needs 12 times the original storage space to "compress" the image. Therefore, the singular value decomposition-based compression may not be of any use in practice.

### 7.3.3 Resizing and rotating images

In Chapter 4, Givens transform was presented. With it, a point in a two-dimensional plane is rotated. Therefore, with Givens transform, each pixel in the image can be manipulated, such that the image is rotated. Low-level rotation like this may be complicated, since there are too many things to consider. For instance, the bounds of the matrix, the color information, and so on. MATLAB Image Processing Toolbox functions are recommended to implement image rotation and geometric transformation. Some of the related functions and their syntaxes are listed below. All these functions can be used to handle color matrices directly:
(1) Image scaling function `imresize()`, with the syntax

$W_1$=imresize($W$,$\alpha$)

The image $W$ is scaled $\alpha$ times. If $\alpha$ is a vector $[n, m]$, the scale of the new image is $n \times m$.
(2) Image translation function `imtranslate()`, with the syntax

$W_1$=imtranslate($W$,$[n,m]$)

which means that the image is shifted to the right by $n$ pixels, and shifted up by $m$ pixels. These parameters can be set to negative numbers.
(3) Image rotation `imrotate()`, with the syntax

$W_1$=imrotate($W$,$\alpha$,methods,'crop')

meaning a rotation of image $W$ in the clockwise direction by $\alpha$ degrees and a transfer to variable $W_1$. The options of rotation algorithms are 'nearest', 'bilinear', and 'bicubic'. If the option 'crop' is used, the image rotated outside the image area will be automatically cut off. The size of the rotated image is kept unchanged.

**Example 7.16.** For the color Lena image, translation and stretching can be performed. The manipulated results can be displayed in the same figure window. See the results. Note that the commands can be nested in the function calls.

**Solutions.** The color Lena image can be loaded into MATLAB, and it can be stretched 3 times. At the upper-left corner, a 0.7 times shrunk image is superimposed, which is superimposed by another rotated, 0.3 times resized, and translated image, as shown in Figure 7.7. Note that, with hold on command, each new image is superimposed on the upper-left corner.

```
>> W=imread('lena512color.tif');
   W1=imresize(W,[512,512*3]); imshow(W1), hold on
   W1=imresize(W,0.7); imshow(W1);
   imshow(imtranslate(imresize(imrotate(W,15),0.3),[15,5]))
```

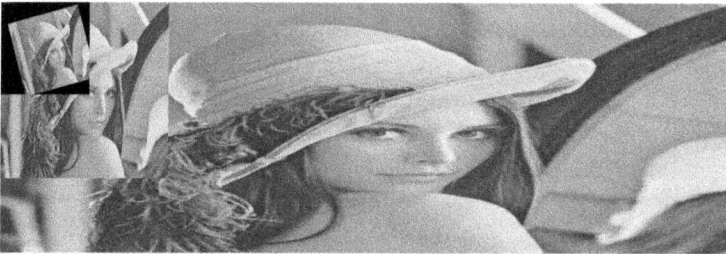

**Figure 7.7:** Translation and extension of Lena image.

**Example 7.17.** For the color Lena image, make clockwise and counterclockwise rotations of the image and observe the results.

**Solutions.** As mentioned before, imrotate() function can be used to rotate color images directly. The original and rotated images can be obtained and displayed as shown in Figure 7.8.

```
>> W=imread('lena512color.tif'); imshow(W);
   figure, W1=imrotate(W,-15); imshow(W1)
   figure, W1=imrotate(W,15,'crop'); imshow(W1)
```

Image rotation may lead to the enlargement of the image size, where more pixels are needed. Some blank regions may appear. The blank ranges are filled in black, since the pixel value is set to zero, corresponding to black in the color image.

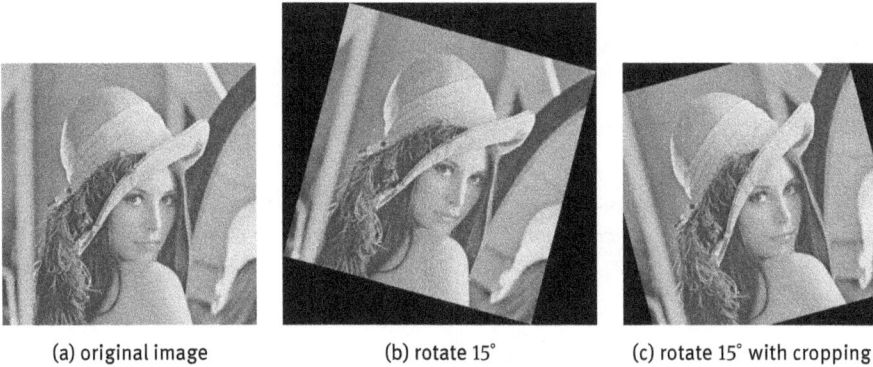

(a) original image        (b) rotate 15°        (c) rotate 15° with cropping

**Figure 7.8:** Lena image rotation.

### 7.3.4 Image enhancement

Many practical image enhancement functions are provided in Image Processing Toolbox. They can be used to extract edges of images, make Gamma correction, and enhance the contrast with the histogram equalization method. Even though the reader might have no knowledge on image processing, the functions can be called to directly complete digital image processing tasks. In this section, several such commonly used functions are introduced.

(1) The image edge detection function `edge()` can be called with $W_1$=`edge(`$W$`,` `methods)`. The input image $W$ is a grayscale one, while the returned $W_1$ is a binary one. The option `methods` is for the methods to use, with the default selection being `'Sobel'` operator. Besides, other operators can be selected, such as `'canny'`, and so on.

(2) The contrast-limited adaptive histogram equalization method `adapthisteq()` function can be invoked with the syntax $W_1$=`adapthisteq(`$W$`)`. Of course, the function can be used with other options. The quality of enhancement will be demonstrated later through examples. Ordinary histogram equalization method is also implemented in function `histeq()`.

(3) Morphological operations on a binary image can be made with the function having the syntaxes `bwmorph()` and $W_1$=`bwmorph(`$W$`,op)`, where the images before and after the conversion are binary. The operation `op` includes `'dilate'`, `'erode'`, `'skel'` (skeleton extraction), `'remove'` (remove interior pixels), and so on.

**Example 7.18.** For the Lena image, extract the edges with different algorithms.

**Solutions.** Since function `edge()` can only be used to handle grayscale images, before manipulating the color Lena image file, it should be converted first to a grayscale image. Different edge detection operators can be tried, and the results obtained are as shown in Figure 7.9.

```
>> W=imread('lena512color.tif'); W=rgb2gray(W);
   figure, W1=edge(W); imshow(W1)
   figure, W1=edge(W,'canny'); imshow(W1)
   figure, W1=edge(W,'roberts'); imshow(W1)
```

(a) default Sobel operator    (b) Canny operator    (c) Roberts operator

**Figure 7.9:** Edge detection of Lena image.

In practical photography, overexposure and underexposure phenomena often happen. In these cases, the histograms of the images may be clustered closely in a condensed interval, and the composition of the image cannot be seen clearly. This phenomenon happens because the histograms may not be well distributed. With the use of histogram equalization methods, the original image can be corrected. Even when the histogram is well distributed, the method can still be used for local processing.

**Example 7.19.** Consider the original image in Figure 7.10(a), which is from [10]. The image is rotated in this book for typesetting purposes. This is the photograph of Martian surface. The right-hand side of the image is dark, and details cannot be observed. Manipulations are needed for image processing.

**Solutions.** This is an underexposed photograph. The adaptive histogram equalization method function can be used to handle the function, with the compensated effect in Figure 7.10(b). It can be seen that through appropriate options, the photograph can be well compensated. The result is satisfactory.

```
>> W=imread('c7fmoon.tif'); imshow(W)
   W2=adapthisteq(W,'Range','original',...
           'Distribution','uniform','clipLimit',0.2);
   figure; imshow(W2);
```

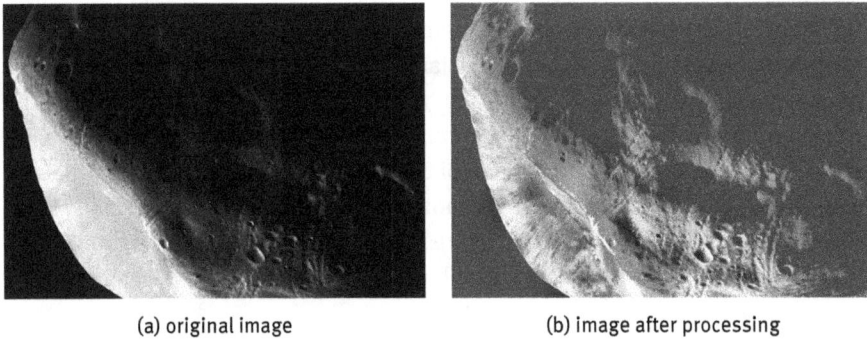

(a) original image                    (b) image after processing

**Figure 7.10:** Processing of the Martian surface image.

**Example 7.20.** The image in Figure 7.11(a) is a binary image stored in file c7ffig.bmp. Detect the edge and extract the skeleton of the image.

**Solutions.** The image can be loaded first, and then with function bwmorph(), the edge of the image can be extracted, as shown in Figure 7.11(b). If the function is used to extract the skeleton of the image, the result obtained is shown in Figure 7.11(c).

```
>> W=imread('c7ffig.bmp'); imshow(W)
   W1=bwmorph(W,'remove'); figure, imshow(~W1)      % edge detection
   W2=bwmorph(~W,'skel',inf); figure, imshow(~W2) % skeleton
```

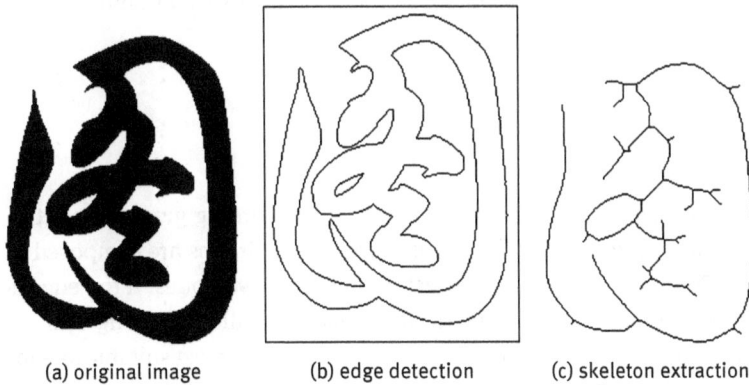

(a) original image          (b) edge detection          (c) skeleton extraction

**Figure 7.11:** Image detection and skeleton extraction.

## 7.4 Graph theory and applications

Graph theory is an important branch in mathematics, and it has wide application field. Graph theory research is originated from Königsberg seven bridge problem, as shown in Figure 7.12(a). The corresponding problem is as follows: in Königsberg, there are seven bridges, joining the banks and two islands. Is it possible to start from one point on the map, and by traveling through each bridge once and only once, return to the starting point? Swiss mathematician Leonhard Euler published a paper in 1 736 with a study of the problem[5]. The north bank of the river is denoted by node C, while the south bank by node B. The left island is denoted by A, and the right island is D. Therefore the seven bridge problem can be expressed directly as shown in Figure 7.12(b). This paper is regarded as the origin of graph theory.

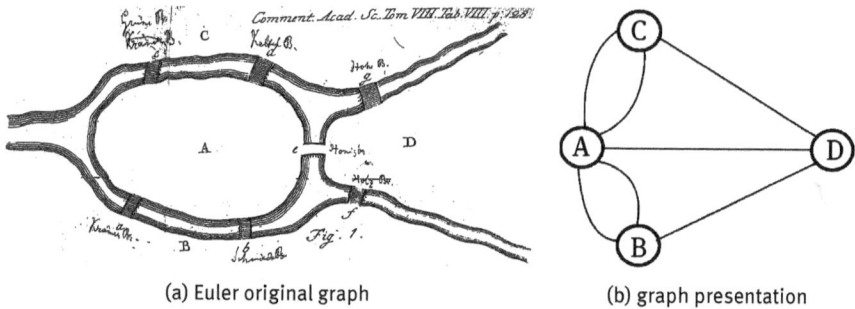

(a) Euler original graph          (b) graph presentation

**Figure 7.12:** Königsberg seven bridge problem.

The fundamental concept of the graph and its representation in MATLAB are introduced. Then the shortest path problem and complicated control system simplification problems are presented.

### 7.4.1 Oriented graph descriptions

Before introducing graphs, some fundamental concepts regarding graphs are introduced, and graph description in MATLAB is also presented. Graphs are composed of nodes and edges. The so-called edge is a direct path joining two nodes. If the edge is oriented, the graph is referred to as oriented. Otherwise, it is called an undigraph.

There are different ways in describing graphs. Of course, the most suitable one for computers is the matrix description. Assuming that there are $n$ nodes in a graph, an $n{\times}n$ matrix $R$ can be used to describe it. Assuming that from node $i$ to node $j$, the weight of the edge is $k$, the corresponding matrix element can be expressed by $R(i, j) = k$. The matrix is referred to as an incidence matrix. If there is no edge from node $i$ to $j$, one may assume that $R(i, j) = 0$. Of course, in certain algorithms, it is required that $R(i, j) = \infty$. Detailed demonstrations are made later.

In MATLAB, incidence matrices can easily be expressed by sparse matrices. Assume that in a graph, there are $n$ nodes and $m$ edges. The edge from node $a_i$ to $b_i$ has a weight of $w_i$, $i = 1, 2, \ldots, m$. Therefore three vectors can be set up, and from them an incidence matrix can be expressed in MATLAB:

$a = [a_1, a_2, \ldots, a_m, n]$; $\quad b = [b_1, b_2, \ldots, b_m, n]$; %start and end nodes

$w = [w_1, w_2, \ldots, w_m, 0]$; %weight vector

$R$=sparse$(a, b, w)$;, %sparse representation of incidence matrix

Note that the last item in each vector makes the incidence matrix a square one. The square matrix is needed in many searching methods. A sparse matrix can be converted into a regular matrix with function full(). A regular matrix can be converted into a sparse one with function sparse(). Of course, the sparse matrix can be created when the $w$ vector is a numeric one. In practice, sometimes the $w$ vector may also be symbolic. The following function is written to implement oriented graph representation:

```
function A=ind2mat(a,b,w)
if size(a,2)==3, b=a(:,2); w=a(:,3); a=a(:,1); end
for i=1:length(a), A(a(i),b(i))=w(i); end
```

The syntax of the function is the same as of function sparse() described above. This function can be used to handle numerical as well as symbolic problems. Therefore if it is possible, when describing oriented graphs, the reader may unify the input method by using function ind2mat(), rather than function sparse().

**Example 7.21.** Consider the oriented graph in Figure 7.13[17]. The numbers on the edges can be regarded as the time needed to travel from the start to the end node. Represent such a graph in MATLAB.

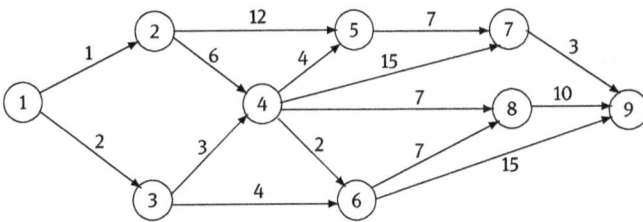

**Figure 7.13:** An oriented graph.

**Solutions.** From Figure 7.13, the information of each edge can be summarized manually in Table 7.1. The start and end nodes, as well as the weight of each edge, are listed in the table.

With the following statements, the incident matrix can be entered into MATLAB first.

**Table 7.1:** Nodal information.

| start node | end node | weight | start node | end node | weight | start node | end node | weight |
|---|---|---|---|---|---|---|---|---|
| 1 | 2 | 1 | 1 | 3 | 2 | 2 | 5 | 12 |
| 2 | 4 | 6 | 3 | 4 | 3 | 3 | 6 | 4 |
| 4 | 5 | 7 | 4 | 7 | 15 | 4 | 8 | 7 |
| 4 | 6 | 2 | 5 | 7 | 7 | 6 | 8 | 7 |
| 6 | 9 | 15 | 7 | 9 | 3 | 8 | 9 | 10 |

```
>> ab=[1 1 2 2 3 3 4 4 4 4 5 6 6 7 8];
   bb=[2 3 5 4 4 6 5 7 8 6 7 8 9 9 9];
   w=[1 2 12 6 3 4 4 15 7 2 7 7 15 3 10];
   R=ind2mat(ab,bb,w); R(9,9)=0  % set incidence matrix to a square one
```

The incidence matrix will be created as below. The matrix can be used to describe uniquely the oriented graph shown in Figure 7.13:

$$R = \begin{bmatrix} 0 & 1 & 2 & 0 & 0 & 0 & 0 & 0 & 0 \\ 0 & 0 & 0 & 6 & 12 & 0 & 0 & 0 & 0 \\ 0 & 0 & 0 & 3 & 0 & 4 & 0 & 0 & 0 \\ 0 & 0 & 0 & 0 & 4 & 2 & 15 & 7 & 0 \\ 0 & 0 & 0 & 0 & 0 & 0 & 7 & 0 & 0 \\ 0 & 0 & 0 & 0 & 0 & 0 & 0 & 7 & 15 \\ 0 & 0 & 0 & 0 & 0 & 0 & 0 & 0 & 3 \\ 0 & 0 & 0 & 0 & 0 & 0 & 0 & 0 & 10 \\ 0 & 0 & 0 & 0 & 0 & 0 & 0 & 0 & 0 \end{bmatrix}.$$

It is worth mentioning that one may try to express a graph such as the seven bridge graph with an incidence matrix. Unfortunately, since there are three or more edges between two nodes, incidence matrices cannot be used.

### 7.4.2 Dijkstra algorithm and applications

The shortest path between two nodes can be directly found with Dijkstra algorithm[3]. In fact, if the start node is specified, the shortest paths to all the other nodes can be found together, without affecting the search speed. In optimum path searching, Dijkstra algorithm is the most effective method. Assume that there are $n$ nodes, with the start node being $s$. The procedure of the searching algorithm is as follows:

(1) Initialization. Three vectors can be used to store the status of the nodes. Vector visited represents each node that has been updated, with initial value of 0; dist0 stores the shortest path from the start node to the end node, with initial

values of $\infty$. Vector parent0 stores the upper-level node, with initial value of 0. Besides, at the start node, one has dist0(s)=0.

(2) Loop solution. Let $i$ represent the $n-1$ loops, updating the upper-level node information. The loops are executed until all the nodes are visited.

(3) Extract the shortest path to node $t$. With parent vector, the optimum path can be traced and extracted.

Based on the Dijkstra algorithm, the following MATLAB function can be written to implement the search:

```
function [d,path0]=dijkstra(W,s,t)
[n,m]=size(W); ix=(W==0); W(ix)=Inf; % unified setting zero to infinity
if n~=m, error('Square W required'); end % nonsquare incidence matrix
visited(1:n)=0; dist0(1:n)=Inf; parent0(1:n)=0;
dist0(s)=0; d=Inf; path0=[];
for i=1:(n-1)   % compute the relation of the start and end nodes
    ix=(visited==0); vec(1:n)=Inf; vec(ix)=dist0(ix);
    [a,u]=min(vec); visited(u)=1;
    for v=1:n, if (W(u,v)+dist0(u)<dist0(v))
        dist0(v)=dist0(u)+W(u,v); parent0(v)=u;
end; end; end
if parent0(t)~=0, path0=t; d=dist0(t); % trace back the shortest path
    while t~=s, p=parent0(t); path0=[p path0]; t=p; end
end
```

The syntax of the function is $[d,p]$=dijkstra($W$,$s$,$t$), where $W$ is the incidence matrix, $s$ and $t$ are respectively the start and end node of the graph. The returned argument $d$ is the shortest path, while $p$ is a vector containing all the nodes on the shortest path. Note that, in the program, the zero elements in $W$ are automatically set to $\infty$, so that Dijkstra algorithm can be executed normally.

**Example 7.22.** Use Dijkstra algorithm to solve Example 7.21, and find the shortest path from node 1 to node 9.

**Solutions.** The following statements can be used to solve the problem. The optimum node vector $p$ is $p = [1, 3, 4, 5, 7, 9]$, and the shortest path is $d = 19$. The meaning is that the optimum path is from node 1, through nodes 3, 4, 5, 7, to node 9.

```
>> ab=[1 1 2 2 3 3 4 4 4 4 5 6 6 7 8];
   bb=[2 3 5 4 4 6 5 7 8 6 7 8 9 9 9];
   w=[1 2 12 6 3 4 4 15 7 2 7 7 15 3 10];
   R=ind2mat(ab,bb,w); R(9,9)=0;   % set up incidence matrix
   [d,p]=dijkstra(R,1,9)           % optimal path
```

A solver is provided in the Bioinformatics Toolbox, which can be used to solve the shortest path problem directly. Function `biograph()` can be used to set up the oriented graph object, and function `view()` can be used to display the graph. The shortest path problem can be solved with function `graphshortestpath()`. The syntaxes of the functions are

$P$=biograph($R$) , % set up the oriented graph $P$

[$d$,$p$]=graphshortestpath($R$,$n_1$,$n_2$) , % solve shortest path problem

where $R$ is the incidence matrix, which can be a regular or a sparse matrix, as demonstrated later in the examples. Function `biograph()` may use other parameters. For the oriented graph in Figure 7.13, the value of $R(i,j)$ is the weight of the edge from node $i$ to $j$. If the oriented graph object $P$ is given, function `graphshortestpath()` can be used to find the shortest path. The input arguments $n_1$ and $n_2$ are respectively the start and end nodes. The returned $d$ is the shortest distance, while vector $p$ returns the node numbers on the shortest path. When the graph is displayed, further decorations can be made. These functions will be demonstrated later through examples.

**Example 7.23.** Solve the problem in Example 7.21 with Bioinformatics Toolbox.

**Solutions.** The following commands can be used to input the incidence matrix as a sparse matrix. Then the oriented graph can be set up and displayed as shown in Figure 7.14(a). Note that when defining the incidence matrix $R$, it is a square matrix.

```
>> ab=[1 1 2 2 3 3 4 4 4 4 5 6 6 7 8];
   bb=[2 3 5 4 4 6 5 7 8 6 7 8 9 9 9];
   w=[1 2 12 6 3 4 4 15 7 2 7 7 15 3 10];
   R=ind2mat(ab,bb,w); R(9,9)=0;              % incidence matrix
   h=view(biograph(R,[],'ShowWeights','on')) % oriented graph handle h
```

With the incidence matrix $R$, function `graphshortestpath()` can be used to compute the shortest path, and the result is shown in Figure 7.14(b). It can be seen that the result is the same as those obtained earlier. With the tools provided in Bioinformatics Toolbox, the result can also be graphically displayed.

```
>> [d,p]=graphshortestpath(R,1,9)   % shortest path from node 1 to 9
   set(h.Nodes(p),'Color',[1 0 0]) % coloring the optimum path in red
   edges=getedgesbynodeid(h,get(h.Nodes(p),'ID'));
                                    % optimum path handles
   set(edges,'LineColor',[1 0 0])  % color the shortest path
```

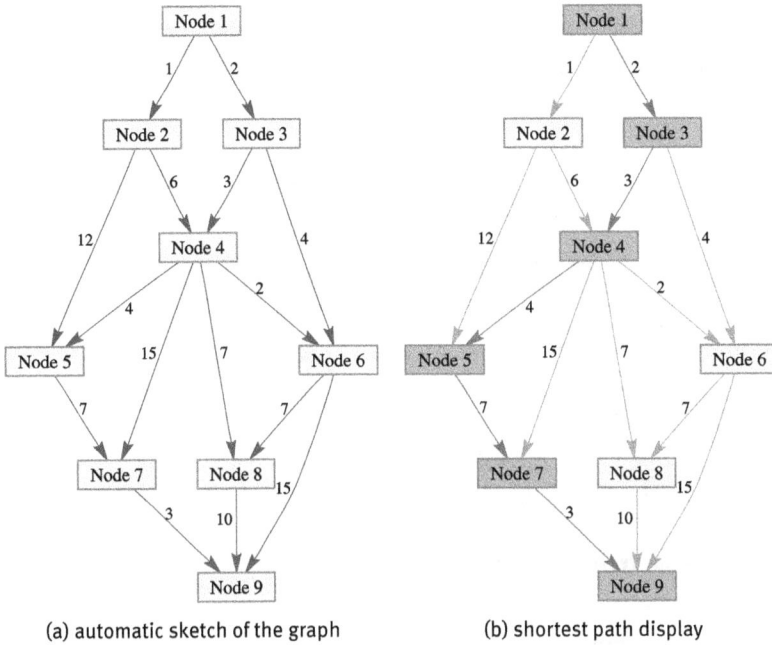

(a) automatic sketch of the graph       (b) shortest path display

**Figure 7.14:** Oriented graph and shortest path.

### 7.4.3 Simplification of control system block diagrams

The mathematical models of control systems can be described as signal flow graphs. Signal flow graphs can be regarded as oriented graphs. In this section, examples are used to describe oriented graph representation of complicated control systems. Incidence matrix construction and simplification methods of signal flow graphs will be illustrated.

**Example 7.24.** Consider a block diagram of a control system shown in Figure 7.15. In the example, if traditional methods are used, the branches in the block diagram

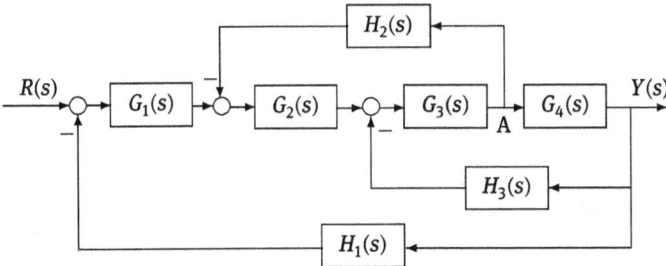

**Figure 7.15:** Block diagram of a control system.

should be moved manually. If there are many overlapped loops, manual manipula-tion of the block diagram may be complicated, and there might be errors if careless manual modifications are encountered. Use signal flow graph to represent the block diagram here.

**Solutions.** Directly manipulating the block diagram of the system, the signal flow graph in Figure 7.16 can be constructed to express the original system. In the signal flow graph, 5 nodes, $x_1 \sim x_5$, are introduced, and an extra input node $u$ is used.

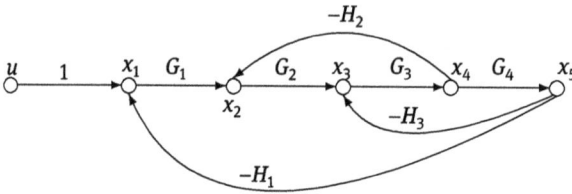

**Figure 7.16:** Signal flow graph representation.

Observing each node, it is not difficult to write out the following formulas:

$$\begin{cases} x_1 = u - H_1 x_5, \\ x_2 = G_1 x_1 - H_2 x_4, \\ x_3 = G_2 x_2 - H_3 x_5, \\ x_4 = G_3 x_3, \\ x_5 = G_4 x_4. \end{cases}$$

Matrix form can be written from the above equations, and the matrix form is the mathematical model needed in system simplifications:

$$\begin{bmatrix} x_1 \\ x_2 \\ x_3 \\ x_4 \\ x_5 \end{bmatrix} = \begin{bmatrix} 0 & 0 & 0 & 0 & -H_1 \\ G_1 & 0 & 0 & -H_2 & 0 \\ 0 & G_2 & 0 & 0 & -H_3 \\ 0 & 0 & G_3 & 0 & 0 \\ 0 & 0 & 0 & G_4 & 0 \end{bmatrix} \begin{bmatrix} x_1 \\ x_2 \\ x_3 \\ x_4 \\ x_5 \end{bmatrix} + \begin{bmatrix} 1 \\ 0 \\ 0 \\ 0 \\ 0 \end{bmatrix} u.$$

**Theorem 7.11.** *It can be seen from the above modeling that the matrix form of the system can be written as*

$$X = QX + PU, \tag{7.4.1}$$

*where $Q$ is the incidence matrix. The transfer functions from the inputs to all the nodes $x_i$ can be expressed as*

$$G = XU^{-1} = (I - Q)^{-1}P. \tag{7.4.2}$$

**Example 7.25.** If we do not want to write down the nodal equations manually, the above mentioned methods can be used to describe the information of each edge, so that the incidence matrix can be set up.

**Solutions.** Each edge in the graph can be expressed by three parameters $(i, j, w)$, where $i$ and $j$ are respectively the start and end nodes of the edge, and $w$ is the weight. For instance, the first edge is from node $x_1$ to node $x_2$, with a weight $G_1$, can be expressed as a vector $[1, 2, G_1]$. The second edge can be expressed as $[2, 3, G_2]$, and so on. The following commands can be used to generate the incidence matrix. It should be noted that the incidence matrix thus created is independent of the orders in the edge descriptions.

```
>> syms G1 G2 G3 G4 H1 H2 H3;
   w=[1,2,G1; 2,3,G2; 3,4,G4; 4,5,G4; 5,1,-H1; 4,2,-H2; 5,3,-H3];
   Q=ind2mat(w)
```

**Example 7.26.** Consider the motor drive system model shown in Figure 7.17. The system has two inputs $r(t)$ and $M(t)$, and one output $n(t)$. Find the equivalent transfer function matrix with Symbolic Math Toolbox in MATLAB.

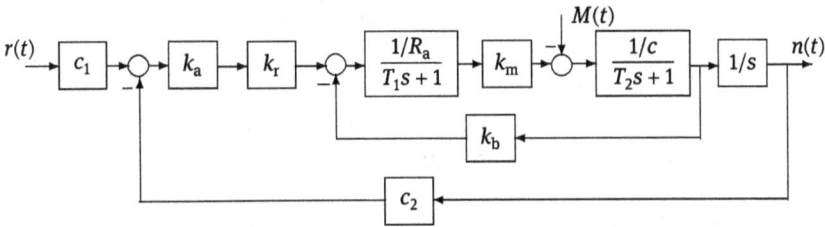

**Figure 7.17:** Motor drive system model.

**Solutions.** For the multivariate system presented here, if one wants to derive the equivalent transfer function manually, it will be a tedious work. Especially, it will be rather complicated if the second input is involved, since the block diagram will have to be redrawn. If the incidence matrix method is used, there is no need to process it manually. From the original block diagram, the signal flow graph in Figure 7.18 can be drawn.

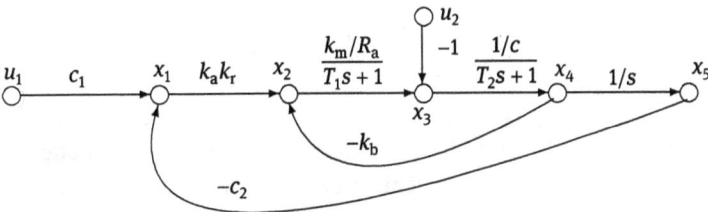

**Figure 7.18:** Signal flow graph of the multivariate system.

From the signal flow graph, the nodal equations can be written as

$$\begin{cases} x_1 = c_1 u_1 - c_2 x_5, \\ x_2 = k_a k_r x_1 - k_b x_4, \\ x_3 = \dfrac{k_m/R_a}{T_1 s + 1} x_2 - u_2, \\ x_4 = \dfrac{1/c}{T_2 s + 1} x_3, \\ x_5 = \dfrac{1}{s} x_4. \end{cases}$$

From the nodal equations, the matrix form can be constructed, and the incidence matrix can be obtained:

$$\begin{bmatrix} x_1 \\ x_2 \\ x_3 \\ x_4 \\ x_5 \end{bmatrix} = \begin{bmatrix} 0 & 0 & 0 & 0 & -c_2 \\ k_a k_r & 0 & 0 & -k_b & 0 \\ 0 & \dfrac{k_m/R_a}{T_1 s + 1} & 0 & 0 & 0 \\ 0 & 0 & \dfrac{1/c}{T_2 s + 1} & 0 & 0 \\ 0 & 0 & 0 & \dfrac{1}{s} & 0 \end{bmatrix} \begin{bmatrix} x_1 \\ x_2 \\ x_3 \\ x_4 \\ x_5 \end{bmatrix} + \begin{bmatrix} c_1 & 0 \\ 0 & 0 \\ 0 & -1 \\ 0 & 0 \\ 0 & 0 \end{bmatrix} \begin{bmatrix} u_1 \\ u_2 \end{bmatrix}.$$

With the following statements, the original multivariate system can be simplified directly. Since $x_5$ is the output node, the following commands may extract the transfer functions from the two inputs to the output node:

```
>> syms Ka Kr c1 c2 c Ra T1 T2 Km Kb s    % declare symbolic variables
   Q=[0 0 0 0 -c2; Ka*Kr 0 0 -Kb 0; 0 Km/Ra/(T1*s+1) 0 0 0
      0 0 1/c/(T2*s+1) 0 0; 0 0 0 1/s 0];
   P=[c1 0; 0 0; 0 -1; 0 0; 0 0]; W=inv(eye(5)-Q)*P;
   W(5,:)
```

The two transfer functions obtained are

$$G^T(s) = \begin{bmatrix} \dfrac{c_1 k_m k_a k_r}{R_a c T_1 T_2 s^3 + (R_a c T_1 + R_a c T_2) s^2 + (k_m k_b + R_a c)s + k_a k_r k_m c_2} \\ -\dfrac{(T_1 s + 1) R_a}{c R_a T_2 T_1 s^3 + (c R_a T_1 + c R_a T_2) s^2 + (k_b k_m + c R_a)s + k_m c_2 k_a k_r} \end{bmatrix}.$$

If the nodal equations are bypassed, the following statements can be used. The incidence matrix can be entered directly with ind2mat () function. The result obtained is exactly the same. When expressing the input matrix $P$, the following statements can also be used, where $P(i, j) = k$ means that $i$ is the input number, $j$ is the node it applied to, and $k$ is the weight:

```
>> w=[1,2,Ka*Kr; 2,3,Km/Ra/(T1*s+1); 3,4,1/c/(T2*s+1);
      4,5,1/s; 4,2,-Kb; 5,1,-c2];
   Q=ind2mat(w), P=zeros(5,2); P(1,1)=1/c; P(3,2)=-1;
```

**Example 7.27.** Consider again the block diagram in Example 7.24. Find the transfer function matrix from the input signal to all the five nodes.

**Solutions.** With the following statements, the incidence matrix $Q$ and input matrix $P$ can be entered, and the transfer functions from the input to each node can be computed directly:

```
>> syms G1 G2 G3 G4 H1 H2 H3
   Q=[0 0 0 0 -H1; G1 0 0 -H2 0; 0 G2 0 0 -H3;
      0 0 G3 0 0; 0 0 0 G4 0];
   P=[1 0 0 0 0]'; G=inv(eye(5)-Q)*P, G(5)
```

The transfer function matrix obtained is as follows:

$$
G = \begin{bmatrix}
(H_3G_3G_4 + 1 + G_3G_2H_2)/(G_4G_3H_3 + G_4G_3G_2G_1H_1 + 1 + G_3G_2H_2) \\
G_1(G_4G_3H_3 + 1)/(G_4G_3H_3 + G_4G_3G_2G_1H_1 + 1 + G_3G_2H_2) \\
G_2G_1/(G_4G_3H_3 + G_4G_3G_2G_1H_1 + 1 + G_3G_2H_2) \\
G_3G_2G_1/(G_4G_3H_3 + G_4G_3G_2G_1H_1 + 1 + G_3G_2H_2) \\
G_4G_3G_2G_1/(G_4G_3H_3 + G_4G_3G_2G_1H_1 + 1 + G_3G_2H_2)
\end{bmatrix}.
$$

Especially, the transfer function from the input to the output (the fifth node) can be obtained as

$$
G_a = \frac{G_4G_3G_2G_1}{G_4G_3H_3 + G_4G_3G_2G_1H_1 + 1 + G_3G_2H_2}.
$$

## 7.5 Solving difference equations

Difference equations are the equations to describe the relationship between the unknown variables and their difference terms.

**Definition 7.4.** The general form of a difference equation is

$$
x(k+1) = f(k, x(k), u(k)), \tag{7.5.1}
$$

where $x(k)$ is referred to as the state vector of the system, $u(k)$ is the input signal, and $k$ is the time instance. Difference equations are the fundamental models used to describe discrete-time systems.

In linear difference equations, linear algebra and the $z$ transform are the fundamental mathematical tools. In this section, the analytical and numerical solutions of linear constant difference equations are explored. Also an attempt is made to solve nonlinear difference equations numerically.

The general form of a linear constant difference equation is

$$y[(k + n)T] + a_1y[(k + n - 1)T] + a_2y[(k + n - 2)T] + \cdots + a_ny(kT)$$
$$= b_1u[(k - d)T] + b_2u[(k - d - 1)T] + \cdots + b_mu[(k - d - m + 1)T], \quad (7.5.2)$$

where $T$ is the sample time. Similar to differential equations, the coefficients $a_i$ and $b_i$ are constants, and the system is referred to as a linear time-invariant discrete system. Besides, the input and output signals are expressed respectively as $u(kT)$ and $y(kT)$, where $u(kT)$ is the $k$th sample of the input signal, $y(kT)$ is the corresponding output sample. For simplicity, denoting $y(t) = y(kT)$ and setting $y[(k + i)T]$ as $y(t + i)$, the difference equation mentioned above can be simply written as

$$y(t + n) + a_1y(t + n - 1) + a_2y(t + n - 2) + \cdots + a_ny(t)$$
$$= b_1u(t + m - d) + b_2u(t + m - d - 1) + \cdots + b_{m+1}u(t - d). \quad (7.5.3)$$

### 7.5.1 Analytical solutions

For the linear constant difference equation presented above, if the initial values $y(0)$, $y(1), \ldots, y(n - 1)$ contain nonzero elements, then taking the $z$ transform of both sides of (7.5.3), we have

$$z^nY(z) - \sum_{i=0}^{n-1} z^{n-i}y(i) + a_1z^{n-1}Y(z) - a_1\sum_{i=0}^{n-2} z^{n-i}y(i) + \cdots + a_nY(z)$$

$$= z^{-d}\left[b_1z^mU(z) - b_1\sum_{i=0}^{m-1} z^{n-i}u(i) + \cdots + b_{m+1}U(z)\right]. \quad (7.5.4)$$

It is found that

$$Y(z) = \frac{(b_1z^m + b_2z^{m-1} + \cdots + b_{m+1})z^{-d}U(z) + E(z)}{z^n + a_1z^{n-1} + a_2z^{n-2} + \cdots + a_n}, \quad (7.5.5)$$

where $E(z)$ is the expression computed from the initial values of the input and output signals, when the $z$ transform is taken:

$$E(z) = \sum_{i=0}^{n-1} z^{n-i}y(i) - a_1\sum_{i=0}^{n-2} z^{n-i}y(i) - a_2\sum_{i=0}^{n-3} z^{n-i}y(i) - \cdots - a_{n-z}y(0) + \hat{u}(n), \quad (7.5.6)$$

where

$$\hat{u}(n) = -b_1\sum_{i=0}^{m-1} z^{n-i}u(i) - \cdots - b_mzu(0). \quad (7.5.7)$$

Taking the inverse $z$ transform of $Y(z)$, the analytical solution $y(t)$ of the difference equation can be found. Based on the above algorithm, a general-purpose MATLAB function can be written to solve the general linear difference equations.

```
function y=diff_eq(A,B,y0,U,d)
E=0; n=length(A)-1; syms z; if nargin==4, d=0; end
m=length(B)-1; u=iztrans(U); u0=subs(u,0:m-1);              % input
for i=1:n, E=E+A(i)*y0(1:n+1-i)*[z.^(n+1-i:-1:1)].'; end   % (7.5.6)
for i=1:m, E=E-B(i)*u0(1:m+1-i)*[z.^(m+1-i:-1:1)].'; end   % (7.5.7)
Y=(poly2sym(B,z)*U*z^(-d)+E)/poly2sym(A,z); y=iztrans(Y); % (7.5.5)
```

The syntax of the function is $y$=diff_eq$(A,B,y_0,U,d)$, where vectors $A$ and $B$ represent the coefficients on the left and right of the difference equation; $U$ is the $z$ transform of the input signal, while $y_0$ is the vector containing the initial values of the output. The argument $d$ is the number of steps for the delay, with a default value of 0. Calling the function, the analytical solution of the difference equation can be obtained directly. The solver can also be used in solving nonmonic difference equations. An example is given below to demonstrate the solution of a general linear difference equation.

**Example 7.28.** For the given difference equation

$$48y(n+4) - 76y(n+3) + 44y(n+2) - 11y(n+1) + y(n)$$
$$= 2u(n+2) + 3u(n+1) + u(n),$$

where $y(0) = 1$, $y(1) = 2$, $y(2) = 0$, and $y(3) = -1$, the input signal is $u(n) = (1/5)^n$. Find the analytical solution of the difference equation.

**Solutions.** From the given problem, the vectors $A$ and $B$ can be extracted. The initial value vector can be sent to the computer, then we call function diff_eq() to solve the difference equation directly:

```
>> syms z n; u=(1/5)^n; U=ztrans(u);          % input z transform
   y=diff_eq([48 -76 44 -11 1],[2 3 1],[1 2 0 -1],U)
                                               % analytical solution
   n0=0:20; y0=subs(y,n,n0); stem(n0,y0)       % graphical display
```

The analytical solution can be found to be

$$y(n) = \frac{432}{5}\left(\frac{1}{3}\right)^n - \frac{26}{5}\left(\frac{1}{2}\right)^n - \frac{752}{5}\left(\frac{1}{4}\right)^n + \frac{175}{3}\left(\frac{1}{5}\right)^n - \frac{42}{5}\left(\frac{1}{2}\right)^n (n-1).$$

The result can be shown graphically in Figure 7.19, and it can be seen that the initial conditions are satisfied.

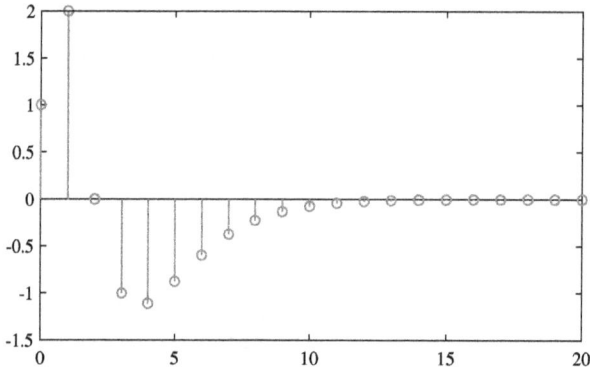

**Figure 7.19:** Solution of the difference equation.

### 7.5.2 Numerical solutions of linear time-varying difference equations

The state space model of linear time-varying difference equation can be expressed as

$$\begin{cases} x(k+1) = F(k)x(k) + G(k)u(k), \\ y(k) = C(k)x(k) + D(k)u(k), \end{cases} \quad x(0) = x_0. \tag{7.5.8}$$

If a recursive method is used, then

$$x(1) = F(0)x_0 + G(0)u(0),$$
$$x(2) = F(1)x(1) + G(1)u(1) = F(1)F(0)x_0 + F(1)G(0)u(0) + G(1)u(1).$$

It can be eventually found that

$$x(k) = F(k-1)F(k-2)\cdots F(0)x_0 + G(k-1)u(k-1)$$
$$+ F(k-1)G(k-2)u(k-2) + \cdots + F(k-1)\cdots F(0)G(0)u(0)$$
$$= \prod_{j=0}^{k-1} F(j)x_0 + \sum_{i=0}^{k-1} \left[ \prod_{j=i+1}^{k-1} F(j) \right] G(i)u(i). \tag{7.5.9}$$

If the matrices $F(i)$ and $G(i)$ are known, the above recursive algorithm can be used to solve the discrete state space model directly. It can be seen from the numerical viewpoint that iterations can be used to solve the difference equations. That is, from the known $x(0)$, we use (7.5.8) to compute $x(1)$, then from $x(1)$ we compute $x(2)$, etc. With this method, the solutions of the equations at each time instance can be computed recursively. It can be seen that iterations are suitable to computer implementation.

**Example 7.29.** Solve the following discrete linear time-varying difference equation[34]:

$$\begin{bmatrix} x_1(k+1) \\ x_2(k+1) \end{bmatrix} = \begin{bmatrix} 0 & 1 \\ 1 & \cos(k\pi) \end{bmatrix} \begin{bmatrix} x_1(k) \\ x_2(k) \end{bmatrix} + \begin{bmatrix} \sin(k\pi/2) \\ 1 \end{bmatrix} u(k),$$

where $x(0) = [1, 1]^T$ and $u(k) = (-1)^k$, $k = 0, 1, 2, 3, \ldots$

**Solutions.** Iterative method can be used to form the following loop structure, from which the state variables at each time instance can be found. The results obtained are shown in Figure 7.20.

```
>> x0=[1; 1]; x=x0; u=-1; % initial values
   for k=0:100, u=-u; F=[0 1; 1 cos(k*pi)]; % alternating signs
      G=[sin(k*pi/2); 1]; x1=F*x0+G*u; x0=x1; x=[x x1]; % updating
   end % compute the output with loop structure
   subplot(211), stairs(x(1,:)), subplot(212), stairs(x(2,:))
```

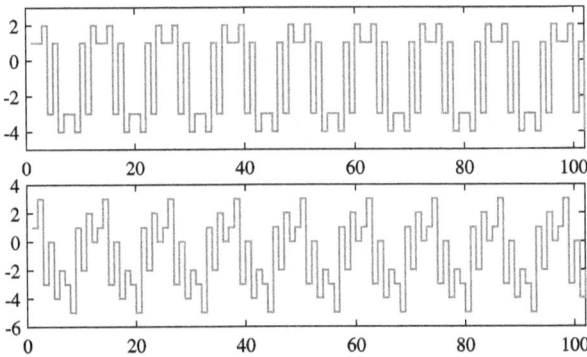

**Figure 7.20:** Time response of the states of the system.

### 7.5.3 Linear time-invariant difference equations

For the linear time invariance difference equation with $F(k) = \cdots = F(0) = F$, $G(k) = \cdots = G(0) = G$, it can be found from (7.5.9) that

$$x(k) = F^k x_0 + \sum_{i=0}^{k-1} F^{k-i-1} Gu(i). \tag{7.5.10}$$

Since with the existing computer languages, it is not likely to find $F^k$ easily, when $k$ is a symbolic variable, the above mentioned formulas cannot be used to find the analytical solutions of the difference equations. Other methods must be considered.

Considering again (7.5.8), the time-invariant form can be written as

$$\begin{cases} x(k+1) = Fx(k) + Gu(k), \\ y(k) = Cx(k) + Du(k), \end{cases} \quad x(0) = x_0. \tag{7.5.11}$$

Taking the $z$ transform of both sides, from the $z$ transform property, it is found that

$$X(z) = (zI - F)^{-1}[zx_0 + GU(z) - Gzu_0].$$  (7.5.12)

It can be found that the analytical solution of the discrete state space equation can be derived as

$$x(k) = \mathcal{Z}^{-1}[(zI - F)^{-1}z]x_0 + \mathcal{Z}^{-1}\{(zI - F)^{-1}[GU(z) - Gzu_0]\}.$$  (7.5.13)

Further observing the above formulas, it can be found that the $k$th power of the constant square matrix $F$ can also be evaluated with the inverse $z$ transform as

$$F^k = \mathcal{Z}^{-1}[z(zI - F)^{-1}].$$  (7.5.14)

**Example 7.30.** For a discrete system given by the following state space model, find the analytical solutions of all the states, when subjected to step inputs:

$$x(k + 1) = \begin{bmatrix} 11/6 & -5/4 & 3/4 & -1/3 \\ 1 & 0 & 0 & 0 \\ 0 & 1/2 & 0 & 0 \\ 0 & 0 & 1/4 & 0 \end{bmatrix} x(k) + \begin{bmatrix} 4 \\ 0 \\ 0 \\ 0 \end{bmatrix} u(k), \quad x_0 = 0.$$

**Solutions.** Using the following statements, the analytical solutions of the state space model can be found directly:

```
>> F=sym([11/6 -5/4 3/4 -1/3; 1 0 0 0; 0 1/2 0 0; 0 0 1/4 0]);
   G=sym([4; 0; 0; 0]); syms z k; U=ztrans(sym(1)); % system and input
   x=iztrans(inv(z*eye(4)-F)*G*U,z,k)                % output signal
```

The analytical solutions of the states are

$$x(k) = \begin{bmatrix} 48(1/3)^k - 48(1/2)^k k - 72(1/2)^k - 24(1/2)^k C_{k-1}^2 + 48 \\ 144(1/3)^k - 48(1/2)^k k - 144(1/2)^k - 48(1/2)^k C_{k-1}^2 + 48 \\ 216(1/3)^k - 192(1/2)^k - 48(1/2)^k C_{k-1}^2 + 24 \\ 24(1/2)^k k - 24(1/2)^k C_{k-1}^2 - 144(1/2)^k + 162(1/3)^k + 6 \end{bmatrix}.$$

In fact, in the results there are nchoosek$(n, k)$ terms, which is a notation for the number of combinations, $C_n^k = n!/[(n - k)!k!]$. The term $C_{k-1}^2$ can further be simplified as $(k - 1)(k - 2)/2$. Therefore the results can be further simplified manually as

$$x(k) = \begin{bmatrix} -12(8 + k + k^2)(1/2)^k + 48(1/3)^k + 48 \\ 24(-8 + k - k^2)(1/2)^k + 144(1/3)^k + 48 \\ 24(-10 + 3k - k^2)(1/2)^k + 216(1/3)^k \\ 12(-14 + 5k - k^2)(1/2)^k + 162(1/3)^k + 6 \end{bmatrix}.$$

Since the result only contains the $C_{k-1}^2$ term, and it can be collected according to $(1/2)^k$ terms, the result can be automatically simplified. The results obtained are the same as those manually simplified above.

```
>> x1=collect(simplify(subs(x,nchoosek(k-1,2),...
      (k-1)*(k-2)/2)),(1/2)^k)
```

**Example 7.31.** Consider the $A^k$ computation problem in Example 6.35. Use the inverse $z$ transform to compute again the matrix power $A^k$.

**Solutions.** It can be seen from (7.5.14) that the matrix power $A^k$ can be computed directly with the following statements. The results are the same as those in Example 6.35.

```
>> A=[-7,2,0,-1; 1,-4,2,1; 2,-1,-6,-1; -1,-1,0,-4];   % matrix input
   syms z k; F1=iztrans(z*inv(z*eye(4)-A),z,k); % inverse z transform
   F2=simplify(subs(F1,nchoosek(k-1,2),(k-1)*(k-2)/2)) % substitution
```

### 7.5.4 Numerical solutions of nonlinear difference equations

Assume that the explicit form of the difference equation is

$$y(t) = f(t, y(t-1), \ldots, y(t-n), u(t), \ldots, u(t-m)). \qquad (7.5.15)$$

With recursive methods, the equations can be solved directly, and the numerical solutions of the equations can be found.

**Example 7.32.** Consider the nonlinear difference equation

$$y(t) = \frac{y(t-1)^2 + 1.1y(t-2)}{1 + y(t-1)^2 + 0.2y(t-2) + 0.4y(t-3)} + 0.1u(t).$$

If the input signal is a sine function $u(t) = \sin t$, and the sample time is $T = 0.05$ seconds, compute its numerical solution.

**Solutions.** Introduce a storing vector $y_0$, whose three components $y_0(1)$, $y_0(2)$, and $y_0(3)$ store respectively $y(t-3)$, $y(t-2)$, and $y(t-1)$. In each recursive step, the vector $y_0$ is updated once. Then the following loop structure can be used to solve the difference equation. The input and output signals can then be drawn, as shown in Figure 7.21. It can be seen that when excited by a sinusoidal signal, the output signal is distorted, due to the actions of the nonlinearities. This is different from linear system responses.

```
>> y0=zeros(1,3); h=0.05; t=0:h:4*pi; u=sin(t); y=[]; % initial values
   for i=1:length(t) % recursively compute the output
      y(i)=(y0(3)^2+1.1*y0(2))/(1+y0(3)^2+0.2*y0(2)+...
```

```
            0.4*y0(1))+0.1*u(i);
    y0=[y0(2:3), y(i)]; % update the stored vector
end
plot(t,y,t,u) % draw the inputs and outputs
```

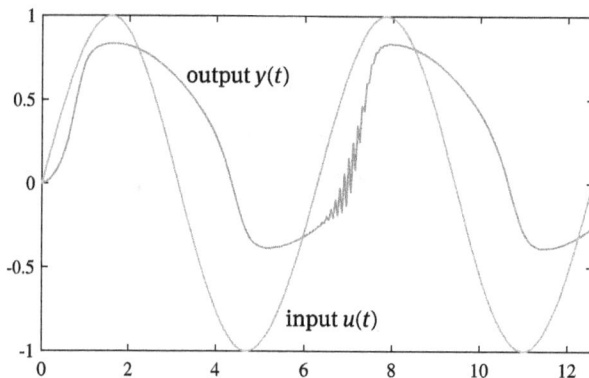

**Figure 7.21:** Numerical solutions of nonlinear discrete systems.

### 7.5.5 Markov chain simulation

A Markov chain is a statistic model, which is named after Russian mathematician Andrey Andreyevich Markov (1856–1922). It can be used to describe a special type of stochastic process.

**Definition 7.5.** A Markov chain $x_0, x_1, x_2, \ldots$ describes a sequence of state vector values. Each state vector only depends upon the state vector values of the previous step. The mathematical model is

$$x_{k+1} = Px_k, \quad k = 0, 1, 2, \ldots, \tag{7.5.16}$$

where $P$ is the state transition probability matrix, whose elements are all nonnegative and the sum of each column is 1.

Markov chain is a special case of a linear time-invariant difference equation. That is, $x_{k+1} = P^k x_0$.

**Definition 7.6.** A steady-state vector $q$ satisfies the following equation:

$$Pq = q, \tag{7.5.17}$$

where $P$ is the state transition probability matrix.

There are two ways in computing the steady-state vector $q$: For small-scale problems, the limit of $P^k$ can be taken when $k \to \infty$. The result is $P_0$, that is, we find $q = P_0 x_0$. An alternative method is to construct the following equation:

$$(P - I)q = 0. \tag{7.5.18}$$

Then for the matrix $[P - I, 0]$, reduced row echelon form can be computed, such that the last row contains all zero elements. The result obtained is then used to find the steady-state matrix elements. Later, examples are given to demonstrate the problem solutions.

**Example 7.33.** In [16], an example was presented to show population transition with a Markov chain. Assume that 5 % of the population in a city transferred to the suburbs, and 3 % of the population transferred to the city from suburbs. In the year 2000, there were 600 000 people in the city, and 400 000 people in the suburbs. How may the population change in the years 2001 and 2002?

**Solutions.** To solve this type of problem, the transition probability matrix $P$ should be created. The transition probability from the city to suburbs is 5 %, that is, 0.05, while the probability for not transferring is 0.95. On the other hand, the transfer probability from suburbs to the city is 0.03, while the probability for not transferring is 0.97. The state transition is shown in Figure 7.22.

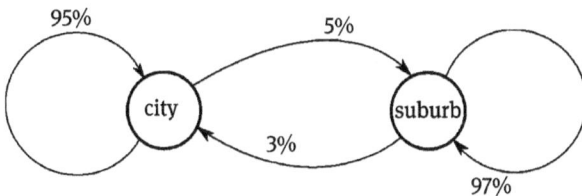

**Figure 7.22:** Illustration of population transition.

The matrix $P$ can be created, and the difference equation of the Markov chain is

$$P = \begin{bmatrix} 0.95 & 0.03 \\ 0.05 & 0.97 \end{bmatrix}, \quad x_{k+1} = \begin{bmatrix} 0.95 & 0.03 \\ 0.05 & 0.97 \end{bmatrix} x_k, \quad x_0 = \begin{bmatrix} 600\,000 \\ 400\,000 \end{bmatrix}, \tag{7.5.19}$$

where $x_0$ is the vector of year 2000. In fact, the matrix obtained here is the same as the incidence matrix considered earlier. The difference equation is the state space model of the Markov chain.

With simple multiplications, the data in years 2001 and 2002 can be computed:

```
>> P=[0.95 0.03; 0.05 0.97]; x0=[600000; 400000];
   x1=P*x0, x2=P*x1
```

In fact, if $P$ is known, it is easy to find $P^k$.

```
>> syms k x; P1=funmsym(P,x^k,x), P0=limit(P1,k,inf)
```

The matrix $P^k$ and steady-state matrix $P_0$ can be found, from which

$$P_1 = \begin{bmatrix} 5(23/25)^k/8 + 3/8 & 3/8 - 3(23/25)^k/8 \\ 5/8 - 5(23/25)^k/8 & 3(23/25)^k/8 + 5/8 \end{bmatrix}, \quad P_0 = \begin{bmatrix} 3/8 & 3/8 \\ 5/8 & 5/8 \end{bmatrix}.$$

Even though it may be difficult to find the steady-state matrix $P_0$, a large $k$ can be selected to compute the matrix power. For instance, selecting $k = 99$ and $k = 100$, the two matrices can be computed:

```
>> P99=P^99, P100=P^100
```

The two matrices obtained are as follows. If can be seen that they are rather close to each other, and close to the theoretical value of $P_0$:

$$P_{99} = \begin{bmatrix} 0.37516 & 0.3749 \\ 0.62484 & 0.6251 \end{bmatrix}, \quad P_{100} = \begin{bmatrix} 0.37515 & 0.37491 \\ 0.62485 & 0.62509 \end{bmatrix}.$$

The reduced row echelon transform is obtained using

```
>> C=[P-eye(2), zeros(2,1)]; D=rref(sym(C))
```

and can be found to be

$$P - I = \begin{bmatrix} 1 & -3/5 & 0 \\ 0 & 0 & 0 \end{bmatrix}.$$

That is, when $x_1 = 3x_2/5$, the system is in steady-state.

## 7.6 Data fitting and analysis

In this section, we concentrate on showing how linear algebra is used in solving data analysis and fitting problems. The linear regression method and its implementation in MATLAB are addressed first. Then, for given data, polynomial and Chebyshev fittings are introduced. Besides, the definition and plotting of Bézier curves are presented, and the theory and solution of principal component analysis problems are discussed in this section.

### 7.6.1 Linear regression

Assume that the output signal $y$ is a linear combination of $n$ channels of inputs $x_1, x_2, \ldots, x_n$, i. e.,

$$y = a_1 x_1 + a_2 x_2 + a_3 x_3 + \cdots + a_n x_n, \tag{7.6.1}$$

where $a_1, a_2, \ldots, a_n$ are undetermined constants. Assume also that there are $m$ experiments carried, so the measured data are:

$$
\begin{aligned}
y_1 &= x_{11} a_1 + x_{12} a_2 + \cdots + x_{1n} a_n + \varepsilon_1, \\
y_2 &= x_{21} a_1 + x_{22} a_2 + \cdots + x_{2n} a_n + \varepsilon_2, \\
&\vdots \\
y_m &= x_{m1} a_1 + x_{m2} a_2 + \cdots + x_{mn} a_n + \varepsilon_m.
\end{aligned}
\tag{7.6.2}
$$

The following matrix equation can be set up:

$$\boldsymbol{y} = \boldsymbol{Xa} + \boldsymbol{\varepsilon}, \tag{7.6.3}$$

where $\boldsymbol{a} = [a_1, a_2, \ldots, a_n]^T$ is the vector of undetermined constants. Since in each experiment, there are errors in the measured data, (7.6.1) is not fully satisfied. In each equation, an error $\varepsilon_k$ is added to each equation, for describing possible measurement noises. The vector $\boldsymbol{\varepsilon} = [\varepsilon_1, \varepsilon_2, \ldots, \varepsilon_m]^T$ is composed of the errors. Vector $\boldsymbol{y} = [y_1, y_2, \ldots, y_m]^T$ stores the measured data, and $\boldsymbol{X}$ is a matrix composed of measured independent variables

$$
\boldsymbol{X} = \begin{bmatrix}
x_{11} & x_{12} & \cdots & x_{1n} \\
x_{21} & x_{22} & \cdots & x_{2n} \\
\vdots & \vdots & \ddots & \vdots \\
x_{m1} & x_{m2} & \cdots & x_{mn}
\end{bmatrix}. \tag{7.6.4}
$$

The objective function is selected such that the sum of squared errors is minimized, that is, $J = \min \boldsymbol{\varepsilon}^T \boldsymbol{\varepsilon}$. The least-squares estimate of undetermined constant vector $\boldsymbol{a}$ of the linear regression model is

$$\hat{\boldsymbol{a}} = (\boldsymbol{X}^T \boldsymbol{X})^{-1} \boldsymbol{X}^T \boldsymbol{y}. \tag{7.6.5}$$

It is known from the matrix analysis in Chapter 3 that the MATLAB command $a=\text{inv}(X'*X)*X'*y$ can be used to find the least-squares solutions, or simply, $a=X\backslash y$. Multivariate linear regression function $\text{regress}()$ is also provided in the Statistics Toolbox in MATLAB to carry out regression parameter and confidence interval estimation tasks. The syntax of the function is $[\hat{a}, a_{ci}]=\text{regress}(y, X, \alpha)$, where $(1 - \alpha)$ is the user-specified confidence level, which can be selected as 0.02, 0.05, or to other values.

**Example 7.34.** From the linear regression model

$$y = x_1 - 1.232x_2 + 2.23x_3 + 2x_4 + 4x_5 + 3.792x_6,$$

generate 120 sets of random input $x_i$ and compute the output vector $y$. Observe whether the undetermined constants $a_i$ can be estimated from the generated data, and find the confidence interval.

**Solutions.** In this example, linear regression and its MATLAB implementation are introduced. In real applications, measured data should be used. With the following statements, the random matrix $X$ and vector $y$ can be generated, and least-squares formula can be used to estimate the undetermined constants $a$:

```
>> a=[1 -1.232 2.23 2 4,3.792]';
   X=0.01*round(100*randn(120,6)); % truncate to the last two digits
   y=0.0001*round(10000*X*a); [a,aint]=regress(y,X,0.02) % regression
```

The estimated values are $a_1 = [1, -1.232, 2.23, 2, 4, 3.792]^T$, which are the same as those given in the model, for generating the samples. The linear regression error is $1.067 \times 10^{-5}$, and it can be neglected. With function regress(), 98 % of confidence can be used, and the estimated values and confidence intervals are respectively:

$$a = \begin{bmatrix} 1 \\ -1.232 \\ 2.23 \\ 2 \\ 4 \\ 3.792 \end{bmatrix}, \quad a_{int} = \begin{bmatrix} 1 & 1 \\ -1.232 & -1.232 \\ 2.23 & 2.23 \\ 2 & 2 \\ 4 & 4 \\ 3.792 & 3.792 \end{bmatrix}.$$

It can be seen that there is no error in the estimation, since the samples are not noised.

Assuming that the measured data are corrupted with noise, and the samples are added to normally distributed noise of $N(0, 0.5)$, the following commands can be used to carry out linear regression, and the undetermined constants and confidence intervals can be found again. Function errorbar() can be used to draw the estimated results with confidence intervals, shown in Figure 7.23:

```
>> yhat=y+sqrt(0.5)*randn(120,1);
   [a,aint]=regress(yhat,X,0.02)           % regression results
   errorbar(1:6,a,aint(:,1)-a,aint(:,2)-a) % error bar plot
```

The new estimation results and confidence interval are

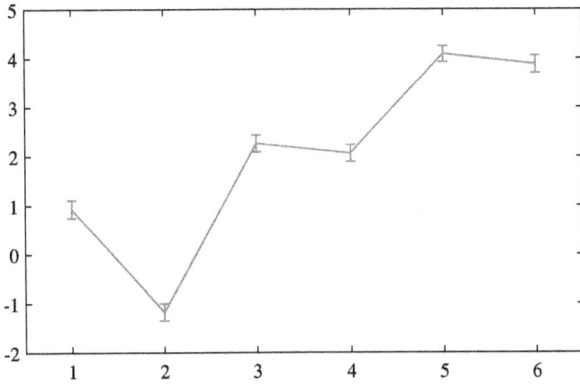

**Figure 7.23:** Parameter estimation when $\sigma^2 = 0.5$.

$$
a = \begin{bmatrix} 0.9296 \\ -1.1392 \\ 2.2328 \\ 1.9965 \\ 4.0942 \\ 3.7160 \end{bmatrix}, \quad
a_{int} = \begin{bmatrix} 0.7882 & 1.0709 \\ -1.2976 & -0.9807 \\ 2.0960 & 2.3695 \\ 1.8752 & 2.1178 \\ 3.9494 & 4.2389 \\ 3.5719 & 3.8602 \end{bmatrix}.
$$

If the variance of the added noise is decreased, for instance, the variance is 0.1, the parameter estimation results under the newly added noise can be obtained as shown in Figure 7.24. It is obvious that the new estimated parameters are more accurate.

```
>> yhat=y+sqrt(0.1)*randn(120,1);
   [a,aint]=regress(yhat,X,0.02);              % regression again
   errorbar(1:6,a,aint(:,1)-a,aint(:,2)-a) % draw error plots
```

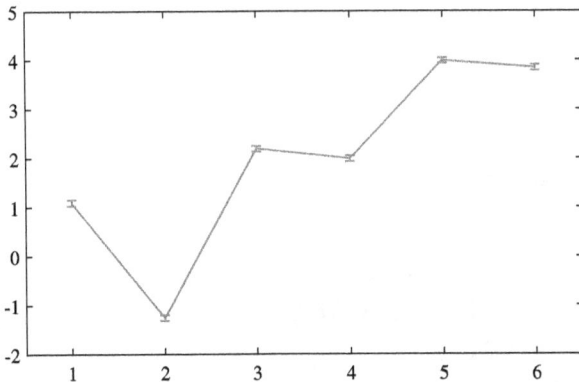

**Figure 7.24:** Parameter estimation after the noise with $\sigma^2 = 0.1$ is added.

### 7.6.2 Polynomial fitting

It is known from calculus that many functions can be approximated by power series. Therefore it might be possible to approximate a given function by polynomials. If the function model is known, MATLAB function `taylor()` can be called to find the approximate polynomials, while if only samples are given, function `polyfit()` can be used to compute the polynomial coefficients. Here, a more generalized method is introduced.

Assume that a function can be written as the following linear combination:

$$g(x) = c_1 f_1(x) + c_2 f_2(x) + c_3 f_3(x) + \cdots + c_n f_n(x), \tag{7.6.6}$$

where $f_1(x), f_2(x), \ldots, f_n(x)$ are known functions and $c_1, c_2, \ldots, c_n$ are undetermined constants. Assuming that the measured data are $(x_1, y_1), (x_2, y_2), \ldots, (x_m, y_m)$, the following linear equation can be set up:

$$Ac = y, \tag{7.6.7}$$

where

$$A = \begin{bmatrix} f_1(x_1) & f_2(x_1) & \cdots & f_n(x_1) \\ f_1(x_2) & f_2(x_2) & \cdots & f_n(x_2) \\ \vdots & \vdots & \ddots & \vdots \\ f_1(x_m) & f_2(x_m) & \cdots & f_n(x_m) \end{bmatrix}, \quad y = \begin{bmatrix} y_1 \\ y_2 \\ \vdots \\ y_m \end{bmatrix}, \tag{7.6.8}$$

and $c = [c_1, c_2, \ldots, c_n]^T$. If $m > n$, then the least-squares solution can be found with left division in MATLAB as $c = A\backslash y$.

If the base functions $f_i(x)$ are selected as $x_i$ in (7.6.1), the problem can then be converted into a linear regression problem. It can be seen that linear regression is only a special case of the problem discussed here.

Assuming that $f_i(x) = x^{i-1}$, the previously introduced method can be used to carry out polynomial fitting, such that the least-squares fitting coefficients can be computed.

**Example 7.35.** Assume that the samples come from the function

$$f(x) = (x^2 - 3x + 5)e^{-5x} \sin x.$$

Use polynomial fitting to the samples and assess the fitting quality.

**Solutions.** The following commands can be used to generate sample vectors $x$ and $y$, and generate matrix $A$. The polynomial fitting of the data can be made with

```
>> x=[0:0.05:1]'; y=(x.^2-3*x+5).*exp(-5*x).*sin(x);
   A=[]; for i=0:7, A=[A x.^(7-i)]; end, c=[A\y]'
   x0=0:0.01:1; y1=polyval(c,x0);
   y0=(x0.^2-3*x0+5).*exp(-5*x0).*sin(x0);
   plot(x0,y0,x0,y1,x,y,'o'), c1=polyfit(x,y,7)
```

The polynomial coefficients in descending order are

$$c = [10.7630, -49.0589, 95.7490, -104.5887, 69.2223, -27.0272, 4.9575, 0.0001].$$

The fitting and the original function are shown in Figure 7.25. It can be seen that when the order is 7, good approximation can be found. Function `polyfit()` yields exactly the same results.

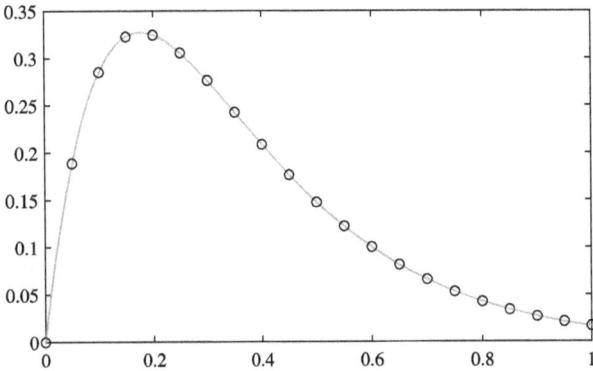

**Figure 7.25:** Polynomial fitting results.

### 7.6.3 Chebyshev polynomials

Chebyshev polynomials are named after Russian mathematician Pafnuty Lvovich Chebyshev (1821–1894). In this section, the definition and properties of Chebyshev polynomials are introduced, followed by examples to illustrate the application and computation.

**Definition 7.7.** The Chebyshev polynomials of the first kind are generated recursively from

$$T_{n+1} = 2xT_n(x) - T_{n-1}(x), \quad n = 1, 2, \dots, \tag{7.6.9}$$

where the initial values are $T_0(x) = 1$ and $T_1(x) = x$.

**Definition 7.8.** The Chebyshev polynomials of the second kind are generated recursively from

$$U_{n+1} = 2xU_n(x) - U_{n-1}(x), \quad n = 1, 2, \dots, \tag{7.6.10}$$

where the initial values are $U_0(x) = 1$ and $U_1(x) = 2x$.

**Definition 7.9.** Two kinds of Chebyshev polynomials are defined as the solutions of the following differential equations:

$$(1 - x^2)y''(x) - xy'(x) + n^2y(x) = 0, \tag{7.6.11}$$
$$(1 - x^2)y''(x) - 3xy'(x) + n(n + 2)y(x) = 0. \tag{7.6.12}$$

**Example 7.36.** Find the first 9 Chebyshev polynomials of the first kind.

**Solutions.** Arbitrary number of Chebyshev polynomials can be found, and here we obtain the first 9 of them

```
>> syms x; T1=1; T2=x; T=[T1 T2];
   for i=1:7, T(i+2)=2*x*T(i+1)-T(i); end
   expand(T.'), fplot(T(1:6),[-1.1,1.1]), ylim([-1.1,1.1])
```

They are as follows, and the curves are shown in Figure 7.26:

$$T_0(x) = 1, \quad T_1(x) = x, \quad T_2(x) = 2x^2 - 1, \quad T_3(x) = 4x^3 - 3x, \quad T_4(x) = 8x^4 - 8x^2 + 1,$$
$$T_5(x) = 16x^5 - 20x^3 + 5x, \quad T_6(x) = 32x^6 - 48x^4 + 18x^2 - 1,$$
$$T_7(x) = 64x^7 - 112x^5 + 56x^3 - 7x, \quad T_8(x) = 128x^8 - 256x^6 + 160x^4 - 32x^2 + 1.$$

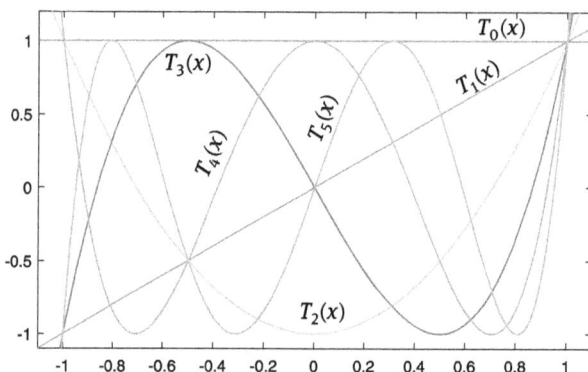

**Figure 7.26:** The curves of the first 6 Chebyshev polynomials.

**Example 7.37.** Use the first 8 Chebyshev polynomials to fit the data in Example 7.35.

**Solutions.** If Chebyshev polynomials are used as basis functions, least-squares fitting can be used again to fit the data.

```
>> x=[0:0.05:1]'; y=(x.^2-3*x+5).*exp(-5*x).*sin(x);
   A=[x.^0 x]; for i=3:8, A=[A 2*x.*A(:,i-1)-A(:,i-2)]; end
```

```
c=A\y, syms x; T1=1; T2=x; T=[T1 T2];
for i=1:6, T(i+2)=2*x*T(i+1)-T(i); end
p=vpa(expand(T*c),4)
```

The coefficient vector obtained is

$$\boldsymbol{c} = [-68.0651, 122.6034, -88.8043, 50.7587, -22.2721, 7.1615, -1.5331, 0.16817].$$

If the Chebyshev polynomials are substituted, the fitting polynomial can be obtained as $p(x) = 10.763x^7 - 49.0589x^6 + 95.749x^5 - 104.589x^4 + 69.2223x^3 - 27.0272x^2 + 4.95751x + 0.00015$. It can be seen that the results are the same as those obtained in Example 7.35.

### 7.6.4 Bézier curves

Bézier curves are commonly used parametric curves in computer graphics and related areas. They are named after French engineer Pierre Étienne Bézier (1910–1999). Vector graphics are used to draw smooth curves. It can also be extended to draw smooth surfaces, known as Bézier surfaces. Bézier used this method in 1960s when designing the bodywork of Renault cars. In this section, Bézier curves of different orders are presented, and MATLAB based Bézier curve drawing is illustrated.

**Definition 7.10.** A quadratic Bézier curve is generated from three given points $\boldsymbol{p}_0$, $\boldsymbol{p}_1$, and $\boldsymbol{p}_2$ as

$$\boldsymbol{B}(t) = \boldsymbol{p}_0(1 - t)^2 + 2\boldsymbol{p}_1(1 - t)t + \boldsymbol{p}_2 t^2, \tag{7.6.13}$$

where $t \in (0, 1)$ and $\boldsymbol{p}_i = [x_i, y_i]^{\mathrm{T}}$. The curve extends from $\boldsymbol{p}_0$ to $\boldsymbol{p}_2$, while $\boldsymbol{p}_1$ is not a point on the curve. It is only a parameter, known as the control point.

**Definition 7.11.** Cubic Bézier curves are generated from four given points, $\boldsymbol{p}_0$, $\boldsymbol{p}_1$, $\boldsymbol{p}_2$, and $\boldsymbol{p}_3$, and are defined as

$$\boldsymbol{B}(t) = \boldsymbol{p}_0(1 - t)^3 + 3\boldsymbol{p}_1(1 - t)^2 t + 3\boldsymbol{p}_2(1 - t)t^2 + \boldsymbol{p}_3 t^3, \tag{7.6.14}$$

where $t \in (0, 1)$. The curve runs from $\boldsymbol{p}_0$ to $\boldsymbol{p}_3$, where $\boldsymbol{p}_1$ and $\boldsymbol{p}_2$ are not points on the curve. They are known as controls.

**Definition 7.12.** High-order Bézier curve is generated from $n + 1$ given points $\boldsymbol{p}_0$, $\boldsymbol{p}_1$, ..., $\boldsymbol{p}_n$ as

$$\boldsymbol{B}(t) = \boldsymbol{p}_0(1 - t)^n + C_n^1 \boldsymbol{p}_1(1 - t)^{n-1} t + \cdots + C_n^{n-1}(1 - t)t^{n-1} + \boldsymbol{p}_n t^n, \tag{7.6.15}$$

where $t \in (0, 1)$ and $C_n^j$ is the binomial coefficient, i. e., the number of combinations of size $j$ from $n$, defined as

$$C_n^j = \frac{n!}{j!(n - j)!}. \tag{7.6.16}$$

It is known from the definition that a first-order Bézier curve also exists, the result is just the segment joining two points. First-order Bézier curves are not discussed in this book.

From the definition of the Bézier curve presented above, it is not hard to write from (7.6.15) a MATLAB function to compute and draw Bézier curves.

```
function [x0,y0]=bezierplot(varargin)
t=varargin{end}; n=nargin-2; x=0; y=0; x1=[]; y1=[];
if length(t)==1, t=linspace(0,1,t); end
for i=1:n+1, C=nchoosek(n,i-1);
    xm=varargin{i}(1); ym=varargin{i}(2);   x1=[x1,xm];
    T=C*(1-t).^(n+1-i).*t.^(i-1); y1=[y1 ym]; x=x+xm*T; y=y+ym*T;
end
if nargout==0, plot(x,y); hold on; plot(x1,y1,'--o'), hold off
else, x0=x; y0=y; end
```

The syntax of the function is $[x,y]$=bezierplot$(p_0,p_1,\ldots,p_n,t)$, where $t$ can be a vector, or the number $m$ of subintervals to generate vector $t$. If no return argument is provided in the function call, the Bézier curve will be drawn automatically.

**Example 7.38.** For the given vectors $p_0 = [82, 91]^T$, $p_1 = [13, 92]^T$, $p_2 = [64, 10]^T$, $p_3 = [28, 55]^T$, $p_4 = [96, 97]^T$, and $p_5 = [16, 98]^T$, draw the Bézier curves of different orders between points $p_0$ and $p_2$.

**Solutions.** Function bezierplot() can be called to draw the Bézier curves. A 1 000-point $t$ vector can be generated first, and this $p$ vector can be sent to the computer to draw the Bézier curves of different orders automatically, as shown in Figure 7.27. The controls are also labeled in the plots.

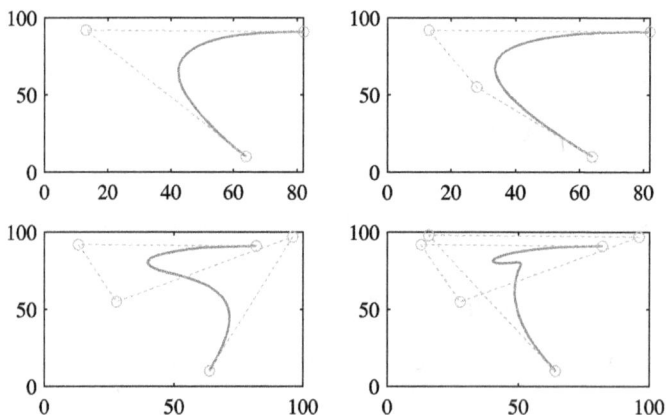

**Figure 7.27:** Different Bézier curves and controls.

```
>> p0=[82; 91]; p1=[13; 92]; p2=[64; 10];
   p3=[28; 55]; p4=[96; 97]; p5=[16; 98]; n=1000;
   subplot(221), bezierplot(p0,p1,p2,n)
   subplot(222), bezierplot(p0,p1,p3,p2,n)
   subplot(223), bezierplot(p0,p1,p3,p4,p2,n)
   subplot(224), bezierplot(p0,p1,p3,p4,p5,p2,n)
```

### 7.6.5 Principal component analysis

Principal components analysis (PCA) is an effective method in modern statistical analysis. Assuming that a phenomenon is affected by several factors, principal component analysis method can be adopted, to recognize which of the factors affect the phenomenon the most. The less important factors can be neglected and the dimension of the problem is reduced, so that the original analysis is simplified.

Assume that an event is affected by $n$ factors $x_1, x_2, \ldots, x_n$, and there are $m$ sets of measured data. Assume that an $m \times n$ matrix $X$ can be used to store the data. The mean value of each column is represented by $\bar{x}_i$, $i = 1, 2, \ldots, n$. The procedure of the principal component analysis method is as follows:

(1) Call function $R$=corr$(X)$ to compute the $n \times n$ covariance matrix $R$ from matrix $X$ such that

$$r_{ij} = \frac{\sqrt{\sum_{k=1}^{m} (x_{ki} - \bar{x}_i)(x_{kj} - \bar{x}_j)}}{\sqrt{\sum_{k=1}^{m} (x_{ki} - \bar{x}_i)^2 \sum_{k=1}^{m} (x_{kj} - \bar{x}_j)^2}}. \tag{7.6.17}$$

(2) From matrix $R$, the eigenvectors $e_i$ and sorted eigenvalues are

$$\lambda_1 \geqslant \lambda_2 \geqslant \cdots \geqslant \lambda_n \geqslant 0. \tag{7.6.18}$$

Each of the eigenvectors of the matrix is normalized such that

$$\|e_i\| = 1, \quad \text{or} \quad \sum_{j=1}^{n} e_{ij}^2 = 1. \tag{7.6.19}$$

The computation can be completed with $[e,d]$=eig$(R)$ function call. The eigenvalues are sorted in the ascending order. Function $e$=fliplr$(e)$ can be used to arrange them in descending order.

(3) The contributions of the principal components and the cumulative contribution can be computed as

$$\gamma_i = \frac{\lambda_i}{\sum_{k=1}^{n} \lambda_k} \quad \text{and} \quad \delta_i = \frac{\sum_{k=1}^{i} \lambda_k}{\sum_{k=1}^{n} \lambda_k}, \quad \text{respectively.} \tag{7.6.20}$$

If the total contribution from the first $s$ eigenvalues is greater than a certain value, for instance, 85 % or 95 %, these $s$ eigenvalues are regarded as the principal components. The original $n$-dimensional problem can be reduced to an $s$-dimensional problem.

(4) Set up new variable coordinates $Z = XL$, i. e.,

$$
\begin{cases}
z_1 = l_{11}x_1 + l_{21}x_2 + \cdots + l_{n1}x_n, \\
z_2 = l_{12}x_1 + l_{22}x_2 + \cdots + l_{n2}x_n, \\
\vdots \\
z_n = l_{1n}x_1 + l_{2n}x_2 + \cdots + l_{nn}x_n,
\end{cases}
\tag{7.6.21}
$$

where the coefficients $l_{ji}$ on the $i$th column of the transformation matrix can be computed from $l_{ji} = \sqrt{\lambda_i}e_{ji}$. Then the principal components can be analyzed with the elements in the matrix $l_{ij}$. Normally, if the first $s$ components are adopted as principal, the values of the elements in the columns after the first $s$ of them in matrix $L$ are close to zero. In equation (7.6.21), the latter $(n - s)$ variables in the $z$ vector can be neglected. A set of $s$ new variables

$$
\begin{cases}
z_1 = l_{11}x_1 + l_{21}x_2 + \cdots + l_{n1}x_n, \\
\vdots \\
z_s = l_{1s}x_1 + l_{2s}x_2 + \cdots + l_{ns}x_n
\end{cases}
\tag{7.6.22}
$$

can be used to describe the original problem. That is, with an appropriate linear transformation, the original $n$-dimensional problem can be reduced to an $s$-dimensional problem.

Assume that a certain physical quantity is affected by several factors, and the values of the factors are measured. In the experimental studies, some of the measured data are redundant. The principal component analysis method can be used to construct a new set of data, so that high-dimensional problems can be reduced to low-dimensional ones. Examples are used to illustrate the use of principal component analysis in certain problems.

**Example 7.39.** Assuming that a set of three-dimensional samples can be generated with $x = t\cos 2t$, $y = t\sin 2t$, and $z = 0.2x + 0.6y$, process the data with the principal component analysis method and reduce the dimension of the problem.

**Solutions.** With the following commands, a set of data can be generated with MATLAB. The three-dimensional curve can be drawn in Figure 7.28.

```
>> t=[0:0.1:3*pi]';
   x=t.*cos(2*t); y=t.*sin(2*t); z=0.2*x+0.6*y; % 3D data
   X=[x y z]; R=corr(X); [e,d]=eig(R)
   d=diag(d), plot3(x,y,z)    % 3D curve plotting
```

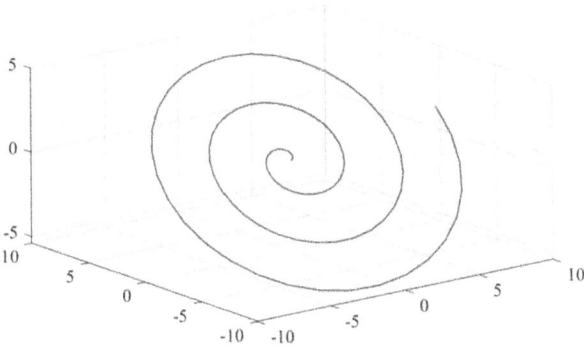

**Figure 7.28:** Three-dimensional curve.

The original three-dimensional curve can be mapped to a certain two-dimensional curve, with the following results found:

$$R = \begin{bmatrix} 1 & -0.0789 & 0.2536 \\ -0.0789 & 1 & 0.9443 \\ 0.2536 & 0.9443 & 1 \end{bmatrix},$$

$$e = \begin{bmatrix} 0.2306 & -0.9641 & 0.1314 \\ 0.6776 & 0.2560 & 0.6894 \\ -0.6983 & -0.0699 & 0.7124 \end{bmatrix}, \quad d = \begin{bmatrix} 0 \\ 1.0393 \\ 1.9607 \end{bmatrix}.$$

It can be seen that the column vector $d$ is arranged in the ascending order, not the expected descending order. Reverse ordering of the vector can be made. Meanwhile, left–right flip to matrix $e$ can be performed. Finally, the matrix $L$ can be constructed. The contributions of the first two eigenvalues are large, while the third is close to zero. It can be seen that two variables are adequate to effectively represent the original problem. With the following statements:

```
>> d=d(end:-1:1); e=fliplr(e); D=[d'; d'; d'];
   L=real(sqrt(D)).*e          % principal component analysis
   Z=X*L; plot(Z(:,1),Z(:,2)) % projection on a 2D plane
```

it can be found that

$$L = \begin{bmatrix} 0.1840 & -0.9829 & 0 \\ 0.9653 & 0.2610 & 0 \\ 0.9975 & -0.0713 & 0 \end{bmatrix}.$$

That is, introducing the new coordinates as

$$\begin{cases} z_1 = 0.1840x + 0.9653y + 0.9975z, \\ z_2 = -0.9829x + 0.2610y - 0.0713z, \end{cases}$$

the original three-dimensional problem can be reduced to a two-dimensional one. After reduction, the two-dimensional curve obtained is shown in Figure 7.29. The two-dimensional plot can be extracted from a certain plane in the three-dimensional space. The reduced curve contains most of the information of the original problem.

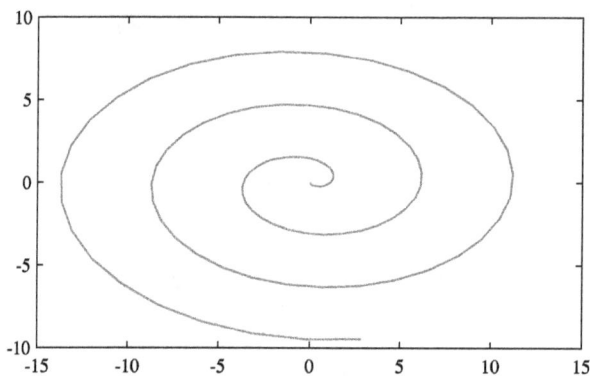

Figure 7.29: Two-dimensional curve after reduction.

## 7.7 Exercises of the chapter

7.1 A resistive network is shown in Figure 7.30. Write down the matrix equation of the $n$-level network, and find the equivalent resistance between A and B, when $n = 12$.

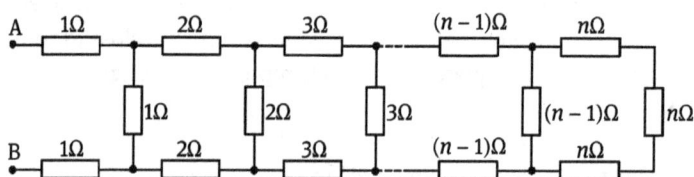

Figure 7.30: Resistive network structure.

7.2 Balance the following chemical reaction equations:
(1) $B_{10}H_{12}CNH_3 + NiCl_2 + NaOH \rightarrow Na_4(B_{10}H_{10}CNH_2)_2Ni + NaCl + H_2O$,
(2) $Pb(N_2)_3 + Cr(MnO_4)_2 \rightarrow Cr_2O_3 + MnO_2 + Pb_3O_4 + NO$,
(3) $Fe_{36}Si_5 + H_3PO_4 + K_2Cr_2O_7 \rightarrow FePO_4 + SiO_2 + K_3PO_4 + CrPO_4 + H_2O$.

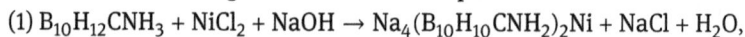

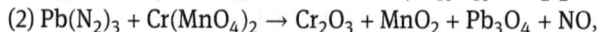

7.3 For the state space model with two inputs and two outputs,

$$\dot{x}(t) = \begin{bmatrix} 2.25 & -5 & -1.25 & -0.5 \\ 2.25 & -4.25 & -1.25 & -0.25 \\ 0.25 & -0.5 & -1.25 & -1 \\ 1.25 & -1.75 & -0.25 & -0.75 \end{bmatrix} x(t) + \begin{bmatrix} 4 & 6 \\ 2 & 4 \\ 2 & 2 \\ 0 & 2 \end{bmatrix} u(t),$$

$$y(t) = \begin{bmatrix} 0 & 0 & 0 & 1 \\ 0 & 2 & 0 & 2 \end{bmatrix} x(t),$$

input the model into MATLAB workspace. Find the equivalent transfer function matrix model, and the transmission zeros of the multivariate system.

7.4 The state space model of the system is

$$\begin{cases} \dot{x}(t) = \begin{bmatrix} -19 & -16 & -16 & -19 \\ 21 & 16 & 17 & 19 \\ 20 & 17 & 16 & 20 \\ -20 & -16 & -16 & -19 \end{bmatrix} x(t) + \begin{bmatrix} 1 \\ 0 \\ 1 \\ 2 \end{bmatrix} u(t), \\ y(t) = [2, 1, 0, 0] x(t), \end{cases}$$

where the initial state vector is $x^T(0) = [0, 1, 1, 2]$, and the input signal is $u(t) = 2 + 2e^{-3t} \sin 2t$. Is it possible to solve directly the differential equation? In [31], a transformation method is provided to convert the equation into the following form:

$$\tilde{x}(t) = \begin{bmatrix} -19 & -16 & -16 & -19 & 0 & 2 & 1 \\ 21 & 16 & 17 & 19 & 0 & 0 & 0 \\ 20 & 17 & 16 & 20 & 0 & 2 & 1 \\ -20 & -16 & -16 & -19 & 0 & 4 & 2 \\ 0 & 0 & 0 & 0 & -3 & -2 & 0 \\ 0 & 0 & 0 & 0 & 2 & -3 & 0 \\ 0 & 0 & 0 & 0 & 0 & 0 & 0 \end{bmatrix} \tilde{x}(t), \quad \tilde{x}(0) = \begin{bmatrix} 0 \\ 1 \\ 1 \\ 2 \\ 1 \\ 0 \\ 2 \end{bmatrix}.$$

Solve the differential equation and validate the results.

7.5 Assume that a matrix-type differential equation is given as

$$X'(t) = AX(t) + X(t)B, \quad X(0) = X_0,$$

where

$$A = \begin{bmatrix} -5 & -2 & -3 & -4 \\ -2 & -1 & -1 & -1 \\ -1 & 2 & -1 & 1 \\ 5 & 0 & 3 & 2 \end{bmatrix}, \quad B = \begin{bmatrix} -2 & -1 & -1 \\ -1 & -2 & 0 \\ 1 & 1 & -1 \end{bmatrix}, \quad X_0 = \begin{bmatrix} -1 & 0 & -1 \\ -1 & -1 & 0 \\ 0 & -1 & 0 \\ 1 & -1 & 0 \end{bmatrix}.$$

Solve the differential equation and validate the results.

7.6 Assuming that the prototype function for the data in Table 7.2 is

$$z(x, y) = a \sin(x^2 y) + b \cos(y^2 x) + cx^2 + dxy + e,$$

use the least-squares method to find the values of $a$, $b$, $c$, $d$, and $e$.

**Table 7.2:** The data for Exercise 7.6.

| $y_i$ | $x_1$ | $x_2$ | $x_3$ | $x_4$ | $x_5$ | $x_6$ | $x_7$ | $x_8$ | $x_9$ | $x_{10}$ | $x_{11}$ |
|---|---|---|---|---|---|---|---|---|---|---|---|
|  | 0.1 | 0.2 | 0.3 | 0.4 | 0.5 | 0.6 | 0.7 | 0.8 | 0.9 | 1 | 1.1 |
| 0.0 | 0.8304 | 0.8273 | 0.8241 | 0.8210 | 0.8182 | 0.8161 | 0.8148 | 0.8146 | 0.8158 | 0.8185 | 0.8230 |
| 0.1 | 0.8317 | 0.8325 | 0.8358 | 0.8420 | 0.8513 | 0.8638 | 0.8798 | 0.8994 | 0.9226 | 0.9496 | 0.9801 |
| 0.2 | 0.8359 | 0.8435 | 0.8563 | 0.8747 | 0.8987 | 0.9284 | 0.9638 | 1.0045 | 1.0502 | 1.1000 | 1.1529 |
| 0.3 | 0.8429 | 0.8601 | 0.8854 | 0.9187 | 0.9599 | 1.0086 | 1.0642 | 1.1253 | 1.1904 | 1.2570 | 1.3222 |
| 0.4 | 0.8527 | 0.8825 | 0.9229 | 0.9735 | 1.0336 | 1.1019 | 1.1764 | 1.2540 | 1.3308 | 1.4017 | 1.4605 |
| 0.5 | 0.8653 | 0.9105 | 0.9685 | 1.0383 | 1.118 | 1.2046 | 1.2937 | 1.3793 | 1.4539 | 1.5086 | 1.5335 |
| 0.6 | 0.8808 | 0.9440 | 1.0217 | 1.1118 | 1.2102 | 1.3110 | 1.4063 | 1.4859 | 1.5377 | 1.5484 | 1.5052 |
| 0.7 | 0.8990 | 0.9828 | 1.0820 | 1.1922 | 1.3061 | 1.4138 | 1.5021 | 1.5555 | 1.5573 | 1.4915 | 1.346 |
| 0.8 | 0.9201 | 1.0266 | 1.1482 | 1.2768 | 1.4005 | 1.5034 | 1.5661 | 1.5678 | 1.4889 | 1.3156 | 1.0454 |
| 0.9 | 0.9438 | 1.0752 | 1.2191 | 1.3624 | 1.4866 | 1.5684 | 1.5821 | 1.5032 | 1.3150 | 1.0155 | 0.6248 |
| 1.0 | 0.9702 | 1.1279 | 1.2929 | 1.4448 | 1.5564 | 1.5964 | 1.5341 | 1.3473 | 1.0321 | 0.6127 | 0.1476 |

7.7 Assume that a person lives in city $C_1$. He often needs to visit other cities $C_2, \ldots, C_8$. The matrix $R_{ij}$ indicates the traveling cost from $C_i$ to $C_j$. Design the cheapest transport map for him to travel from city $C_1$ to other cities if

$$
R = \begin{bmatrix}
0 & 364 & 314 & 334 & 330 & \infty & 253 & 287 \\
364 & 0 & 396 & 366 & 351 & 267 & 454 & 581 \\
314 & 396 & 0 & 232 & 332 & 247 & 159 & 250 \\
334 & 300 & 232 & 0 & 470 & 50 & 57 & \infty \\
330 & 351 & 332 & 470 & 0 & 252 & 273 & 156 \\
\infty & 267 & 247 & 50 & 252 & 0 & \infty & 198 \\
253 & 454 & 159 & 57 & 273 & \infty & 0 & 48 \\
260 & 581 & 220 & \infty & 156 & 198 & 48 & 0
\end{bmatrix}.
$$

7.8 Find the shortest path from nodes A to B in the graphs shown in Figure 7.31, (a) and (b).

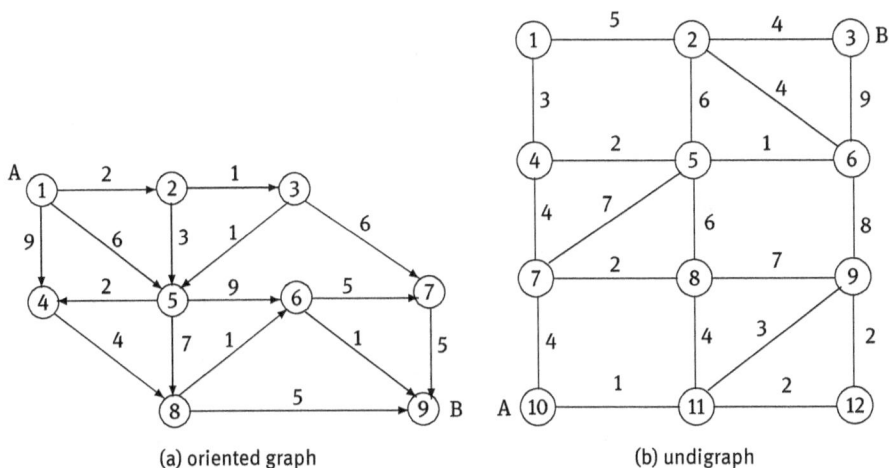

(a) oriented graph    (b) undigraph

**Figure 7.31:** Shortest path problems.

7.9 For the block diagram of the system shown in Figure 7.32, derive the overall model from the input signal $r(t)$ to the output signal $y(t)$.

7.10 The block diagram of a control system is shown in Figure 7.33. Derive the overall model from the input signal $r(t)$ to the output signal $y(t)$.

7.11 Solve the following difference equations:

(1) $72y(t) + 102y(t-1) + 53y(t-2) + 12y(t-3) + y(t-4) = 12u(t) + 7u(t-1)$, where $u(t)$ is step input, and $y(-3) = 1$, $y(-2) = -1$, $y(-1) = y(0) = 0$;

(2) $y(t) - 0.6y(t-1) + 0.12y(t-2) + 0.008y(t-3) = u(t)$, $u(t) = e^{-0.1t}$. The initial values of $y(t)$ are zeros.

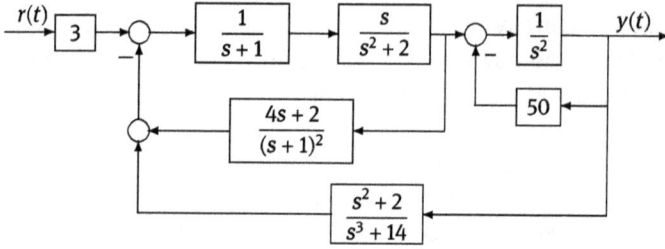

**Figure 7.32:** Block diagram of the system in Exercise 7.9.

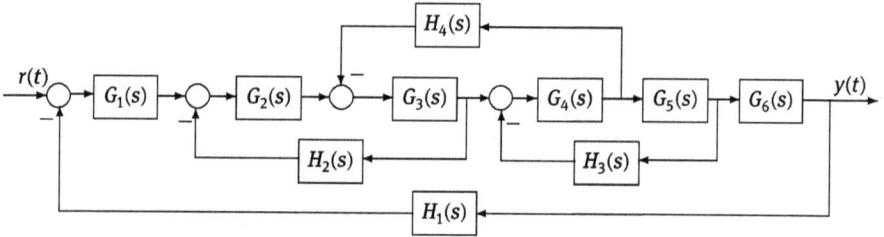

**Figure 7.33:** Block diagram of the system in Exercise 7.10.

7.12 Solve the following nonlinear difference equation:

$$y(t) = u(t) + y(t-2) + 3y^2(t-1) + \frac{y(t-2) + 4y(t-1) + 2u(t)}{1 + y^2(t-2) + y^2(t-1)},$$

and when $t \leqslant 0$, $y(t) = 0$, $u(t) = e^{-0.2t}$.

7.13 The Fibonacci sequence $a(1) = a(2) = 1$, $a(t+2) = a(t) + a(t+1)$, $t = 1, 2, \ldots$, in fact, is a linear difference equation. Find the analytical expression of the general term $a(t)$.

7.14 For a certain discrete state space model of the system given below, and the initial value of $x^T(0) = [1, -1]$, find the analytical solution of the step response, and compare it with the numerical solutions:

$$(1) \; x(t+1) = \begin{bmatrix} 0 & 1 \\ -0.16 & -1 \end{bmatrix} x(t) + \begin{bmatrix} 1 \\ 1 \end{bmatrix} u(t);$$

$$(2) \; x(t+1) = \begin{bmatrix} 11/6 & -1/4 & 25/24 & -2 \\ 1 & 1 & -1 & -1 \\ 0 & 1 & -1 & 0 \\ 0 & 1 & -3/4 & 0 \end{bmatrix} x(t) + \begin{bmatrix} 2 \\ 1/2 \\ -3/8 \\ 1/4 \end{bmatrix} u(t).$$

7.15 Assume that the state transition probability matrix of a Markov chain is given below[16]. Write the mathematical model of the Markov chain. Compute the steady-state transition matrix if

$$P = \begin{bmatrix} 0.50 & 0.70 & 0.30 \\ 0.20 & 0.80 & 0.30 \\ 0.10 & 0.10 & 0.40 \end{bmatrix}.$$

7.16  Design a Bézier curve from $(0, 0)$ to $(2, 0)$, and see how to select the controls such that the curve approaches a certain arc.

# Bibliography

[1]    Axler S. Linear Algebra Done Right. New York: Springer, 2nd edition, 1997.
[2]    Beezer R A. A First Course in Linear Algebra, Version 2.99. Washington: Department of
       Mathematics and Computer Science University of Puget Sound, 1500 North Warner, Tacoma,
       Washington, 98416-1043, http://linear.ups.edu/, 2012.
[3]    Dijkstra E W. A note on two problems in connexion with graphs. Numerische Mathematik, 1959,
       1: 269–271.
[4]    Dongarra J J, Bunsh J R, Molor C B. LINPACK User's Guide. Philadelphia: Society of Industrial and
       Applied Mathematics, 1979.
[5]    Euler L. Solutio problematis ad geometriam situs pertinentis. Commentarii Academiae
       Scientiarum Imperialis Petropolitanae, 1736, 8: 128–140.
[6]    Forsythe G, Malcolm M A, Moler C B. Computer Methods for Mathematical Computations.
       Englewood Cliffs: Prentice-Hall, 1977.
[7]    Forsythe G, Moler C B. Computer Solution of Linear Algebraic Systems. Englewood Cliffs:
       Prentice-Hall, 1965.
[8]    Garbow B S, Boyle J M, Dongarra J J, Moler C B. Matrix Eigensystem Routines – EISPACK Guide
       Extension. Lecture Notes in Computer Sciences, vol. 51. New York: Springer-Verlag, 1977.
[9]    Golub G H, Reinsch C. Singular value decomposition and least squares solutions. Numerische
       Mathematik, 1970, 14(7): 403–420.
[10]   Gongzalez R C, Woods R E. Digital Image Processing. Englewood Cliffs: Prentice-Hall,
       2nd edition, 2002.
[11]   Grassmann H. Extension Theory. History of Mathematics, vol. 19. Translated from the German
       by Lloyd C Kannenberg. Rhode Island: American Mathematical Society, 2000.
[12]   Higham N J. Functions of Matrices: Theory and Application. Philadelphia: SIAM Press, 2008.
[13]   Huang L. Linear Algebra in System and Control Theory (in Chinese). Beijing: Science Press,
       1984.
[14]   Klema V, Laub A. The singular value decomposition: its computation and some applications.
       IEEE Transactions on Automatic Control, 1980, AC-25(2): 164–176.
[15]   Laub A J. Matrix Analysis for Scientists and Engineers. Philadelphia: SIAM Press, 2005.
[16]   Lay D C, Lay S R, McDonald J J. Linear Algebra and Its Applications . Boston: Pearson, 5th
       edition, 2018.
[17]   Lin Y X. Dynamic Programming and Sequential Optimization (in Chinese). Zhengzhou: Henan
       University Press, 1997.
[18]   Mathematics Handbook Editorial Group. Mathematics Handbook (in Chinese). Beijing: People's
       Education Press, 1979.
[19]   Moler C. Experiment with MATLAB. Beijing: BUAA Press, 2014.
[20]   Moler C B, Stewar G W. An algorithm for generalized matrix eigenvalue problems. SIAM Journal
       of Numerical Analysis, 1973, 10: 241–256.
[21]   Moler C B, Van Loan C F. Nineteen dubious ways to compute the exponential of a matrix. SIAM
       Review, 1979, 20: 801–836.
[22]   Smith B T, Boyle J M, Dongarra J J, Moler C B. Matrix Eigensystem Routines – EISPACK Guide.
       Lecture Notes in Computer Sciences. New York: Springer-Verlag, 2nd edition, 1976.
[23]   Suda N (translated by Cao C X). Matrix Theory in Automatic Control (in Chinese). Beijing:
       Science Press, 1979.
[24]   Vlach J, Singhal K. Computer Methods for Circuit Analysis and Design. New York: Van Nostrand
       Reinhold Company, 1983.
[25]   Wilkinson J H. Rounding Errors in Algebraic Processes. Englewood Cliffs: Prentice-Hall, 1963.
[26]   Wilkinson J H. The Algebraic Eigenvalue Problem. Oxford: Oxford University Press, 1965.

https://doi.org/10.1515/9783110666991-008

[27] Wilkinson J H, Reinsch C. Handbook for Automatic Computation. Volume II, Linear Algebra. Berlin: Springer-Verlag, 1971.

[28] Xue D Y. Fractional-order Control Systems – Fundamentals and Numerical Implementations. Berlin: de Gruyter, 2017.

[29] Xue D Y, Chen Y Q. MATLAB Solutions of Problems in Advanced Applied Mathematics (in Chinese). Beijing: Tsinghua University Press, 2004.

[30] Xue D Y, Chen Y Q. Solving Applied Mathematical Problems with MATLAB. Boca Raton: CRC Press, 2008.

[31] Xue D Y, Chen Y Q. Modeling, Analysis and Design of Control Systems Using MATLAB and Simulink. Singapore: World Scientific, 2014.

[32] Xue D Y, Chen Y Q. Scientific Computing with MATLAB. Boca Radon: CRC Press, 2nd edition, 2016.

[33] Xue D Y, Ren X Q. Simulation and analytical algorithm for linear continuous system simulation (in Chinese). ACTA Automatica Sinica, 1992, 18(6): 694–701.

[34] Zheng D Z. Linear System Theory (in Chinese). Beijing: Tsinghua University Press, 2nd edition, 2002.

# MATLAB function index

Bold page numbers indicate where to find the syntax explanation of the function. The function or model name marked by * are the ones developed by the authors. The items marked with ‡ are those downloadable freely from Internet.

# Index